LOEB CLASSICAL LIBRARY

FOUNDED BY JAMES LOEB 1911

EDITED BY

JEFFREY HENDERSON

HIPPOCRATES

VII

LCL 477

HIPPOCRATES

VOLUME VII

WITH AN ENGLISH TRANSLATION BY

WESLEY D. SMITH

HARVARD UNIVERSITY PRESS
CAMBRIDGE, MASSACHUSETTS
LONDON, ENGLAND

First published 1994

LOEB CLASSICAL LIBRARY® is a registered trademark
of the President and Fellows of Harvard College

Library of Congress Control Number 93–19601
CIP data available from the Library of Congress

ISBN 978-0-674-99526-0

Printed on acid-free paper and bound by
The Maple-Vail Book Manufacturing Group

CONTENTS

EPIDEMICS

2,4-7

INTRODUCTION

The five books of *Epidemics*[1] presented here have generally been less well known and studied than books 1 and 3.
In Roman imperial times they were judged by Galen and
his predecessors to be "less genuine" than books 1 and 3.
Since then they have been less frequently copied, edited,
translated and interpreted. W. H. S. Jones was following
that tradition in volume I of the Loeb Hippocrates when
he printed 1 and 3 only, and praised them as "the most re-
markable product of Greek science." I hope to make some
amends for that here, and, by making all seven books of
Epidemics available, to help to restore these unique and
interesting works to their proper place.

Books 1 and 3 were distinguished from the others at the
time of the formal publication of the Corpus, in the first or
second century A.D., about five centuries after the time
when the works were probably composed. The editors who
made the judgments were ignorant of the origin and authorship of the miscellany of works attributed to Hippocrates, as we still are, but from reading the seven books of

[1] The word *epidemics* means "visits," and may refer to the itin-
erant physician's visits to the towns in which he practices, or more
likely to the visitations of diseases in those communities. (This
latter was Galen's interpretation.)

1

Epidemics they easily judged that 1 and 3 were better, more finished, more ready for publication than the others, and more unified in style. Further judgments and conjectures followed: that *Epid.* 2 and 6 are notes made in preparation for revising them for publication, and that either 4 is such notes or 4 was composed by the grandson of the great Hippocrates. But 5 and 7 were judged to lack the reserved, theoretical bent of the others and to be more rhetorically elaborate. Hence they must have been written after the great Hippocrates wrote, and perhaps by his descendants. These conjectures by the editors reflected their own training and predilections, but strongly affected the way in which the various books were received and treated by Galen and thence by those who transmitted them to us. We are fortunate that these primitive works were copied and transmitted to us at all. But we must realize that antiquity's inferences from style and substance are not better than our own—in fact not as good in some respects.

A large part of their attraction is their freshness, one might even say innocence. They are technical prose from the time when prose was coming into being and authors were realizing its potential; unique jottings by medical people in the process of creating the science of medicine. In reading the Epidemics one seems to be present while they are first formulating their descriptions of the way the body is put together, the way it responds to disease, the things that make a difference for good or ill, the ways in which the medical men should intervene. One finds the authors musing about the nature of their experience, and planning how to extend and evaluate it, admonishing themselves, "study this," "think about that," and explaining "this is what I observed, and this is what I made of it." This intense intellectual activity is carried forward in primitive, simple ways: the works have no developed language of science, no sophisticated methodology, no protocols for testing theories or correcting the inferences drawn from them. The *Epidemics* are also a unique genre. We know of nothing like them written before or after.[2] But because they differ from one another, it is not easy, especially if we include all seven books, to say what we mean when we speak of their genre.

In language and style they are simple, and at the risk of some awkwardness I have often tried to mimic them in my English rendering, though English is not well adapted to some of the effects of the Greek. To get a sense of the mind and the prose style of the *Epidemics* we need to recall how prose style was developing into a powerful tool of reflection and persuasion such as Plato and Demosthenes, for example, exhibit. They are opposites, one a self-conscious rhetorician, the other a philosopher who scorned rhetoric. But both as artists are in some ways at the opposite extreme from the writers of the *Epidemics*: both of them developed sentence structures into profound dramatic media for conveying complex thought and manipulating the audience, each of them working with long, leisurely sentences, sometimes difficult to understand, but whose individual elements or clauses are of a length to be readily comprehensible and are closely related grammatically to what precedes and follows; these clauses all lead the hearer from beginning through middle to end, using a

[2] They were in part revived in the seventeenth century by Guillaume de Baillou and Sydenham, who systematically recorded catastases in hopes of establishing statistical epidemiology.

series of promises and fulfillments whose effect is to mes-
merize and to convince of their inevitable rightness just
as their syntax comes clear and their ambiguities are resolved. At their best, such sentences, along with brief con-
necting ones, cumulatively produce increased confidence
that they are part of a worthy comprehensive design. The
Epidemics do not exhibit such conscious prose style,
whether from deliberate choice, or because their writers
are unaware of it.[3] Some stylistic tendencies are particularly striking. The *Epidemics* generally deal in bursts of
observation and judgment, reports of cases, statements
of ideas, posing of questions. What they report or ask
will have profound significance, but often it is stated simply, without indication of how it relates to a larger design,
theory, or observation. "The patient's extremities were
cool, his center burning hot." "Tongue peripneumonic."
"Bilious excrement." Syntax is often only juxtaposition.
An abstraction is a major achievement, e.g., "apostasis," a
term that describes the movement of the noxious material
of the disease towards deposit or excretion. Attaching appropriate verbs and adjectives to the abstraction is the test
of professional competence as well as of compositional
skill: "they [apostases] are best when they go down from
the disease, like [meaning 'as in the case of'] varicose
veins."[4] Much of the search for method is a search for pat-

[3] On these as on many other questions it is better to reserve
judgment in consideration of our ignorance of date, authorship,
and intended audience.

[4] The whole section, 2.1.7, is instructive in the studied attempt
to attach the right evaluative adjectives to various phenomena.

terns that will permit analogy, as in this passage testing how many notions can be transferred from the waning of the day to that of the year: "In autumn there are worms and cardialgic ailments (heartburn), shivering, melancholy. One should watch for paroxysms at the onset; also in the whole disease: as is the exacerbation at evening, so is the year at its evening. Intestinal worms also" (*Epid*. 6.1.11).

Often the authors seek to reduce general structures and principles to aphorisms, "opposites cure opposites," "purge after crises," giving an air of confident knowledge, and making the principles memorable. Satisfactory presentational structure appears to be most easily achieved by offering a general truth followed by illustrations, some of which simply illustrate, but some of which qualify the statement. Sometimes a writer will venture a judgment that seems naive, e.g., "Intestinal gas is contributory to protruding shoulder blades, for such people are flatulent." Sometimes apparently hard-won inferences seem banal or tautological. But for the most part, however fumbling the expression is, the *Epidemics* give the impression of sincere, intense, and productive intellection. The rare methodological formulations confirm our impression that the particulars are being pursued in the hope of successful generalities, e.g., 6.3.12: "The summary conclusion comes from the origin and the going forth, and from very many accounts and things learned little by little, when one gathers them together and studies them thoroughly, whether the things are like one another; again whether the dissimilarities in them are like each other, so that from dissimilarities there arises one similarity. This would be the road (i.e., method). In this way develop verification of correct accounts and refutation of erroneous ones."

The freshness and directness of these works have produced various outrageous claims for them. In Roman imperial times they were embraced, most eloquently by Galen, as early monuments of dogmatism, i.e., deductive rationalism like Plato's, which started from principles like "opposites cure opposites" and deduced the rest; by others they were embraced as models of empiricism, science based on observation of phenomena without preconceptions, and for this view one could adduce such methodological statements as the above, and reiterated statements of what "we must seek." It is easy to demonstrate that Galen and others were wrong to read sophisticated dogmatic theories into the *Epidemics*.[5] And yet there is a kernel of truth there: in part they aspire to the kinds of answers that dogmatism later produced. Equally, calling them Empirical is an anachronism. They are not "empirical" in the proper sense because they do not have the sophistication that empiricism developed when it was formulated in the centuries after these works, namely sceptical critique of dogmatism and systematic methods for dealing with observation and for evaluating hypotheses drawn from it. The *Epidemics* show great concern for developing effective method, but their concept of method is at the beginning. Their attention is on extending their theory, not on methods of testing and refining it. The Empiric's question, "How many observations make theory?" is far in the future, as is the terminology that developed along with Empirical analyses. Hence the *Epidemics* ex-

[5] My book *The Hippocratic Tradition*, Cornell Univ. Press 1979, gives an account of ancient and modern interpretations of Hippocratic medicine.

hibit many wild leaps from observation to finished theory, of the sort that the method of the ancient Empirics was developed to avoid. Yet, indeed, the *Epidemics* are full of reports of actual observations, and they show concern with the problems of creating a method, like the following: "For good physicians similarities cause wanderings and uncertainty, but so do opposites. It has to be considered what kind of explanation one can give, and that reasoning is difficult even if one knows the method"[6] (*Epid.* 6.8.26).

Besides sharing a general outlook about what medicine is and what the physician concerns himself with, the individual groups of *Epidemics* have their own personalities.

As has often been observed, *Epidemics* 1 and 3 are most finished in composition, though still structurally very loose. Primarily they present *catastases*[7] with accounts of the illnesses they produced, and individual case histories, along with a few methodological observations.

[6] Galen indicates that there were many interventions by editors and commentators in the text of this section of *Epid.* 6, which indicates to us both great interest in Hippocratic methodology on the part of the ancients, and great confusion. For example, Galen tells us that he is not even reporting Capito's reading of one sentence, since no one else knows of it. In the physiognomic example that follows the theoretical statement Galen gives us the "plausible" reading and interpretation of Rufus of Ephesus but says that it was Rufus' own, different from those of the other texts and interpreters. For modern discussions, see Manetti-Roselli ad loc., and Volker Langholf, *Medical Theories in Hippocrates*, Berlin and N.Y.: Walter de Gruyter, 1990, p. 206.

[7] Catastasis means condition or situation. In medicine it became a technical term for a description of the dominant weather and characteristic diseases of a period of time, usually a year.

7

INTRODUCTION

Epidemics 5 and 7 are collections of case histories, often grouped by type and subject matter to illustrate various subjects of interest. For example, the series 7.64ff and
5.7ff aim at evaluating therapeutic procedures. Some, especially in 5, express indignation or remorse at the fate
of patients who could have been helped.[8] The group beginning at 7.35 is prognostic (what will happen in head
wounds with denuded bone, what will happen in tetanus),
as are others, e.g., 7.56 and 58. Many of the cases in 5 and 7
seem to be sorting out the course of disease in relation to
critical periods and sequences of symptoms. It is worth
notice that the author of 7 leans towards drama in his case
histories.[9]

Epidemics 2 and 6 show a preoccupation with the way
in which the body is organized and part communicates
with part. Evidence of that concern is shown by *Epidemics*
2's unique anatomy of veins and nerves (*Epid.* 2.4.1–2).
Both works concern themselves with the ways in which the
various parts affect one another: sympathy between lungs
and testicles (2.1.6 and 7), between breast, womb, and
consciousness (6.5.11; 2.6.32), between mucus and semen
(6.6.8). Similarly the works pay much attention to pains
and flows on the same side of the body as the disease (the

[8] 5.15 and 17 report deaths of patients from medicine, 5.27–31
report deaths from failures to treat properly.

[9] Tendencies are shown by the vocabulary; e.g., *Epidemics* 7
uses ἐπιεικῶς ("reasonably," or "somewhat") 18 times, while *Epidemics* 5 uses it 4 times and the other *Epidemics* never use it. It
uses σφόδρα ("extremely") 39 times, while the other *Epidemics*
use it very rarely (it does not occur in 2, is used once in 4 and 6, and
5 times in 5).

catch phrase for it is κατ᾽ ἴξιν, 6.2.5 etc.) and they are also concerned with exits from the body which the physician can exploit, including the skin (2.4.22). *Epidemics* 2 and 6 are not simply interested in mapping such things, they want to create a technical medicine that will take control of them. A clear example is given by 2.3.8, in which the author builds on the assumption that a disease progresses towards an apostasis, a deposit or excretion of the noxious disease material, and he considers that his medical craft should learn how to control it and make it happen: "Create apostases, leading the material yourself. Turn aside apostases that have already started, accept them if they come where they should and are of the right kind and quantity, but do not offer assistance. Turn some aside if they are wholly inappropriate, but especially those that are about to commence or are just begun."[10]

This urge to a strong, invasive approach to therapy was congenial to Galen and to many others, including Empirics, in later antiquity. It has generally been discounted in post-renaissance times, when, in accord with contemporary movements in medicine, a picture was developed of Hippocrates as an advocate of restrained, expectative therapy, who trusted in the healing power of Nature. It is important to appreciate both tendencies in these works. *Epidemics* 6 reaches a climax of listing all the kinds of things that need investigation (6.8.7ff). Galen, the voluminous writer, read that section as a list of topics to be expanded in rewriting. Modern interpreters, all academics

[10] Apostasis is used only once in *Epidemics* 5 and once in 7. From this we can infer that their authors know the subject but are not preoccupied with it.

by profession, tend to read it as a list of lecture topics. Thus, we read ourselves into these works.

Epidemics 4 is closely related to 2 and 6. It mentions some of the same cases and discusses some of the same material, but it has its own personality and style, different from theirs. Its author seems to emphasize prognosis especially, collecting numbers of similar cases that differ in small ways. One of his fascinations is chlorotic coloring. And he reports how his predictions of the outcomes of cases fared (note, e.g., 4.35). In 4.25 the author tries to worry out the variations in tooth and gum infections as related to sex, age, and differences in timing. He manages to articulate questions, but he is not explicit about conclusions.

Overall, we get from *Epidemics* 2, 4, and 6 the impression of numbers of physicians working in proximity and communicating with one another. Similarly from *Epidemics* 5's comments on other physicians' errors, we get the sense of the author in a medical community. But the relations among, and the dating of, the various groups of Epidemics remain doubtful. Apparent coincidences between the patients of *Epidemics* 1 and the names of magistrates in documents on stone found on Thasos make it seem reasonable to date *Epidemics* 1 around 410 B.C. The other books of *Epidemics* could be earlier or later, though their points of view and assumptions are so similar that one assumes that they were composed close in time to 1 and 3. There is nothing except later unreliable tradition to associate the writing of the *Epidemics* with Cos and Hippocrates.

Manuscripts

	Symbol	Date	Contents
Marcianus graecus 269[11]	M	saec. X	*Epid.* 5.14–7
Vaticanus graecus 276	V	saec. XII	*Epid.* 2–7
Parisinus graecus 2140	I	saec. XIII	*Epid.* 2–7
Parisinus graecus 2142	H	saec. XIV	*Epid.* 2–7
Vaticanus graecus 277	R	saec. XIV	*Epid.* 2–7

The oldest and best, the only independent manuscripts which contain the *Epidemics*, are M and V. M has many descendants. For the text of M in *Epidemics* 2, 4, and 5, where M itself is defective, I use HIR, recentiores of the M tradition.[12] For *Epidemics* 2 and 6 there are richly infor-

---

[11] M is mutilated, and after folio 408 has lost all of the *Epidemics* preceding 5.14.

[12] For descriptions of MIVH, see *Ippocrate, Epidemie Libro Sesto*, a cura di Daniela Manetti e Amneris Roselli (Florence 1982, *Biblioteca di Studi Superiori* LXVI) xxv–xxxviii. Cay Lienau, ed., *Hippocratis De Superfetatione, CMG* I.2.2 (Berlin 1973) distinguishes older and younger parts of manuscript V, to the younger part of which (V[b]) the *Epid.* belong, and similarly older and younger parts of H (H[a] and H[b]) are distinguished by Hermann Grensemann, *Über Achtmonatskinder, Über das Siebenmonatskind (unecht) CMG* I 2, 1 (Berlin 1968). For a study of the relations of the recentiores to M, and the scribal corrections and conjectures that they exhibit, see J. Irigoin, "Le rôle des recentiores dans l'établissement du texte hippocratique," *Corpus Hippocraticum, Colloque de Mons*, ed. R. Joly (Mons 1977) 9–17, and S. Byl, "Les recentiores du traité pseudo-hippocratique *Du Régime*; quelques problèmes," *Hippocratica, Actes du Colloque Hippocratique de Paris*, ed. M. Grmek (Paris 1980) 73–83. Jacques Jouanna, *Scriptorium* 38 (1984) 56–9, establishes that H[b] and G were copied from ms. I after it had lost a number of folios.

11

mative commentaries by Galen which tell us much about the ancient textual tradition and interpretations. The commentary on *Epid*. 2 and parts of the *Epid*. 6 commentary are lost in Greek but preserved in Arabic. I sometimes quote the German translation of the Arabic from the Corpus Medicorum Graecorum. For *Epid*. 6 we have commentaries by Palladius and John of Alexandria, useful more for establishing the text of Galen's commentaries, on which they are based, than that of the Hippocratic mss. I cite them sparingly.

Dialect

The Ionic dialect of the Hippocratic Corpus is not a spoken language, but a book language whose original forms cannot be known because of successive corruptions and corrections of the words in the process of the transmission of the texts. All the sources of the text, including the earliest papyrus manuscripts, exhibit considerable inconsistency in their use of the dialect.[13] The process of adjusting the dialect seems to have been repeated in each generation as the texts were copied, and recovery of the original

[13] There were discussions of, and disagreements about, the dialect in the ancient commentators and editors, concerning which we get some, but not much, information from Galen. See his commentary on *Fractures* 18B 322 Kühn, in which he says that some commentators treated Hippocrates' dialect as Old Attic, and his commentary on Epidemics 6, CMG 5.10.2.2, p. 483, where he tells us that the editors Dioscurides and Capito presented all Hippocrates' works in the Coan dialect (as they understood it). We cannot attain the truth.

has been put beyond our reach.[14] A scholarly consensus, with which I am in accord, has developed whereby, in addition to following the tendencies of the best manuscript or manuscripts, editors introduce corrections for a general consistency of inflectional form, such as αὐτοῖσι, not αὐτοῖς, ποιέουσι, not ποιοῦσι, etc., and remove the clear hyper-Ionisms, those erroneous forms that have apparently been introduced into the text by scribes as elegant "corrections." I have admitted some inconsistency into my text. I have only rarely reported manuscript variations in orthography, when they seemed interesting or where they might be preferable to the text I print.

Enclitics

For accentuation of enclitics I have followed the principles enunciated by W. S. Barrett, *Euripides' Hippolytus* (Oxford 1964) 424–427. In the few cases in which enclitics come in series I treat the group as an accentual unit, e.g., ἦγόν τε με.

[14] See the discussions of medical Ionic by Hugo Kühlewein, *Hippocratis opera quae feruntur omnia*, vol. 1 (Teubner, Leipzig 1894), pp. lxv–ccxxviii, E. Schwyzer in Karl Deichgräber, ed., *Hippocratis de carnibus* (Leipzig and Berlin 1935) 62–70, Hans Diller, ed., *Hippocratis De aere aquis locis*, CMG 1.1.2 (Berlin 1970) 13–17, and Jacques Jouanna, ed., *Hippocratis De natura hominis*, CMG 1.1.3 (Berlin 1975) 133–155.

Editions

Asul.   *Omnia opera Hippocratis*, Venetiis in aedibus Aldi et Andreae Asulani 1526

Corn.   *Hippocratis Coi libri omnes*, Froben, Basileae 1538 (edited by Janus Cornarius)

Merc.   *Hippocratis Coi opera quae existunt*, ed. Hieronymus Mercurialis, Venetiis 1588

Foës   *Magni Hippocratis opera omnia quae exstant*, ed. Anutius Foesius, Francofurti 1595

Lind.   *Magni Hippocratis Coi opera omnia Graece et Latine*, ed. Ioan. Anton. van der Linden, Leiden 1665

Li.   *Oeuvres complètes d'Hippocrate* par Émile Littré, vol. 5, Paris 1846

Erm.   *Hippocratis et aliorum medicorum veterum reliquiae*, ed. Franciscus Z. Ermerins, Traiecti ad Rhen. 1859–64.

Foës used a number of manuscripts and much medical insight and philological ingenuity to improve the text of the early editions, which were based on manuscripts of the M tradition. Littré extended the manuscript evidence greatly by using the manuscripts in Paris, including his C (Parisinus 2146), which is a copy of V.

Sigla

M	Marcianus graecus 269	saec. X
V	Vaticanus graecus 276	saec. XII
I	Parisinus graecus 2140	saec. XIII
H	Parisinus graecus 2142	saec. XIV
R	Vaticanus graecus 277	saec. XIV

Gal.	Galen's Hippocratic Commentaries, Glossary, and discussions of individual passages in his other works.
(v.l.)Gal.	An ancient variant reading reported but rejected by Galen
Erot.	Erotianus, ed. Ernst Nachmanson, Uppsala 1918.
Man.-Ros.	*Ippocrate*, *Epidemie Libro Sesto*, ed., tr., comm., by Daniela Manetti and Amneris Roselli, Florence 1982.
Pall.	Palladii *Commentarii in Hipp. librum VI de morbis popularibus*, ed. F. R. Dietz in *Scholia in Hipp. et Galenum*, II 1–204, Königsberg 1834 (Repr. 1966).

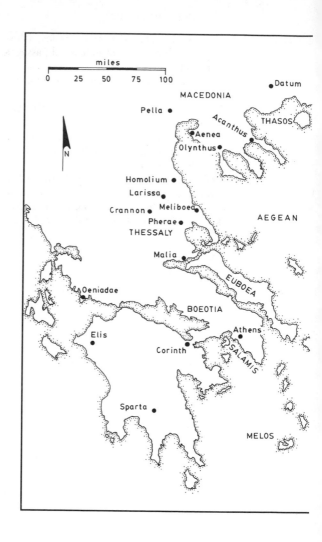

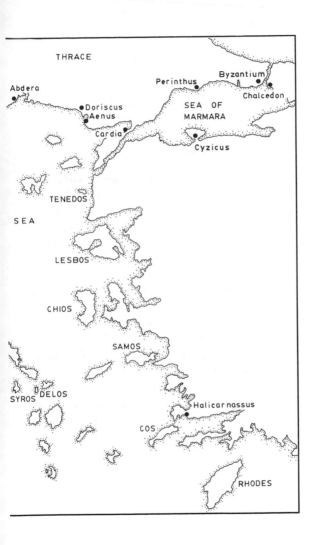

THRACE

Abdera

Doriscus
Aenus

Cardia

Perinthus

Byzantium

Chalcedon

SEA OF
MARMARA

Cyzicus

TENEDOS

SEA

LESBOS

CHIOS

SAMOS

SYROS DELOS

Halicarnassus

COS

RHODES

ΕΠΙΔΗΜΙΩΝ ΤΟ ΔΕΥΤΕΡΟΝ

ΤΜΗΜΑ ΠΡΩΤΟΝ

V 72
Littré
1. Ἄνθρακες ἐν Κρανῶνι θερινοί· ὗεν ἐν καύμασιν
ὕδατι λάβρῳ δι᾽ ὅλου· ἐγένετο καὶ μᾶλλον νότῳ καὶ
ὑπεγίνοντο μὲν ἐν τῷ δέρματι ἰχῶρες· ἐγκαταλαμ-
βανόμενοι δὲ ἐθερμαίνοντο καὶ κνησμὸν ἐνεποίεον·
εἶτα φλυκταινίδες ὥσπερ πυρίκαυστοι διανίσταντο
καὶ ὑπὸ τὸ δέρμα καίεσθαι ἐδόκεον.

2. Ἐν καύμασιν ἀνυδρίης οἱ πυρετοὶ ἀνίδρωτες τὰ
πλεῖστα· ἐν τούτοισι δὲ ἢν ἐπιψεκάσῃ ἱδρωτικώτεροι
γίνονται κατ᾽ ἀρχάς· ταῦτα δυσκριτώτερα μένει ἢ[1]
ἄλλως· ἀτὰρ ἧσσον εἰ μὴ εἴη διὰ ταῦτα ἀλλὰ διὰ τῆς
νούσου τὸν τρόπον. οἱ καῦσοι ἐν τῇσι θερινῇσι μᾶλλον γίνονται· καὶ ἐν τῇσιν ἄλλῃσιν ὥρῃσιν, ἐπιξηραί-
νονται δὲ μᾶλλον θέρεος.

3. Φθινοπώρου μάλιστα τὸ θηριῶδες καὶ ἡ καρ-
διαλγία· καίτοι καὶ αὐτὴ ἧσσον κακουργοίη[2] ἂν ἢ[3]
αὐτοῦ τοῦ νοσήματος τοιούτου[4] ἐόντος. αἱ ἀσκαρίδες
δείλης ὁμοίως τούτῳ καὶ ἐκεῖναι τηνικαῦτα ὀχλέουσι

mss. HIRV   [1] εἰ V   [2] κακουργέοι HIR
[3] om. HIR   [4] τουτέου HIR

18

EPIDEMICS 2

SECTION 1

1. In Crannon[1] in summer: anthrax. During the hot weather there was continuous violent rain. It occurred more with wind from the south. There were watery gatherings in the skin. When formed, they grew hot and caused itching, and then small blisters as though from burns rose up. They seemed like burns on the skin beneath.

2. In hot weather when it is dry fevers are mostly free from sweat. But if there is any rain during them there is more sweat at the outset. Then their crises are more difficult than otherwise, but less so if it come from the nature of the disease and not from that. Causus (burning fever) occurs more in summer; it occurs in other seasons, but is drier in summer.

3. Mostly in fall worms and cardialgia. This too is less harmful than when the disease itself is of that kind. Round worms similarly are worse in the afternoon and give most

[1] See map, pages 16–17.

τῆς ἡμέρης τὰ πλεῖστα, οὐ μόνον διὰ τὸ μᾶλλον
πονεῖν ἀλλὰ καὶ αὐταὶ διὰ σφᾶς ἑωυτάς.

4. Ἐν φθινοπώρῳ ὀξύταται αἱ[5] νοῦσοι καὶ θανατω-
δέσταται. τὸ ἐπίπαν ὅμοιον τῷ δείλης παροξύνεσθαι·
ὡς τοῦ ἐνιαυτοῦ περίοδον ἔχοντος τῶν νούσων οἷον[6] ἡ
ἡμέρη τῆς νούσου· οἷον τὸ δείλης παροξύνεσθαι,
τοιοῦτον τῆς νούσου καὶ ἑκάστης καταστάσιος πρὸς
ἀλλήλας ὅταν μή τι νεωτεροποιηθῇ ἐν τῷ ἄνω εἴδει· εἰ
δὲ μή, ἄλλης ταῦτα ἂν ἄρχοι ὥστε καὶ τὸν ἐνιαυτὸν
πρὸς ἑωυτὸν οὕτως ἔχειν.

5. Ἐν τοῖσι καθεστεῶσι καιροῖσι καὶ ὡραίως τὰ
ὡραῖα ἀποδιδοῦ ἔτεσιν εὐσταθεῖς καὶ εὐκρινέσταται
αἱ νοῦσοι, ἐν δὲ τοῖσιν ἀκαταστάτοισιν ἀκατάστατοι
καὶ δύσκριτοι· ἐν γοῦν Περίνθῳ ὅταν τι ἐκλίπῃ ἢ
πλεονάσῃ ἢ πνευμάτων ἢ ἀπνοίων[7] ἢ ὕδατι ἢ αὐχμῷ ἢ
καύματι ἢ ψύχει.[8] τὸ δὲ ἔαρ τὸ ἐπίπαν ὑγιεινότατον
καὶ ἥκιστα θανατῶδες.

6. Πρὸς τὰς ἀρχὰς σκεπτέον τῶν νούσων εἰ αὐτίκα
ἀνθεῖ· δῆλον δὲ τῇ ἐπιδόσει· τὰς δὲ ἐπιδόσιας τῇσι
περιόδοισιν· καὶ αἱ κρίσιες ἐντεῦθεν δῆλοι, καὶ τοῖσιν
ἐν τῇσι περιόδοισι παροξυσμοῖσιν εἰ πρωϊαίτερον ἢ
οὔ, καὶ εἰ πλείονα χρόνον ἢ οὔ, καὶ εἰ μᾶλλον ἢ οὔ.[9]
πάντων δὲ τῶν ξυνεχέων ἢ διαλειπόντων τῶν χρονίων
καὶ τρωμάτων καὶ πτυέλων ὀδυνωδέων καὶ φυμάτων,
φλεγμοναὶ καὶ ὅσα ἄλλα ἐπιφαίνεται[10] ὕστερον, ἴσως
δὲ καὶ ἄλλων πρηγμάτων κοινῶν, τὰ μὲν θᾶσσον

---

[5] om. HIR   [6] οἴην V   [7] μὴ πνευμάτων IR

trouble in that part of the day, not simply because pain is generally greater then, but of themselves.

4. In fall diseases are most acute and most deadly; generally similar to exacerbation in the afternoon. The year has a cycle of diseases, just as the day has of one disease. As is the exacerbation of the disease in the afternoon, so is the worsening of each disease and weather pattern relative to each other if there be no change in the upper air. If there is, these things would commence from the new weather pattern. Hence, the year, too, is related to itself in this way.

5. In stable times and years which produce seasonal things at their proper times, diseases are dependable and have proper crises, but in unstable years they are unstable and have bad crises. So in Perinthos when there is deficiency or excess of wind or calm, with rain or drought, heat or cold. Spring is generally very healthy and minimally deadly there.

6. One must examine the beginning of a disease, whether it comes to full flower immediately—it is clear in the advancement—and examine the advancements according to their periods. From this the crises too are clear, as well as from the exacerbations within the periods (i.e., whether or not they are earlier and whether or not they continue a longer time, and whether or not they are more severe). In all continuous or intermittent chronic diseases, wounds, painful salivation, and swellings, it is true of inflammation and all other things that appear later, and probably this applies to all common events, that the ones

---

8 ὑδάτων . . . αὐχμῶν . . . καυμάτων . . . ψυξέων HIR
9 καὶ εἰ πλείονα . . . μᾶλλον ἢ οὐ om. V
10 ἐπιφαίνονται HIR

76 βραχύτερα, τὰ δὲ βραδύτερα | μακρότερα· καὶ ἐν
περιόδοισι τὸ ἐπὶ πρωῖτερον καὶ ἄλλης ἐπιδόσιος
ἀπαυδώσης τῆς νούσου· καὶ γὰρ τῶν παραχρῆμα
ἀπολλυμένων ταχύτεραι αἱ κρίσιες ὅτι ταχεῖς[11] οἱ
πόνοι καὶ ξυνεχεῖς καὶ ἰσχυροί. τὰ δὲ κρίνοντα ἐπὶ τὸ
βέλτιον μὴ αὐτίκα ἐπιφαινέσθω. τὰ κρίσιμα μὴ κρίνοντα, τὰ μὲν θανατώδεα, τὰ δὲ δύσκριτα. τὰ προκρι-
νόμενα ἢν ὅμως κριθῇ, ὑποστροφαί· ἢν δὲ μή, ἀκρι-
σίαι· γένοιτο δ᾽ ἂν καὶ ὀλέθρια τὰ μὴ σμικρά. ὅσα
κρίσιμα σημεῖα γινόμενα, τὰ αὐτὰ ταῦτα μὴ γινόμενα
δύσκριτα· τὰ ἐναντία δὲ σημαίνοντα κακὸν οὐ μόνον
ἢν παλινδρομέῃ ἀλλὰ καὶ τῆς ἀρχαίης φύσιος τὰ
ἐναντία ῥέποντα, ὥσπερ καὶ τῶν κακῶν σημεῖον ἐπὶ
τὰ ἐναντία ῥέποντα. θεωρεῖν δὲ οὕτω δεῖ· χρωμάτων,[12]
ξυμπτωσίων φλεβῶν, ὄγκων[13] ὑποχονδρίων, ἀναρρο-
πιῶν,[14] καταρροπιῶν·[15] πολλὰ δὲ καὶ τῶν τοιούτων,
οἷον ἀποφθειρουσέων οἱ τιτθοὶ προσισχναίνονται·
οὐδὲ γὰρ ἐναντίον. οὐδὲ βῆχες χρόνιαι ὅτι ὄρχιος
οἰδήσαντος παύονται· ὄρχις οἰδήσας ὑπὸ βηχωδέων[16]
ὑπόμνημα κοινωνίης στηθέων, μαζῶν, γονῆς, φωνῆς.

7. Ἀποστάσιες ἢ διὰ φλεβῶν ἢ τόνων ἢ δι᾽ ὀστέων
ἢ νεύρων ἢ δέρματος ἢ ἐκτροπέων ἑτέρων· χρησταὶ δὲ
αἱ κάτω τῆς νούσου οἷον κιρσοί, ὀσφύος βάρεα. ἐκ
78 τῶν ἄνω ἄρισται δὲ μάλιστα | κάτω,[17] καὶ αἱ κατω-
τάτω κοιλίης, καὶ προσωτάτω ἀπὸ τῆς νούσου, καὶ αἱ

---

[11] ταχέως V   [12] βρωμάτων HIR
[13] ὄγκοι VI   [14] ἀναρροπίαι V   [15] καταρροπίαι V

that come quicker are shorter in duration, the slower ones last longer; also within periods, what appears earlier indicates another subtle progression of the disease: those who die suddenly have swifter crises, because the suffering is quick, continuous and strong. But if things are going to turn to the better at crisis, let them not appear straight away. Critical symptoms without a crisis: some are fatal, some indicate a bad crisis. Those that appear in advance of a crisis: if there is a crisis, there is relapse; otherwise, lack of a crisis. These too may be deadly if they are not small. All signs which are critical when they occur also indicate bad crises when they are absent. Things that give contrary signs are bad, not only when they run backwards, but when they incline to the opposite of the original nature; just so, things inclining the opposite way are a sign of trouble. One should observe in this way: colors, collapse of blood vessels, swelling of the hypochondria, upward tendencies, downward tendencies. There are many things of the kind, such as withering of the breasts in women who are going to abort. This is not contrary, nor that chronic coughs stop when a testicle swells. When a testicle swells from a cough it is a reminder of the relationship of chest, breasts, genitals, voice.

7. Apostases, through the blood vessels, the nerves, the bones, the tendons, the skin or other diversions. They are best when they go down from the disease, like varicose veins, heaviness of the loins. From the upper parts the best are the farthest below, those of the lowest intestine and farthest from the disease; also the ones that come by outflow,

---

κατ᾿ ἔκρουν, οἷον αἷμα ἐκ ῥινέων, πύον ἐξ ὠτός,
πτύαλον, οὖρον κατ᾿ ἔκρουν. οἷσι μὴ ταῦτα, ἀποστάσιες, οἷον ὀδόντες, ὀφθαλμοί, ῥίς, ἱδρώς. ἀτὰρ καὶ
τὰ ὑπὸ δέρμα ἐς τὸ ἔξω ἀφιστάμενα[18] φύματα οἷον
ταγγαί,[19] καὶ τὰ ἐκπυοῦντα οἷον ἕλκος καὶ τὰ τοιαῦτα
ἐξανθήματα, ἢ λόποι ἢ μάδησις τριχῶν, ἀλφοί, λέ-
πραι ἢ τὰ τοιαῦτα. ὅσα ἀποστάσιες μέν εἰσιν ἀθρόως
ῥέψασαι[20] καὶ μὴ ἡμιρρόπως[21] καὶ ὅσα ἄλλα εἴρηται
κακὰ ἢν ἀναξίως[22] τῆς περιβολῆς τῆς νούσου, οἷον τῇ
Τιμένεω ἀδελφιδῇ ἐκ νούσου ἰσχυρῆς ἐς δάκτυλον
ἀπεστήριξεν, οὐχ ἱκανὸν δέξασθαι τὴν νοῦσον·
ἐπαλινδρόμησεν, ἀπέθανεν. ἀποστάσιες ἢ διὰ φλεβῶν
ἢ διὰ κοιλίης ἢ διὰ νεύρων ἢ διὰ δέρματος ἢ κατὰ
ὀστέα ἢ κατὰ τὸν νωτιαῖον ἢ κατὰ τὰς ἄλλας ἐκροάς,
στόμα, αἰδοῖον, ὦτα,[23] ῥῖνας. ἐξ ὑστέρης ὀκταμήνῳ τὰ
τῶν κρίσεων[24] τῇ ὑστεραίῃ ὡς ἂν ἐς τὴν ὀσφῦν ἢ ἐς
τὸν μηρόν. καὶ ἐς ὄρχιας ἔστι δ᾿ ὅτε ἐκ βηχέων, καὶ
ὄρχις αὐτὸς ἀφ᾿ ἑωυτοῦ. βηχώδεις ἀποστάσιες αἱ μὲν
ἀνωτέρω τῆς κοιλίης οὐχ ὁμοίως τελέως ῥύονται. αἱμορραγίαι λάβροι ἐκ ῥινῶν ῥύονται πολλά, οἷον τὸ
Ἡραγόρεω· οὐκ ἐγίνωσκον οἱ ἰητροί. |

80   8. Τὰς φωνὰς οἱ τρηχέας φύσει ἔχοντες, καὶ αἱ
γλῶσσαι ὑποτρηχεῖς, καὶ ὅσαι τραχύτητες ὑπὸ νούσων ὡσαύτως· αἱ οὖν ἐοῦσαι σκληραὶ τῇ φύσει καὶ

[18] ἀφιστ. ἐς τὰ ἔξω HIR
[19] γαγγαὶ V
[20] ῥεύσασαι HIR

24

as blood from nostrils, pus from ear, expectoration, urine in its outflow. For those who lack these, expect apostases, for example, in teeth, eyes, nose, sweat. Also swellings under the skin which push out, for example scrofulous tumors and suppurations like ulcers, and similar eruptions, or peeling skin, loss of hair, leprous skin, scaly skin, or the like. All abscessions inclining in a mass, not gradually, and the others that have been described, are bad if inappropriate for the compass of the disease, as with Timenes' niece: from a strong disease it settled in her toe, which was not adequate to receive the disease. It ran back up, and she died.[2] Apostases are either through the blood vessels, or the intestine, or the tendons, or the skin, or along the bones or along the spine, or through other exits, mouth, genitals, ears, nostrils. From the womb, in the case of the eighth-month child, the critical flows are on the next day, as it were to the loins or the thigh. Also to the testicles sometimes from coughs. And the testicle itself from itself. Apostases from coughs, when they are above the intestines, do not cure completely in the same way. Violent nosebleeds cure many ills, as in the case of Heragoras. The physicians were unaware of it.

8. There are people who by nature have rough voices, and whose tongues are somewhat rough, and there is roughness from disease just so. Voices rough by nature are

---

[2] This case is described at *Epid.* 4.26.

---

[21] edd.: ἡμίρροπος mss.　　　[22] Smith from Gal. ("schädlich, wenn sie nicht entsprechend"): καὶ ἦν μὴ ἀναξίως mss.
[23] μάζους Gal.　　　[24] κίρσων (v.l.) Gal.

ἄνοσοι τοῦτ' ἔχουσιν· αἱ[25] δὲ μαλακαὶ καὶ βραδύτεροι
ἐς ἀμαρτωλίην· ἢ[26] χρηστὸν ἀρχαίη φύσις. σκεπτέον
καὶ τὰ ἀπὸ τῶν διαιτέων τὰ μακροκέφαλα καὶ μακραύ-
χενα ἀπὸ τῶν ἐπικυψίων· καὶ τῶν φλεβῶν ἡ εὐρύτης
καὶ παχύτης ἀπὸ τοῦ αὐτοῦ, καὶ στενότητες[27] καὶ
βραχύτητες καὶ λεπτότητες ἀπὸ τῶν ἐναντίων· ὧν αἱ
φλέβες εὐρεῖαι, καὶ αἱ κοιλίαι, καὶ τὰ ὀστέα εὐρέα·
εἰσὶ δὲ οὗτοι λεπτοί,[28] οἱ δὲ πίονες τἀναντία τούτων·
καὶ ἐν τοῖσι λιμαγχικοῖσιν αἱ μετριότητες ἀπὸ τούτων
σκεπτέαι. αἱ[29] προαυξήσιες ἑκάστῳ ἃ μειοῦσι, καὶ αἱ
μειώσιες ἃ προαυξοῦσι, καὶ τῇσι προαυξήσεσιν[30]
ὁποῖα ξυμπροαύξεται καὶ ὁποῖα συγκρατύνεται, καὶ
διασφύξιες[31] ποῖαι κοιναὶ τῶν φλεβῶν.

9. Αἱ τῶν ἤτρων ῥήξιες αἱ μὲν περὶ ἥβην τὰ
πλεῖστα ἀσινεῖς τὸ[32] παραυτίκα, αἱ δὲ σμικρὸν ἄνωθεν τοῦ ὀμφαλοῦ ἐν δεξιᾷ αὗται ὀδυνώδεις[33] καὶ
ἀσώδεις καὶ κοπριήμετοι, οἷον καὶ τὸ Πιττακοῦ. |
γίνονται δὲ αὗται ἢ ἀπὸ πληγῆς ἢ σπάσιος ἢ ἐμπη-
δήσιος ἑτέρου,[34] οἷσι τὸ μεταξὺ τοῦ ἤτρου καὶ τοῦ
δέρματος ἐμφυσᾶται καὶ οὐ καθίσταται.

10. Τὸ τῶν χροιῶν οἷον τὸ Πολυχάρου,[35] < . . . > τό
τε ἐκ λευκοχρόου ὅτι ἀπὸ τοῦ ἥπατος πᾶν τὸ τοιοῦτον,

---

25 Gal.: οἶσι mss.   26 Smith from Gal.: εἰ (v.l.) Gal.: ἢ
mss.   27 add. καὶ πλατύτητες HIR
28 οἱ λεπτοί HIR   29 αἷ V
30 προαύξεσιν mss.: corr. Corn.   31 Smith from Gal.:
-σφαξ- mss.: σφιγξ- (v.l.) Gal.   32 τὰ V
33 ὀδυν. αὗται HIR   34 ἐντέρου (v.l.) Gal.

so inclined even without disease. Those with soft voices are slower to show problems; the original nature is the effective thing. One must consider also what comes from way of life, e.g., the long heads and the long necks from bending. The breadth of the blood vessels and their thickness come from the same causes, and their narrowness, shortness, thinness from the opposite ones. People whose vessels and intestines are wide also have broad bones. These people are thin, and fat ones have opposite characteristics. In those who are wasted with hunger, normalcy must be thought about on the basis of this. For each thing consider what its growth reduces and what its reduction increases; and for the increases what kinds of things are increased with them and what kinds are suppressed.[3] And throbbing, what kind is common for the blood vessels.

9. Breaks in the peritoneum: around the genitals, they are mostly harmless for the time being; slightly above the navel to the right they are painful and produce nausea and vomiting of feces, as in the case of Pittacus. They come from blows, stretching, or someone jumping on the patient. Those cases have swelling between peritoneum and skin which is not stable.

10. Colors, like those of Polychares: ⟨yellowish, greenish,⟩ dead white are to be observed, since everything of

[3] Galen suggests that this refers to castration: if the testicles flourish, other things are suppressed, and vice versa. But perhaps it refers only to the bones and vessels mentioned just above.

---

[35] πουλοχάριον mss.: I correct this on the basis of Galen's commentary, and indicate a hiatus for his "weisse, grüne und gelbe," i.e., something like ὑπόχλωρον, ὑπόξανθον

27

καὶ ἀπὸ τούτου ἡπατικὰ νοσήματα, ἐν τούτοισι καὶ
ἴκτεροι οἱ ἀπὸ ἥπατος ἐς τὸ ὑπόλευκον, καὶ οἱ ὑδαται-
νόμενοι καὶ οἱ λευκοφλέγματοι· οἱ δὲ ἀπὸ σπληνὸς
μελάντεροι. καὶ ὕδρωπες καὶ οἱ ἴκτεροι καὶ αἱ δυσελ-
κίαι τῶν ἐκλεύκων τῶν ὑποφακωδέων, καὶ τὸ δέρμα
καταρρήγνυται καὶ τὰ χείλεα, οἷος Ἀντίλοχος καὶ
Ἀλεύας· τὸ ἀπὸ τῶν χυμῶν τῶν ἐκ τοῦ σώματος τοῦ
ἁλμώδεος· ὅτι ὑπὸ τὸ δέρμα μάλιστα καὶ ἀπὸ τῆς[36]
κεφαλῆς ὅταν ἀπὸ τοῦ πλεύμονος διαθερμαίνηται.

11. Τὰς ἀφορμὰς ὁπόθεν ἤρξατο κάμνειν σκεπτέον,
εἴτε κεφαλῆς ὀδύνη εἴτε ὠτὸς εἴτε πλευροῦ. σημεῖον δὲ
ἐφ᾽ οἷσιν ὀδόντες καὶ ἐφ᾽ οἷσι βουβῶνες. τὰ γενόμενα
ἕλκεα καὶ κρίνοντα πυρετούς, καὶ φύματα· οἷσι ταῦτα
μὴ παραγίνεται ἀκρισίη· οἷσιν ἐγκαταλείπεται βεβαιόταται ὑποστροφαὶ καὶ τάχισται.

12. Τὰ ὠμὰ διαχωρήματα καὶ ὑγρὰ κέγχρος στε-
ρεὸς ἐν ἐλαίῳ ἑφθὸς ἵστησιν, οἷον τὸ ναυτοπαίδιον καὶ
ἡ μυριοχαύνη.|

ΤΜΗΜΑ ΔΕΥΤΕΡΟΝ

84   1. Γυνὴ ἐκαρδιάλγει καὶ οὐδὲν καθίστατο πλὴν[37] ἐς
ῥοιῆς χυλὸν ἄλφιτον ἐπιπάσσουσα. καὶ μονοσιτίη
ἤρκεσε, καὶ οὐκ ἀνήμει οἷα τὰ[38] Χαρίωνος.

[36] om. HIR
[37] Gal. comm. Gal. *Al. Fac.*: πάλαιον V: πάλην HIR
[38] οἷον τὰ V

this kind comes from the liver. Thence come hepatic diseases and among them are the jaundices from the liver that tend to whiteness, and dropsical diseases and leucophlegmatic ones. Jaundices from the spleen are darker. Dropsy, jaundice, and a tendency toward sores appear in those of a whitish lentil color: the skin cracks, and the lips, as in Antilochus and Aleuas. What comes from the humors of the body that is salty is to be observed: this mostly comes under the skin and down from the head when it is heated from the lung.

11. The point of departure should be studied whence the patient began to be ill, whether pain is of head or ear or side. It is a sign in those who have tooth problems and those who have swellings in the groin. Developing sores and swellings bring crises to fevers; those in which they do not occur are without crisis, those in which they persist have the surest and quickest relapses.[4]

12. Raw and liquid feces are fixed by hard millet boiled in olive oil, as in the case of the sailor boy and the silly woman.[5]

SECTION 2

1. A woman had heartburn. Nothing relieved it except sprinkling barley meal in pomegranate broth. She survived on one meal daily. Her vomitus was unlike that of Charion.

[4] Cf. *Epid.* 6.3.21.
[5] Possibly a proper name, Myriochaune.

ΕΠΙΔΗΜΙΑΙ

2. Αἱ μεταβολαὶ ὠφελέουσιν ἢν μὴ ἐς πονηρὰ μεταβάλλῃ, οἷον ἀπὸ φαρμάκων ἐμέουσι πυρετῶν ἕνεκα· αἱ ἐς ἀκρητέστερα τελευτὴν[39] σῆψιν σημαίνουσιν, οἷον Δεξίππῳ.

3. Ἡ Σεράπις[40] ἐξ ὑγρῆς κοιλίης ᾤδησεν· κνησμοὶ[41] δ' οὐκ οἶδα ποσταίῃ· οὐ πρόσω· ἔσχε δ' ἔτι[42] καὶ ἀπόστημα ἐν κενεῶνι, ὅπερ μελανθὲν ἀπέκτεινεν.

4. Καὶ ἡ Στυμάργεω ἐκ ταραχῆς ὀλιγημέρου πολλὰ
86 στήσασα | καὶ παιδίου μετὰ στάσιν θήλεος ἀποφθορῆς, τετράμηνον ὑγιήνασα ᾤδησεν.

5. Μόσχῳ λιθιῶντι ἰσχυρῶς ἐπὶ τῷ βλεφάρῳ τῷ ἄνω κριθὴ ἐγένετο πρὸς τοῦ ὠτὸς μᾶλλον, ἔπειτα ἐξηλκώθη ἔσω· πέμπτῃ καὶ ἕκτῃ ἔσωθεν πύον ἐρράγη, τὰ κάτωθεν ἔλυσεν βουβὼν παρ' οὖς· ἦν καὶ κάτω ἐπὶ τῷ τραχήλῳ κατ' ἴξιν τοῦ ἄνω βουβῶνος.

6. Ὁ τῆς Ἀρισταίου γυναικὸς ἀδελφὸς χλιαινόμενος ἐταλαιπώρει ὁδῷ, κἄπειτα ἐν κνήμῃ τέρμινθοι ἐγένοντο· ἔπειτα ξυνεχὴς πυρετὸς ἐγένετο, καὶ τῇ ὑστεραίῃ ἱδρὼς ἐγένετο, καὶ τὰς ἄλλας τὰς ἀρτίους ἐγένετο αἰεί. ἔτι δὲ ὁ πυρετὸς εἶχεν· ἦν δὲ ὑπόσπληνος, ἡμορράγει ἐξ ἀριστεροῦ πυκνὰ κατ' ὀλίγον· ἐκρίθη. τῇ ὑστεραίῃ ἀριστερὸν παρ'[43] οὖς οἴδημα· τῇ δὲ ὑστεραίῃ καὶ παρὰ δεξιόν, ἧσσον δὲ τοῦτο· καὶ ἐπεχλιαίνετο· ταῦτα κατεμωλύνθη καὶ οὐκ ἀπεπύησεν.

[39] Gal.: τελευταὶ mss.
[40] VR Gal.: -ες HI: Σεράπους Gal.Gloss.
[41] σκνησμοὶ V

30

2. Changes help unless there is a change toward something bad. As in those who vomit from drugs in fever: if the alteration at the end tends towards undigested matter it indicates sepsis, as with Dexippus.

3. Serapis swelled up from a moist intestine. Itching after I don't know how many days; no further development. She also had an abscess in the flank which became black, and she died.

4. The wife of Stymarges was constipated after an upset of a few days' duration. Having, after the constipation, aborted a girl child, she was well for four months, then suffered from swelling.

5. Moschus, who was severely ill with stone, got a sty on the upper eyelid towards the ear. Then it ulcerated inside. In the fifth and sixth days pus broke forth inside. The matter beneath was relieved by a swollen gland near the ear, and there was one on the neck in line with the swollen gland above.

6. The brother of Aristaeus' wife, who had a mild fever, grew fatigued on a journey. He developed pustules called terebinths on his calf. Then there was a continuous fever and sweating on the next day, and it returned on all subsequent even days. The fever continued. His spleen swelled. There was hemorrhage from the left nostril, frequent but not of great quantity. Then a crisis. On the following day swelling by the left ear. On the following by the right also, but smaller. The fever increased. The swellings withered away, and did not suppurate.

---

42 Langholf: δέ τι mss.
43 ἀριστερὸν παρ' om. V

7. Ὁ παρ᾽ Ἀλκιβιάδεω ἐλθὼν ἐκ[44] πυρετῶν ὀλίγον[45] πρὸ κρίσιος ὄρχις ἀριστερὸς ᾤδησεν· ἦν δὲ σπλῆνα μέγαν ἔχων καὶ ἀεὶ ἔχων. καὶ δὴ τότε ἐκρίθη | ὁ πυρετὸς εἰκοσταῖος· κἄπειτα ὑπεχλιαίνετο ἄλλοτε καὶ ἄλλοτε καὶ ἔπτυεν ὑπάνθηρον.

88

8. Ἦι ἡ χεὶρ ἡ δεξιὴ σκέλος δὲ ἀριστερὸν ἐκ τῶν βηχωδέων, βραχὺ οὐκ ἄξιον λόγου βηξάσῃ, παρελύθη παραπληγικῶς, ἄλλο δὲ οὐδὲν ἠλλοιώθη οὔτε πρόσωπον οὔτε γνώμην, οὐ μὴν ἰσχυρῶς· ταύτῃ ἐπὶ τὸ βέλτιον ἤρξατο χωρεῖν περὶ εἰκοστὴν ἡμέρην· σχεδὸν ἐγένετό οἱ περὶ γυναικείων κατάρρηξιν, καὶ ἴσως τότε πρῶτα γινόμενα, παρθένος γὰρ ἦν.

9. Ἀπήμαντος καὶ ὁ τοῦ τέκτονος πατὴρ τοῦ τὴν κεφαλὴν κατεαγέντος καὶ Νικόστρατος οὐκ ἐξέβησσον· ἦν δ᾽ ἑτέρωθι κατὰ νεφροὺς ἀλγήματα. 9b. Ἐρωτήματα· εἰ ῥήιον[46] ἀεὶ πληροῦσθαι ποτοῦ ἢ σίτου.

10. Ὀδύνας τὰς ἰσχυροτάτας ὅτῳ τρόπῳ γνοίη[47] ἄν τις· ἴδιος[48] φόβος, αἱ εὐπορίαι,[49] αἱ ἐμπειρίαι, καὶ αἱ δειλίαι.[50]

11. Ὕδωρ τὸ ταχέως θερμαινόμενον καὶ ταχέως ψυχόμενον αἰεὶ κουφότερον. τὰ βρώματα καὶ τὰ πόματα πείρης δεῖ εἰ ἐπὶ τὸ ἴσον μένει.

---

[44] ὁ ἐκ V
[45] Smith from Gal. ("kurz"): ὀλίγων mss.
[46] εἰ ῥήιον Gal: εἴρεον ἀεὶ V: ἤρεον γὰρ αὐτοὺς ἀεὶ HIR: ῥᾶον *Aph.*
[47] διαγνοίη HIR
[48] Gal.: ἴδων mss.

7. The man who came from Alcibiades, after fever, swelled in the left testicle shortly before the crisis. He had an enlarged spleen which he had had always, and which was then troublesome. The fever reached a crisis on the twentieth day. Afterwards there was low fever from time to time and slightly colored expectoration.

8. The girl whose right arm and left leg were paralyzed after a cough (the cough was brief and insignificant) had no other change in aspect or intelligence, at any rate nothing extreme. She began to change for the better around the twentieth day. It occurred about the time her menses broke forth, which perhaps occurred then for the first time, since she was a maiden.

9. Neither the father of Apamas (the carpenter whose head was cracked) nor Nicostratus came down with coughing; but they had pains by the kidneys on both sides. 9b. Question: is it easier always to satiate with food or with drink?

10. How can one recognize very serious pains? Peculiar fear, simple treatments, experiences, cowardice.[6]

11. Water: the quickest to heat and cool is always lighter.[7] We need experience as to whether food and drink have equal staying power.

[6] The meaning here is not clear. The two parts of ch. 10 may not be related.

[7] *Aph.* 5.26 virtually repeats this.

[49] Gal. Heraclides (Gal.): εὐφορίαι mss. Capito (Gal.)
[50] after δειλίαι Gal. repeats "ἐρωτήματα . . . σίτου" (= 9b)

12. Ῥητέον ὅτι αἵματος ῥυέντος ἐκχλοιοῦνται καὶ ὅσα ἄλλα τοιαῦτα, ὅτι πρὸς τὸ ὑγραίνειν καὶ ξηραίνειν καὶ θερμαίνειν καὶ ψύχειν πολλὰ ἄν τις τοιαῦτα[51] εἴποι. |

90   13. Τὸ ἐξηκονθήμερον ἀπόφθαρμα ἄρσεν, τόκων μετ᾽ ἐπίσχεσιν,[52] ὑγιηρόν.

14–15. Ἡράκλεια[53] ᾧδε τοῦ κακοῦ ὀγδοαίῳ, δυσεντεριώδης. ἆρ᾽ ὅτι ἦν καὶ τεινεσμώδης[54]

16. Θηλάζουσα εἶτα ἐκθύματα ἀνὰ τὸ σῶμα πάντη εἶχεν, ἐπεὶ ἐπαύσατο θηλάζουσα κατέστη, θέρεος.

17. Τῇ τοῦ σκυτέως ὃς τὰ σκύτινα ἐποίησε τεκούσῃ, καὶ ἀπολυθείσῃ τελέως ἐδόκει, τοῦ μὲν χορίου τι[55] τὸ ὑμενοειδὲς ἄπεσχεν·[56] ἀπῆλθε τεταρταίη κακῶς·[57] στραγγουριώδης ἐγένετο[58] αὐτίκα συλλαμβάνουσα. ἔτεκεν ἄρσεν·[59] πολλὰ δὲ ἔτεα ἤδη εἶχε· τὰ ὕστατα οὐδὲ[60] ἐπιμήνια ἤει· ὅτε δὲ τέκοι διέλιπεν ἐπ᾽ ὀλίγον[61] ἡ στραγγουρίη.

18. Ἰσχίον δέ τις ἤλγει πρὶν ἴσχειν· ἐπεὶ δὲ ἔσχεν οὐκ ἔτι ἤλγει. ἐπεὶ δὲ ἔτεκεν, εἰκοσταίη ἐοῦσα, αὖθις ἤλγησεν· ἔτεκεν μέντοι[62] ἄρσεν. 18b. Ἐν γαστρὶ ἐχούσῃ ἐν κνήμῃ κάτω δεξιῇ ἢ τρίτῳ ἢ τετάρτῳ μηνὶ

[51] HIR omit ὅτι . . . τοιαῦτα, which V and Galen have.

[52] μετ᾽ ἐπίσχεσιν Smith: "nach Verhinderung" Gal.: ἐπίσχεσις Heraclides (Gal.): ἐν ἐπισχέσεσιν mss.

[53] Ἡρακλείδης (v.l.) Gal.: Ἡρακλείδες V: Ἡρακλεῖ HIR

[54] ὅστις ἄρα καὶ τεινεσ. HIR   [55] om. V

[56] ἔπεσχεν V

[57] "so wie es sein soll" (καλῶς?) Gal.

12. One must say that in hemorrhage patients develop a greenish color, and one can find many other such things related to wetness and dryness, to hotness and cold.

13. The sixty-day-old male fetus, aborted after the delay of childbirth, was healthy.

14–15 Heracleia: in the eighth month (her abortion occurred) with an evil odor; mostly she was dysenteric—because there was tenesmus too?[8]

16. The nursing woman got swellings all over the body. When she stopped nursing they disappeared, in summer.

17. The wife of the leatherworker who made my shoes, having given birth, thought she had been completely delivered. She retained some tissues of the afterbirth; they came out on the fourth day with difficulty. As soon as she conceived she developed strangury. She delivered a male child. She was of rather advanced age; she had not even menstruated in the most recent period. But when she gave birth, the strangury was relieved for a time.

18. A woman was pained in the hips before she conceived. When she conceived the pain disappeared. But when she gave birth (at age 20) it again commenced. The baby was a boy. 18b. A woman carrying in the third or fourth month had eruptions on the lower left leg and on

---

8 This passage was quite obscure and corrupt before Galen's time. I offer his interpretation except at the end, where, instead of the question, he says she did not have hemorrhaging.

---

58 γὰρ ἐγένετο HIR   59 δ' ἄρσεν V
60 καὶ οὐδ' HIR
61 ὀλίγον χρόνον V
62 οὖν HIR

ἐξανθήματα πρὸς ἃ τῇ μάννῃ χρώμεθα, καὶ ἐν χειρὶ
92 δεξιῇ παρὰ μέγαν | δάκτυλον· οὐκ οἶδ᾽ ὅτι ἔτεκε,
κατέλιπον γὰρ ἐξάμηνον· ᾤκει, ὡς[63] ἐγῷμαι, ὡς τὰ
Ἀρχελάου[64] πρὸς τῷ κρημνῷ.

19. Ἡ Ἀντιγένεος ἡ τῶν περὶ Νικόμαχον ἔτεκε
παιδίον, σαρκῶδες μέν, ἔχον δὲ τὰ μέγιστα διακεκρι-
μένα, μέγεθος δὲ ὡς τετραδάκτυλον, ἀνόστεον, ὕστερον δὲ παχὺ στρογγύλον· αὕτη δὲ ἀσθματώδης ἐγένετο πρὸ τοῦ τόκου· ἔπειτα ἅμα τῷ τόκῳ πύον ἀνήμεσεν
ὀλίγον, οἷον ἐκ δοθιῆνος.

20. Θυγατέρας τεκούσης διδύμους, καὶ δυστοκη-
σάσης καὶ οὐ πάνυ[65] καθαρθείσης, ἐξῴδησεν ὅλη·
ἔπειτα ἡ γαστὴρ μεγάλη ἐγένετο, τὰ δ᾽ ἄλλα ἐταπει-
νώθη· καὶ ἐρυθρὰ ᾔει μέχρι τοῦ ἕκτου μηνός, ἔπειτα
λευκὰ[66] κάρτα πάντα ἤδη τὸν χρόνον· πρὸς δὲ τἀφροδίσια οἱ ῥόοι ἔβλαπτον,[67] καὶ οἱ ἄκρητα ἐρυθρὰ ἱκνευμένως ᾔει.[68]

21. Τῇσι χρονίῃσι λειεντερίῃσιν ὀξυρεγμίη γενομένη, πρόσθεν μὴ[69] γενομένη, σημεῖον χρηστόν, οἷον
Δημαινέτῃ ἐγένετο· ἴσως δ᾽ ἐστὶ καὶ τεχνήσασθαι· καὶ

[63] δὲ HIR
[64] Ἀχελώου V
[65] πάντῃ HIR
[66] λεπτὰ Gal.
[67] πρὸς δὲ τἀφροδίσιον αἱ οὐραὶ ἔβλεπον Gal.
[68] καὶ . . . ᾔει om. Gal.: ἱκνεύμενα HIR: ἱκνευμένως I (above the line), V
[69] μηδέποτε HIR

the right hand next to the thumb, eruptions of the sort for which we give frankincense. I do not know what her baby was. I left her six months pregnant. She lived, I believe, in Archelaus' property near the cliff.

19. The wife of Antigenes, of Nicomachus' house, produced a child that was flesh with the largest limbs distinguished, about eight centimeters in breadth, boneless, a thick globular exterior. She became asthmatic before the birth. At the time of the birth she brought up a small quantity of pus, as though from a small abscess.

20. A woman gave birth to twin daughters. She had trouble at the birth, the purgation was incomplete. She swelled up all over. Then her belly became large and the rest of her went down. The discharge came out red up to the sixth month, then very white continuously from then on. The flows were harmful to sexual relations.[9] Pure red flows of the proper sort did come.

21. When women who are chronically ill with lientery develop acid belching which was not present before, it is a very favorable sign, as it was for Demaenete. It might be

---

[9] Galen says in his commentary (how he claims to know this is unclear, but perhaps through Heraclides) that he is reproducing the original old reading of the manuscripts: πρὸς δὲ τἀφροδίσιον αἱ οὐραὶ ἔβλεπον ("The woman's tail pointed toward the Temple of Aphrodite"), but that commentators changed it to read πρὸς δὲ τἀφροδίσια αἱ οὐραὶ ἔβλεπον ("Her tail inclined toward sexual intercourse"), and Artemidorus Capito altered the text to πρὸς δὲ τἀφροδίσια οἱ ῥόοι ἔβλαπτον ("The flows were harmful to sexual relations"), which is the reading of all the medieval manuscripts. Galen praises the conjecture offered by Heraclides, πρὸς δὲ τἀφροδίσιον αἱ θύραι ἔβλεπον ("The door of her house opened in the direction of the Temple of Aphrodite").

γὰρ αἱ ταραχαὶ αἱ τοιαῦται ἀλλοιοῦσιν·[70] ἴσως δὲ καὶ
ὀξυρεγμίαι λειεντερίην λύουσιν. |

94   22. Ἰήθη ἐλλεβόρου πόσει Λυκίη· τὰ ὕστατα
σπλὴν μέγας καὶ ὀδύναι καὶ πυρετὸς καὶ ἐς ὦμον
ὀδύναι· καὶ ἡ φλὲψ ἡ κατὰ σπλῆνα ἐπ᾽ ἀγκῶνι ἐτέτατο·
καὶ ἔσφυζε μὲν πολλάκις, ἔστι δ᾽ ὅτε καὶ οὔ.[71] οὐκ
ἐτμήθη, ἀλλ᾽ ἅμα ἱδρῶτι διῆλθεν ἢ[72] αὐτόματον, ἔξω[73]
διόντων·[74] ὁ σπλὴν τὰ δεξιὰ ἐνετείνετο, πνεῦμα ἐνεδι-
πλασιάζετο, οὐ μὴν μέγα· παρεφέρετο, περιεστέλλετο·
φῦσα ἐνεοῦσα οὐ διῄει. κάτω οὐδέν, οὐδὲ οὔρει· ἀπέθανε.

23. Πρὸ τοῦ τόκου τὰ ἀμφὶ φάρυγγα ἑτερόρροπα
ὁρμήσαντα οὐκ ἐφηλκώθη. ἐπὶ τὰ ἀριστερὰ μετῆλθεν,
ἐς σπλῆνα ὀδύνη ἦλθεν ἀκρίτως. 23b. Ἱέρωνι ἐκρίθη
πεντεκαιδεκαταίῳ. 23c. Τῇ Κῴου ἀδελφεῇ ἧπαρ ἐπήρθη σπληνικὸν τρόπον· ἀπέθανε.[75] 23d. Βίων ἅμα οὔρει
τε ὑπέρπολυ ἀνυπόστατον καὶ αἷμα ἐξ ἀριστεροῦ· ἦν
γὰρ καὶ ὁ[76] σπλὴν κυρτὸς καὶ σκληρὸς κατ᾽[77] ἄνω·
περιεγένετο· ὑποστροφή.

24. Ἦν δὲ τῶν κυναγχικῶν τὰ παθήματα τάδε·[78]
τοῦ τραχήλου οἱ σπόνδυλοι ἔσω ἔρρεπον τοῖσι μὲν ἐπὶ
96   πλέον τοῖσι δ᾽ ἐπ᾽ | ἔλασσον· καὶ ἔξωθεν[79] ἔνδηλος

---

[70] ἀλύουσιν V   [71] Gal.: om. mss.
[72] om. Gal.   [73] Diller from Gal.: ἐς οὗ mss.
[74] δὲ ἰόντων HIR
[75] ἀπέθανε δευτεραίῃ HIR
[76] om. V
[77] Smith: καὶ mss.

possible to induce it. Such upsets produce change. It may be that acid belching gets rid of lientery.[10]

22. Lycie was treated with a potion of hellebore. Towards the end she had an enlarged spleen, pains, fever, pains towards the shoulder. The blood vessel from the spleen was tense at her elbow. It throbbed frequently, but sometimes did not. No phlebotomy, but it passed with the sweat or spontaneously, the matter passing to the outside. The spleen on its right was stretched tight; breathing doubled in frequency, but without great depth. She became delirious. She was wrapped up.[11] She was full of wind which did not pass. No feces and no urine. She died.

23. Prior to giving birth, affections in the area of the pharynx, inclining to one side or the other, were not ulcerous when they commenced. They moved to the left side. Pain came on the spleen, without crisis. 23b. Hieron's disease reached climax on the fifteenth day. 23c. The sister of the man from Cos had her liver elevated in a splenic manner. She died. 23d. Bion urinated excessively without sediment, and bled from the left nostril. And indeed his spleen was swollen and hard on the top. He survived. There was a relapse.

24. People with cynancus[12] had the following affections: neck vertebrae inclined inward, some severely, some less so; on the outside the neck had a conspicuous hollow

[10] *Aph.* 6.1 seems to have been drawn from this passage.

[11] Or "she suffered general contraction."

[12] The term refers to a severe sore throat that feels like a dog's choke collar.

---

[78] τοιάδε Gal. comm. *Prog.*   [79] ἔσωθεν HIR

ἔγκοιλον[80] ἔχων ὁ τράχηλος· καὶ ἤλγει ταύτῃ ψαυόμε-
νος· ἦν δὲ καὶ κατωτέρω τινὶ τοῦ ὀδόντος καλεομένου,[81] ὃ οὐχ ὁμοίως ὀξύ ἐστιν· ἔστι δ' οἷσι καὶ πάνυ
περιφερὲς μέζονι περιφερείῃ ἢ[82] εἰ μὴ ξὺν τῷ ὀδόντι
καλεομένῳ. φάρυγξ οὐ φλεγμαίνουσα, κειμένη δέ. τὰ
ὑπὸ γνάθους ὀγκηρά, οὐ φλεγμαίνουσιν εἴκελα· οὐδὲ
βουβῶνες[83] οὐδενὶ ἀλλὰ φύσει ἔμενον·[84] καὶ γλῶσσαν
οὐ ῥηϊδίως στρέφοντες, ᾤδησαν ἀλλὰ μέζων τε αὐτοῖ-
σιν ἐδόκει εἶναι καὶ προπετεστέρη· καὶ ὑπὸ γλώσσῃ
φλέβες ἐκφανεῖς. καταπίνειν οὐκ ἐδύναντο, ἢ πάνυ
χαλεπῶς, ἀλλ' ἐς τὰς ῥῖνας ἔφευγεν εἰ πάνυ ἐβιῶντο·[85]
καὶ διὰ τῶν ῥινῶν διελέγοντο. πνεῦμα δὲ τούτοισιν οὐ
πάνυ μετέωρον. ἔστι δ' οἷσι φλέβες αἱ ἐν κροτάφοισι
καὶ ἐν κεφαλῇσι καὶ ἐπ' αὐχένι ἐπηρμέναι. βραχὺ δὲ
τούτων τοῖσι παλιγκοτωτάτοισι[86] κρόταφοι θερμοί,
εἰ[87] καὶ τἆλλα μὴ πυρεταίνοιεν. οὐ πνιγόμενοι οἱ
πλεῖστοι εἰ μὴ καταπίνειν προθυμέοιντο ἢ σίαλον[88] ἢ
ἄλλο τι· οὐδ' οἱ ὀφθαλμοὶ ἐγκαθήμενοι. οἷσι μὲν οὖν
ἦν ἐς ὀρθὸν ἐξόγκωμα μήτε ἑτερόροπον οὗτοι οὔτε
παραπληκτικοὶ ἐγένοντο· ἀπολόμενον[89] δὲ[90] εἴ τινα |
98   εἶδον ἀναμνήσομαι· οὓς δὲ οἶδα νῦν περιεγένοντο. ἦν
δὲ τὰ μὲν τάχιστα ῥηΐζοντα, τὰ δὲ πλεῖστα καὶ ἐς
τεσσαράκοντα ἡμέρας περιήιει· τοῦτο δὲ οἱ πλεῖστοι
καὶ ἄπυροι· πολλοὶ δὲ καὶ πάνυ ἐπὶ πολὺν χρόνον
εἶχον[91] ἔχοντές τι μέρος τοῦ ἐξογκώματος· καὶ κατά-
ποσις καὶ φωνὴ ἐνσημαίνουσα· κίονές τε τηκόμενοι

---

[80] ἦν δῆλος ἐγκοίλως Gal. comm. *Prorrh.*

which was painful to the touch. Sometimes it was lower than the so-called odontoid process (second vertebra), and was less acute, but in some cases very rounded, with a greater circumference than if it was not at the odontoid process. The pharynx was not inflamed, but quiet. The area under the jaw was swollen, but without similar inflammation. The glands were not swollen, but normal. Sufferers could not easily move their tongues, which felt larger and somewhat protruding. And blood vessels under their tongues were very obvious. They could not drink, or only with difficulty. Drink passed into the nostrils if they forced it. And they talked through the nose. But breathing was not excessively shallow. Some had elevated blood vessels in their temples and in the head and neck. To an extent, in those with exceptional malignancy, their temples were hot, though they were not otherwise feverish. Most did not choke, save when they too eagerly swallowed down their saliva or something else. Their eyes were not fixed. People whose swelling was straight on, not on one side, did not develop paralysis, and as for dying, if I saw anyone I will recall it. All I can think of now survived. Some cases were quickly eased, but most extended to forty days. Many had it for a long time, retaining part of the swelling; and there were drinking symptoms and the symptomatic voice,

---

81 καλεομένου ὀστοῦ HIR   82 Gal.: om. mss.
83 βουβῶνας HIR   84 Gal.: μὲν mss.
85 ἐβίαζον HIR: "würgten" Gal.
86 παλιγκοτάτοισι mss.: corr. Li.   87 εἰ Gal.: om. mss.
88 πτύελον HIR   89 Gal.: ἀπολλόμενον mss.
90 τε V   91 om. HIR

μινύθησίν[92] τινα παρεῖχον, πονηρὸν οὐδὲν δοκέοντες κακὸν ἔχειν. οἱ δὲ ἑτερόρροπα ἔχοντες, οὗτοι ὁκόθεν ἂν[93] ἐγκλιθείησαν οἱ σπόνδυλοι αὐτῇ[94] παρελύοντο, τὰ δ' ἐπὶ θάτερα εἵλκετο.[95] ἦν δὲ ταῦτα ἐν προσώπῳ καταφανέα μάλιστα καὶ τῷ στόματι καὶ τῷ κατὰ γαργαρεῶνα διαφράγματι· ἀτὰρ καὶ γνάθοι αἱ κάτω παρήλλασσον κατὰ λόγον. αἱ δὲ παραπληγίαι[96] οὐ διὰ παντὸς τοῦ σώματος ἐγίνοντο οἷον ἐξ ἄλλων, ἀλλὰ μέχρι χειρός. τὰ ὑπὸ τοῦ κυναγχικοῦ οὗτοι καὶ πέπονα ἀναπτύοντες καὶ βραχυμογεῖς ἦσαν· οἷσι δ' ἐς ὀρθὸν καὶ ἀπέπτυον· οἷσι δὲ καὶ ξὺν πυρετῷ οὗτοι πολλῷ μᾶλλον καὶ δύσπνοοι καὶ διαλεγόμενοι σιαλοχόοι[97] καὶ φλέβες τούτοισι μᾶλλον ἐπηρμέναι· καὶ πόδες πάντων μὲν ψυχρότατοι τούτων δὲ μάλιστα· καὶ ὀρθοστατεῖν οὗτοι ἀδυνατώτεροι καὶ οἵτινες[98] μὴ αὐτίκα ἔθνησκον· οὓς δὲ ἐγὼ οἶδα πάντες ἔθνησκον. |

ΤΜΗΜΑ ΤΡΙΤΟΝ

100   1. Ἐς Πέρινθον περὶ ἡλίου τροπὰς ὀλίγον τὰς θερινὰς ἤλθομεν. ἐγεγόνει δὲ ὁ χειμὼν εὔδιος νότιος· τὸ δὲ ἔαρ καὶ τὸ θέρος πᾶν ἄνυδρον μέχρι Πληϊάδων δύσιος· εἰ γάρ τι καὶ ἐγένετο, ὅσον[99] ψεκάς· καὶ ἐτησίαι οὐ κάρτα ἔπνευσαν, καὶ οἱ πνεύσαντες διεσπασμένως. τοῦ θέρεος καῦσοι ἐπεδήμησαν πολλοί· ἦσαν δὲ ἀνήμετοι· κοιλίαι[100] ταραχώδεις λεπτοῖσιν, ὑδατώδε-

----

[92] μινυθεῖσι V   [93] ἄλλη V

42

and the uvula shriveled and showed withering, though it appeared to have no ill effects. Those whose swellings were on one side were paralyzed on whichever side the vertebrae inclined to, and were drawn up on the other. This was evident in the face especially, and in the mouth, and the division below the uvula. The lower jaw was deviant proportionately. The paralysis did not occur over the whole body, as in other diseases, but only as far as the hand. When expectorating even ripened material from the cynancus these produced little with difficulty, while those with the affection in the center spat it right out. Those with fever had much more difficulty with breathing and drooled when they talked, and their blood vessels were more elevated. All had very cold feet, but these last especially. They had more difficulty standing erect, even when they did not die immediately. All I am aware of did die.

SECTION 3

1. We arrived in Perinthus in summer, near the solstice. The winter had been mild, southerly. Spring and summer were quite dry until the setting of the Pleiades. If there was rain it was showers. The Etesian winds hardly blew. The winds that did were intermittent. Causus (burning fever) was epidemic in the summer. Patients were without vomiting, intestines were upset, with light, watery, unbilious,

---

94 HI: αὐτοὶ R: αὕτη V   95 εἵλκοντο HIR

96 παράπληγαι mss.: corr. Li.   97 Gal. cit. in Loc. Aff.: διαλεγομένοισιν ἀλλοχόοι mss.: "war bitter" (v.l.) Gal. comm. (= σίαλα ὄξεα?)   98 εἴ τινες HR   99 ἦν ὅσον HIR

100 καὶ κοιλίαι HIR

σιν, ἀχόλοισιν ἐπάφροισι πολλοῖσιν, ἴσχοντα[101]
ἔστιν ὅτε καὶ ὑπόστασιν τεθέντα, ἐξ οἵων δὴ καὶ
ἐξαιθριαζομένων τὸ εἴκελον ἰσατώδει διαχωρήματι,
διὰ παντὸς κακῶν.[102] ἐν τούτοισι πολλοὶ κωματώδεις
ἦσαν καὶ παράφοροι· οἱ δὲ ἐξ ὕπνων τοιοῦτοι ἐγίνοντο, ὅτε δὲ ἐγερθεῖεν κατενόουν πάντα. πνεύματα
μετέωρα, οὐ μὴν πάνυ· οὖρα λεπτὰ μὲν τοῖσι πλείστοισι καὶ ὀλίγα, ἄλλως δὲ οὐκ ἄχροα. αἱμορραγίαι
ἐκ ῥινῶν οὐκ ἐγένοντο εἰ μὴ ὀλίγοισιν, οὐδὲ παρ' ὦτα
εἰ μή τισι περὶ ὧν ὕστερον γράψω.[103] οὐδὲ σπλῆνες
ἐπήροντο, οὐδὲ δεξιὸν ὑποχόνδριον οὐδὲ ἐπώδυνον
κάρτα οὐδὲ ἐντεταμένον ἰσχυρῶς· ἦν δέ τι ἐνσημαῖ-
νον. καὶ μάλιστα ἐκρίνετο πάντα τὰ πολλὰ περὶ τεσσαρεσκαίδεκα, ὀλίγα σὺν ἱδρῶτι, ὀλίγα σὺν ῥίγει, καὶ
πάνυ ὀλίγοισιν ὑποστροφαὶ ἐγίνοντο. ὑπὸ δὲ τὰς
ψεκάδας τὰς γενομένας ἐν τῷ θέρει ἐπεφαίνετο ἱδρώς·
ἐν τοῖσι πυρέττουσι[104] καί τινες αὐτίκα ἱδρῶτες ἐπ'
ἀρχῆς ἐγίνοντο, οὐ μὴν κακοήθως· καὶ | τοῖσιν ὑπὸ
102 τοῦτον τὸν χρόνον καὶ ἐκρίθη ξὺν ἱδρῶτι. ἐγένοντο δὲ
ἐν τοῖσι θερινοῖσι πυρετοῖσι περὶ ἑβδόμην καὶ ὀγδόην
καὶ ἐνάτην τρηχύσματα ἐν τῷ χρωτὶ κεγχρώδεα, τοῖσιν ὑπὸ κωνώπων μάλιστα[105] εἴκελα ἀναδήγμασιν, οὐ
πάνυ κνησμώδεα· ταῦτα διετέλει μέχρι κρίσιος· ἄρσενι δὲ οὐδενὶ εἶδον ταῦτα ἐξανθήσαντα, γυνὴ δὲ
οὐδεμία ἀπέθανεν ᾗ ταῦτα ἐγένετο. ὅτε δὲ ταῦτα ἐγένετο[106] βαρυήκοοί τε ἦσαν καὶ κωματώδεις, πρόσθεν δὲ

[101] ἴσχον τάδ' V

frothy stools frequently; when left to stand they often separated, if they were exposed to the air there was something
blue like the woadlike excrement; they were always foul.
Many patients suffered coma and delirium, some developed it during sleep, but when they woke they were entirely sane. Their breathing was shallow, but not very much
so. Most had thin urine, and little of it, but still not color-
less. There were no hemorrhages from the nose save in
a few cases, no swollen glands by the ears, save for some
that I will describe later. Spleens not elevated, nor was
the right hypochondrium very painful nor stretched very
tight, but it was indicative. Most reached complete crisis at
about fourteen days, a few with sweat, a few with shivering,
and there were relapses for a very few. When there were
showers in the summer, sweat appeared on the fevered patients, and some produced sweat initially—but not in a bad
way. And for those ill at that time the crisis, too, came
through sweat. In the summer fevers there appeared on
the seventh, eighth, and ninth days roughness on the skin,
granulous, very like bites of gnats, not very itchy. They persisted until the crisis. I did not see these eruptions on any
male, and none of the women who had them died. But
when they appeared the patients became hard of hearing and comatose. Women who were going to develop

---

102 ἐξαιθριαζόμενον τὸ ἴκελλον ἴσα τῷ εἴδει διαχωρήματα
κακόν (κακῶν H) HIR

103 παρ' ὦτα . . . γράψω Li.: παρωτάτοισιν ὕστερον γράψω
V: παρωτάτοισιν περὶ ὧν ὕστερον γράψω HIR: "Ebenso war es
mit den Anschwellungen an den Ohren" (om. εἰ μὴ κτλ.) Gal.

104 πυρετοῖσι V   105 γιγνομένοις μάλιστα HIR
106 ὅτε . . . ἐγένετο om. HIR

οὐ κάρτα ἦσαν κωματώδεις ᾗσιν ἔμελλε ταῦτα ἔσε-
σθαι· οὐ μὴν τὸ σύμπαν διετέλεον, κωματώδεις δὲ καὶ
ὑπνώδεις τὸ θέρος καὶ μέχρι Πληϊάδων δύσιος, ἔπειτα
μὴν ἀγρυπνίαι μᾶλλον. ἀτὰρ οὐδὲ τὸ σύμπαν ὑπὸ
τῆς καταστάσιος ταύτης ἔθνησκον. κοιλίην μὲν οὖν
οὐκ ἐνεδέχετο οὐδὲ[107] τοῖσι γεύμασιν ἱστάναι ἀλλὰ
παρὰ λόγον[108] ᾤετο ἄν τις ἰήσασθαι ξυμφέρειν, καίτοι
ὑπέρπολλα ἔστιν οἷσι τὰ διόντα[109] ἦν.

1b. Τὸ[110] ἐν ψύχει κεῖσθαι ἐπιβεβλημένον[111] ὡς
ἕλκη μὲν τὸ ψυχρόν, θάλπῃ δέ, τὸ τοιοῦτον εἶδος ἐκ
προσαγωγῆς ἐστι μᾶλλον. 1c. Καὶ τὸ μηδὲν[112] τῇ
φύσει πάθῃ[113] γίνεσθαι· ἐφ᾽ οἷσί τε[114] καὶ ὁκοῖα τὰ
σημεῖα καὶ πλείω ἢ μείω γινόμενα, χάσμη, βήξ,
πταρμός, σκορδίνημα, ἔρευξις φῦσα· πάντα τὰ τοιαῦτα. 1d. Διαφθείρουσιν ᾗσιν ἐν πυρετοῖσιν, ἀσώδε-
σιν, φρικώδεσιν, ἐρεύθονται πρόσωπα. 1e. Κοπιώδεις,
ὀμμάτων ὀδυνώδεις, καρηβαρίαι, παραπληγίαι· καὶ
γυναικεῖα ἢν ἐπιφαί|νηται, μάλιστα δὲ ᾗσι πρῶτον,
ἀτὰρ καὶ παρθένοισι καὶ γυναιξὶν ᾗσι διὰ χρόνου,
ἀτὰρ καὶ ᾗσι μὴ ἐν ᾧ εἴθισται χρόνῳ ἢ ὡς δεῖ
ἐπιφαίνονται, ἔπειτα ἔξωχροι γίνονται. μέγα δ᾽ ἐν
ἅπασι τὸ καὶ ἑξῆς[115] καὶ ἐν ᾧ χρόνῳ καὶ ἐφ᾽ οἷσιν. 1f.
Τοῖσι πάνυ χολώδεσιν ἐν πυρετοῖσι μάλιστα ὅλως[116]
ἐπὶ σκέλεα ἡ κάθαρσις.

104

---

107 οὐδ᾽ ἐν V   108 παράλογον V
109 τάδ᾽ ἰόντα HIR   110 τῷ HIR
111 ὑποβεβλημένον HIR   112 μηδ᾽ ἐν V

them were not comatose beforehand. And it did not persist throughout. But they were comatose and sleepy through the summer and to the setting of the Plciades, whereupon they were sleepless instead. But generally no one died in this condition. It was not possible to stabilize the intestine even with small amounts of food; it would have been irrational to think treatment helpful; and yet some passed excessive amounts of excrement.

1b.[13] Lying in the cold covered, in order to breathe cool air but to be warmed: this sort of treatment is more gradual.[14] 1c. When the affections do not accord with the patient's nature: look at the circumstances and type of the signs, increasing or becoming fewer: yawning, cough, sneezing, stretching, belching, flatulence, everything of that sort. 1d. Women abort who, along with fevers, nausea, shivering, develop red faces. 1e. Fatigue, eye pains, heaviness in the head, paralyses, and whether menstruation appears, and especially when it is the first occurrence. And both for maidens and women, those for whom it lasts long, those for whom it is at a time different than it ought to be, after which they become very pale. Important in each case: the sequence and the time and circumstances. 1f. For the very bilious, especially in fevers, the purge is generally to the legs.[15]

---

[13] 1b–f are notes not obviously connected to the above.
[14] *Epid.* 6.4.14 seems to give a brief version of this.
[15] Cf. *Epid.* 4.20f, which this passage may refer to.

---

[113] πάθει HIR   [114] ἐφ' οἷς εἴτε V
[115] ἐξ ἧς V
[116] ὅλως οἷσιν mss.: οἷσιν om. Gal., del. Li.

2. Φαρμάκων[117] τρόπους ἴσμεν ἐξ ὧν γίνεται[118] ὁκοῖα ἄσσα· οὐ γὰρ πάντες ὁμοίως[119] ἀλλ᾽ ἄλλοι ἄλλως εὖ κέονται· καὶ ἄλλοθι[120] πρωϊαίτερον ἢ ὀψιαίτερον ληφθέντα· καὶ οἱ διαχειρισμοὶ οἷον ἢ[121] ξηρᾶναι ἢ κόψαι ἢ ἑψῆσαι καὶ τὰ τοιαῦτα, ἐῶ τὰ πλεῖστα·[122] καὶ[123] ὁκόσα[124] ἑκάστῳ, καὶ ἐφ᾽ οἷσι νοσήμασι, καὶ ὁπότε τοῦ νοσήματος, ἡλικίην, εἴδεα,[125] δίαιταν, ὁκοίη ὥρη ἔτεος, καὶ ἥτις καὶ ὁκοίως ἀγομένη, καὶ τὰ τοιαῦτα.[126]

3. Ζωΐλῳ τῷ παρὰ τὸ τεῖχος ἐκ βηχὸς πεπείρης πυρετὸς ὀξὺς καὶ ἔρευθος προσώπου,[127] καὶ κοιλίη ἀπολελαμμένη πλὴν πρὸς ἀνάγκην. πλευροῦ ὀδύνη ἀριστεροῦ καὶ οὖς κατ᾽ ἴξιν ὀδυνῶδες[128] πάνυ καὶ κεφαλῆς. οὗτος οὕτω πτύων[129] διὰ παντὸς ὑπόπυον ἐνόσει. ἀλλὰ τὰ ἄλλα ἐκρίθη, καὶ κατὰ οὖς ἐρράγη 106 πύον πολὺ περὶ | ὀγδόην ἢ ἐνάτην. αἱ δ᾽ ἀρχαὶ τῆς[130] ὀδύνης τοῦ ὠτὸς οὐκ οἶδ᾽ ὅπως ἄνευ ῥίγεος· ἐκρίθη·[131] ἵδρωσε κεφαλὴν κάρτα καὶ οὗτος.

---

117 οὕτως φαρμάκων (v.l.) Gal. comm. Gal. Ther. Pis. (K.14.229f.): φαρμάκων τε mss. (v.l.) Gal.: om. Gal. comm.: φαρμάκων δὲ Gal. Th.P.   118 γεγένηται Gal. Th.P.

119 Gal. comm. Gal. Th.P.: om. mss.

120 συγκεῖνται καὶ ἄλλα ὅσα Gal. comm. Gal. Th.P.

121 om. Gal. Th.P.

122 ἕως τὰ πλεῖστα μειώσει πλείω Gal. Th.P.

123 om. HIR   124 ὁκοῖα Gal. Th.P.

125 εἰδέαν HIR   126 καὶ ὁπότε . . . τοιαῦτα] καὶ ἐφ᾽ ᾗτε τοῦ νουσήματος ἡλικίᾳ ἰδέᾳ, καὶ διαίτῃ ὁκοίᾳ ἢ ὥρῃ ἔτεος ὁκοίως ἄγωμεν, καὶ τὰ τοιαῦτα Gal. Th.P.

2. We know the characteristics of drugs, from what ones come what kinds of things. For they are not all equally good, but different characteristics are good in different circumstances. In different places medicinal drugs are gathered earlier or later; also the preparations differ, such as drying, crushing, boiling, and so on (I pass over most things); and how much for each person and in what diseases and when in the disease, in relation to age, appearance, regimen, what kind of season, what season and how it is developing, and the like.[16]

3. Zoilus, who lived by the wall, after a ripe cough, had acute fever and redness of face, stoppage of the intestines save when constrained. Pain in the left ribs, and the ear and head were very painful on the corresponding side. Thus he continued ill throughout, spitting up somewhat purulent matter. But other symptoms reached a crisis, and much pus broke through by the ear on the eighth or ninth day. The beginning of the pain in the ear was somehow without shivering. He reached crisis. He, too, sweated much about the head.

---

[16] Galen's commentary on this passage is close to our text, but his citation of the passage in *Theriac to Piso* differs considerably in its words, though not in import. It appears that this passage became famous and developed a special text through repeated citations in drug books.

---

127 προσώπου ἔρευθος HIR
128 ὀδυνώδεες mss.: corr. Corn.
129 Gal.: πτύον mss.
130 Galen: τῆς ἐννάτης mss.
131 ἡ κρίσις HIR

4. Ἐμπεδοτίμῃ ξύγκαυσις καὶ ἀριστεροῦ πλευροῦ ἄνω ἅμα ᾠτὶ ὀδύνη,[132] μάλιστα κατ᾽ ὠμοπλάτην ἀτὰρ καὶ ἔμπροσθεν. πτύαλα πολλὰ κατ᾽ ἀρχὰς ἔπτυεν[133] ἀνθηρά, καὶ ἀμφὶ ἑβδόμην ἢ ὀγδόην ἐπὶ τὰ ἐπιπέπονα.[134] κοιλίη ἑστήκει μέχρι ἀμφὶ ἐνάτην καὶ δεκάτην. ἡ ὀδύνη ἀπέσβη, οἴδημα ἐνῆν,[135] καὶ ἱδρώτια ἐγένετο· οὐ μὴν ἔκρινεν· δῆλα δὲ ἦν καὶ ἄλλοισι καὶ τῇ ἐξόδῳ· περὶ γὰρ ἀρχομένην τὴν τοῦ ὠτὸς ὀδύνην καὶ ἡ

108 γαστὴρ | ἐπεταράχθη. ἐρράγη δὲ ἐκ τοῦ ὠτὸς ἐνάτῃ καὶ ἐκρίθη τεσσαρεσκαιδεκάτῃ ἄνευ ῥίγεος ἡ νοῦσος. τῇ αὐτῇ ἡμέρῃ ἀτὰρ καὶ τὸ πτύελον λαυρότερον ᾔει. ἐπὶ[136] τὸ οὖς ἐρράγη καὶ πεπειρότερον. ἱδρῶτες δὲ καὶ ἔπειτα[137] ἐπὶ πολὺν χρόνον τῆς κεφαλῆς ἐγίνοντο· ἐξηράνθη. 4b. Ὡς τρίτῃ ὁπόσα ἀσήμως ἀφανίζεται δύσκριτα, οἷον τῇ τοῦ Πολεμάρχου ἐρυσίπελας τῇ παιδίσκῃ.

5. Οἱ ἐπὶ βουβῶσι πυρετοὶ κακὸν πλὴν τῶν ἐφημέρων, καὶ οἱ ἐπὶ πυρετοῖσι βουβῶνες κακίονες ἐν τοῖσιν ὀξέσιν ἐξ ἀρχῆς παρακμάσαντες.[138]

6. Τὰ πνεύματα ἐν[139] ὑποχονδρίοισιν[140] ἔπαρσις μαλακὴ καὶ ἔντασις οὐδετέρη. ἐπ᾽ αὐτῶν ἄνω στρογγύλον ἐν τοῖσι δεξιοῖσιν οἷον περιφέρεια ἀποπυητική· ἄλλο πρόμακρον[141] ἐπὶ πλέον· ἄλλο κεχυμένον· ἄλλο κάτω ῥέπον καὶ ἔνθεν καὶ ἔνθεν ξύντασις μέχρι τοῦ

---

[132] ᾠτὶ ὀδύνη Li.: ὅτι ἀνοδύνη V: ᾠτὶ ἂν ὀδύνη HR: ᾠτὶ ἀνοδύνη I   [133] πτύεντα HIR
[134] ἔπειτα ἐπίπονα HIR

4. Empedotime had intense heat and pain in the upper part of the left torso, together with the ear, especially at the shoulder blade, but also in front. Much expectoration, florid at the beginning, tending to be concocted about the seventh or eighth day. The bowels were stopped, to about the ninth and tenth day. The pain was stopped. There was swelling inside, and sweat came on. But there was no crisis. That was clear from other factors and from the excrement. At the beginning of the ear pain the stomach was upset. There was a breaking forth from the ear on the ninth day and the disease reached crisis without shivering on the fourteenth day. On the same day the expectoration came more vigorously. Riper matter broke forth towards the ear. Subsequently there were sweats on the head for a long time. There was drying. 4b. Whatever disappears about the third day without signs indicates an unfavorable crisis, like the erysipelas of the young daughter of Polemarchus.

5. Fevers following swollen glands are a bad thing unless they last one day, and swollen glands following fevers are worse in acute diseases if they pass their prime near the beginning.

6. Wind in the hypochondria: there is a soft swelling stretching neither way. After that a rounding on the right side like a suppurated round area or another that is more elongated, another that is broken up, another that goes downward, with stretching this way and that as far as the

---

135 οἴδημα ἐνήει HIR: "on the eighth day" (om. οἴδημα) Gal.: Langholf conj. ἐν ηʹ   136 ἐπεὶ HIR   137 ἔρπειτα HIR
138 ἢ παρακμάσασι Gal.   139 om. HIR
140 add. ἐν τοῖσι λαπαροῖσιν V
141 μακρότερον HIR

ὀμφαλοῦ ἐν πάσῃ τῇ ἄνω ἴξυϊ[142] ἦν ἐπανειλεῖται[143] καὶ
ἐπείληπται[144] ἐς τὸ περιφερές. ἢν μὲν πνεῦμα ᾖ, ἀκρίτως λεπτύνεται θέρμῃ· ἢν δὲ τοῦτο διαφύγῃ, ἐς ἐμπύησιν ὁρμᾷ.

7. Πυκνὰ πνεύματα, σμικρά, μεγάλα, ἀραιά,[145]
ἔξεισιν· ἔξω μέγα, ἔσω σμικρόν· τὸ μὲν ἐκτεῖνον, τὸ δὲ
110   κατεπεῖγον· διπλῆ ἔσω | ἐπανάκλησις, οἷον ἐπεισπνέ-
ουσιν θέρμῳ ψυχρόν.[146] ἰητήριον ξυνεχέων χασμέων
μακρόπνους· τοῖσιν ἀπότοισι καὶ μόγις πίνουσι
μικρόπνους.[147]

8. Κατ' ἴξιν καὶ πλευρῶν ἔντασις ὀδυνώδης καὶ
ἐντάσιες ὑποχονδρίων καὶ σπληνὸς ἐπάρσιες, ἐκ ῥινῶν ῥήξιες. 8b. Τὰ ἐγκαταλιμπανόμενα μετὰ κρίσιν
ὑποστροφώδεα· τὸ γοῦν πρῶτον σπληνῶν ἐπάρσιες,
ἢν μὴ ἐς ἄρθρα τελευτήσῃ ἢ[148] αἱμορραγίη γίνηται,
ὑποχονδρίου δεξιοῦ ἔντασις ἢν μὴ διεξοδεύσῃ οὖρα,
112   αὕτη γὰρ ἡ | κατάληψις ἀμφοῖν καὶ ὑποστροφαί.
ἀποστάσιας οὖν ποιεῖσθαι αὐτὸν ἡγεύμενον· τὰς δὲ
παρακλίνειν ἤδη γινομένας ἀποδέχεσθαι ἢν ἴωσιν ᾗ
δεῖ[149] καὶ ὁποῖα δεῖ καὶ ὁκόσα, μὴ ξυνδρᾶν δέ· τὰς δ'

142 Gal.: ἴξει mss.
143 ἐπανείληται HIR
144 ἐπίχρημπται Ι: ἐπίλημπται Η: ἐπίπληκται R
145 πνεύματα μικρὰ πυκνά, μεγάλα ἀραιά, μικρὰ ἀραιά,
μεγάλα πυκνά Gal. comm. Gal. Diff. resp. (K.895.18): πνεῦμα
σμικρὸν κτλ. (i.e., all singular) Gal. Diff. resp. (K.891.13)
146 θερμὸν ψυχρόν Gal. comm. Gal. Diff. resp.: ἐν θερμῷ
ψυχρόν HIR: θερμῷ ψυχρῷ Artem. (Gal.)
147 μακρόπνους (-πους Ι) HIR

navel. In the upper waist generally if there is bulging it is
trapped in that region. If it is wind it is relieved without crisis by heat. But if it evades that it will move toward suppuration.

7.[17] Exhalation is frequent, small, large, infrequent.
Large exhalation, small inhalation; one stretched out, the
other hurried. Double inspiration, like people breathing in
a cold breath after a hot. Cure for continuous yawning,
long breather; for inability to drink and difficulty drinking,
short breather.[18]

8. On the same side occur painful stretching of the
pleura and stretching of the hypochondria and elevation of
the spleen, breaking forth from the nostrils.[19] 8b. Material
left after a crisis tends towards relapses. First, elevation of
the spleen, unless it terminates in the joints, or hemorrhage occurs; there is stretching of the hypochondria on
the right if the urine does not go out, for there is blockage
in both places, and relapses. Create apostases, leading the
material yourself. Turn aside apostases that have already
started, accept them if they come where they should and
are of the right kind and quantity, but do not offer assis-

17 Galen quotes this passage in two ways, both different from
this, in *De diff. resp.* (K 7.891.13 and 895.18). It had been refined,
perhaps, by later medical theorists, but our manuscripts seem
credible.

18 Galen comments on the solecism, breather for breathing.
Epid. 6.2.3 deals with the same material.

19 Cf. *Epid.* 6.2.5.

---

148 om. HIR
149 Gal.: ἤδη mss.

ἀποτρέπειν ἦν πάντη ἀσύμφοροι ἔωσι, μάλιστα δὲ
ταύτας μελλούσας, εἰ δὲ μὴ ἄρτι ἀρχομένας.[150]

9. Αἱ τεταρταῖαι αἱμορραγίαι δύσκριτοι.

10. Οἱ διαλείποντες μίαν τῇ ἑτέρῃ ἐπιρριγεῦσιν
ἅμα κρίσει ἐς ἑβδόμην.[151]

11. Σκόπα[152] ἐκ κορυζωδέων χολωδέων καὶ φάρυγγος φλεγμονῆς, φλαύρως διαιτηθέντι ἡ κοιλίη ἀπελήφθη καὶ πυρετὸς ξυνεχὴς ἐγένετο, καὶ γλῶσσα
εὐανθὴς καὶ ἄγρυπνος· ἤτρου ἔντασις ἰσχυρῶς ὁμαλῶς κατὰ σμικρὸν ἐς τὸ κάτω ἐν τοῖσι δεξιοῖσιν·
πνεῦμα ὑπόπυκνον· ὑποχόνδριον[153] ἤλγει καὶ ἀνα-
πνέων καὶ στρεφόμενος· ἄνευ δὲ βηχὸς ἀνεχρέμπτετο
ὑποπάχεα.[154] ὀγδοαίῳ[155] πέπλος δοθεῖσα ἀπὸ τοῦ ὑποχονδρίου μὲν ἀπῶσεν, ἐπεραιώθη δὲ οὐδέν. τῇ δὲ
ὑστεραίῃ βάλανοι δύο προστεθεῖσαι, οὐκ ἐφάνησαν,
οὖρον δὲ παχὺ καὶ θολερὸν λείη καὶ ὁμαλῇ καὶ ἑστη-
κυίῃ[156] θολερότητι· εἶτα[157] γαστὴρ μαλακωτέρη ἦν καὶ
σπλὴν ἐπηρμένος καὶ κατάρροπος ἐγένετο· ποτῷ
ἐχρῆτο ὀξυγλύκει. δεκάτη αἷμα ἐξ ἀριστεροῦ ὑδα|ρὲς
114 ὀλίγον ἦλθεν· οὐ πάνυ δὲ ἢ ἄρρωστον[158] αὐτὸ τότε,[159]
καὶ οὖρον ὑπόστασιν ἔχον, ὑπὸ τῇ ὑποστάσει ὑπό-
λευκόν τι προσεχόμενον πρὸς τῷ ἀγγείῳ λεπτόν, οὔτε

150 ἐρχομένας V   151 ἐς ἑβδόμην mss. Artem. (Gal.):
ἐκ τῶν πέντε εἰς τὰς ἑπτά Gal. (cf. Epid. 6.2.9)

152 σκοπαί V: σκοπῶ HIR: Skopas Gal.

153 ὑποχόνδρια HIR   154 ὑποπάχεος mss.: corr. Lind.:
ἀνεχρέμπτετο ὑποπάχεα om. Gal.

155 om. HIR   156 ἑστηκυίη V: ἔστη κοιλίη ἐν HIR

tance. Turn some aside if they are wholly inappropriate, but especially those that are about to commence or are just begun.[20]

9. Fourth day hemorrhages indicate unfavorable crises.[21]

10. The fevers that intermit one day have shivering on the other along with a crisis towards the seventh day.

11. Scopas: after bilious running at the nose and phlegm in the pharynx, from a poor diet his bowels were seized and a continuous fever came on; florid tongue; sleepless; stretching of the lower abdomen strongly, evenly, gradually downward on the right. Fairly rapid breathing. Pain in the hypochondrium when he breathed or twisted. With no cough he brought up fairly thick material. On the eighth day wartweed which was administered expelled matter from the hypochondria, but nothing was brought through. The next day two suppositories were given; feces were not exhibited, but the urine was thick and turbid with smooth and even and stable sediment. Afterwards his belly was softer, his spleen became elevated and tended to hang down. As drink he took honey and vinegar water. On the tenth day a small amount of watery blood from the left nostril, not a very small amount, nor sickly, and urine containing a suspension; and under the suspension it put a light deposit on the vessel, not like se-

---

20 Cf. *Epid.* 6.2.7.   21 = *Epid.* 6.2.8.

---

157 Smith: ἥτε mss.: "gegen Abend" Gal.
158 ἢ ἄρρωστον Smith: δέ τι ἄρρωστος mss.
159 Nikitas: τοῦτο mss.: "noch schwach damals" Gal.

οἷον γονοειδὲς οὔτε ἀνόμοιον, ἐρρύη τοῦτο βραχύ. τῇ
δὲ ὑστεραίῃ κριθείς, ἀπύρετος· καὶ ὑπῆλθεν ὑπό-
γλισχρον τῇ ἑνδεκάτῃ, τὸ δέ τι περιρροῦν χολῶδες.
οὔρου δὲ κάθαρσις πολλὴ καὶ πλήθει καὶ ὑποστάσει
καὶ πρὶν μὲν οἰνοποτεῖν ἤρξατο σμικροῦ < . . . >[160]
λάπῃ ὁμοίη. διῆλθε δὲ τῇ ἑνδεκάτῃ ὡς ὀλίγων ἐόντων,[161] γλίσχρα δὲ καὶ κοπρώδεα θολερά· τὸ τοιοῦτον
ἢ[162] κρίσιμον,[163] ὅτι καὶ τὸ[164] Ἀντιγένεος;

12. Ἐν Περίνθῳ τὰ περὶ τὰς γλώσσας αἱρόμενα
ξυστρέμματα, καὶ ταπεινὰ ἐόντα, λιθίδια. καὶ τὰ τοῖσι
ποδαγρικοῖσιν τὰ ἀσθενέα παρ᾽ ἄρθρα ἐκείνων ἐστίν·
καὶ γὰρ ἡ ὀστέων φύσις καὶ τοῦ[165] σκληρύνεσθαι
τοῦτο αἴτιον καὶ τοῦ ξυντείνεσθαι.

13. Τὸ τῆς Ἱπποστράτου ἐκ τεταρταίου ἐνιαυσίου
ἀπεκορύφου· ὑπόψυχρος φανερῶς δοκέουσα ἔφοδος
ἐπὶ πᾶν τὸ σῶμα καὶ ἱδρώς· ἐκρίθη ταύτῃ· καὶ μετὰ
ταῦτα γυναικεῖα πλείω πλήθει καὶ χρόνῳ, τότε γὰρ
ἐπεῖχεν· μὴ[166] ἑστάναι[167] ἔδοξεν ἀπόστασις.

14. Ἐν τῇσι σφυζούσῃσιν[168] αἱμορραγίῃσι σχῆμα
116 εὑρητέον[169] καὶ | τὸ ξύμπαν εἰ ἐκ τοῦ πάνυ[170] κατάν-

---

160 I have indicated a lacuna here because the Galenic commentary says: "war der Bodensatz weiss; dann nachdem er Wein
getrunken hatte . . . "   161 ὡς ὀλίγον ἐόντα (v.l.) Gal.
162 ἢ Smith: εἰ Langholf: ἤει mss.: "Man muss untersuchen,
ob" Gal.   163 κρισίμως HIR
164 τῷ V   165 τὸ HIR
166 ἢ Ps. Gal. Caus. Aff.
167 ἱστάνειν (sic) V

men but not unlike it. This disappeared shortly. On the next day a crisis, and he was without fever. He passed sticky material on the eleventh day, and the little surrounding fluid was bilious. Much purging of urine in quantity and in sediment, and before he began to drink some wine ⟨there was white sediment, afterwards⟩ a kind of scum. He passed, on the eleventh day, although there was no great quantity, slimy, feces-like turbid matter. Does such excrement indicate crises, as did that of Antigenes?

12. In Perinthus the gatherings that arose around the tongue, even small ones, were small stones (concretions). And in those with podagra the illness around the joints is of that sort. This is the nature of bones and the cause of the hardening and the contraction.

13. The disease of Hippostratus' wife, after a year-long quartan, reached the apex. There was an accession of visible chill all over the body, and sweat. She reached crisis on that day. Afterwards, there were menstrual flows rather great in quantity and time. For then she had retained them. There seemed to be no apostasis that stayed.[22]

14. In hemorrhages that throb, one must find a good posture. In general it is preferable if the hemorrhaging

---

[22] In his commentary Galen has a different text for the last two sentences: "She let it run and considered it not good to stop the expulsion." The text of the paraphrase in the Pseudo-Galenic *De causa affectionum* (Helmreich p. 18) might be taken to mean "she persisted rather than decide to stop it."

---

168 σφύζουσιν HIR
169 εὐρύτεον V: εὐρύ τε ὂν HIR: "muss man sorgen" Gal.
170 πάντη HIR

τεος ἄναντες ποιοῖτο. διὸ καὶ αἱ ἀποδέσιες αἱ ἐν τῇσι φλεβοτομίῃσιν ὁρμῶσιν, αἱ δὲ ἰσχυραὶ κωλύουσιν.

15. Οἶμαι τὸ ἔναιμον καὶ τὸ ὑπόχολον ὀξυρεγμιῶδες, ἴσως δὲ ἐς μέλαιναν τούτοισι τελευτᾶν.

16. Ῥίγη ἄρχεται γυναιξὶ μὲν μᾶλλον ἀπὸ ὀσφύος διὰ νώτου ἐς κεφαλήν· ἀτὰρ καὶ ἀνδράσιν ὄπισθεν μᾶλλον. φρίσσουσι τὰ ἔνδοθεν μᾶλλον[171] ἢ τὰ ἔξωθεν τοῦ σώματος οἷον πήχεων, μηρῶν· ἀτὰρ καὶ τὸ δέρμα ἀραιόν· δηλοῖ δὲ ἡ θρὶξ τῶν ζώων.[172]

17. Ἧισιν οὐδὲν ἔσω τοῦ τεταγμένου χρόνου, ἑκάστῃσι τὰ τικτόμενα ἀπόγονα γίνεται. τὰ ἐπιφαινόμενα ἐν οἷσι μησὶ γίνεται. οἱ πόνοι ἐν περιόδοισιν· ὅτι ἐν ἑβδομήκοντα κινεῖται, ἐν τριπλασίῃσι τελειοῦται. ὅτι μετὰ γυναικεῖα δεξιὰ[173] τὰ δὲ ἀριστερὰ χάσκων·[174] ὑγρότης διὰ τῶν ἐπιέοντων·[175] διαίτης ξηρότης.[176] ὅτι θᾶσσον κινηθέν, διακριθέν, αὖτις αὔξεται βραδύτερον ἐπὶ πλείονα χρόνον. οἱ πόνοι περὶ τρίτην ἡμέρην πρὸς τῆσι πεντήκοντα καὶ ἕκτην πρὸς τῆσιν[177] ἑκατόν,

[171] φρίσσουσι . . . μᾶλλον om. HIR
[172] τῶν ζώων om. Gal.   [173] om. (v.l.) Gal.
[174] χασκῶν Gal. *Gloss.* Dioscurides' text
[175] ἀπιόντων Gal.   [176] ξηρότητι HIR
[177] τοῖσιν HIR

---

[23] Cf. *Epid.* 6.6.14.
[24] Cf. *Epid.* 6.3.11 and the note there.
[25] Chapters 17 and 18 are loosely organized aphorisms. While there is unique material here some of these statements occur, of-

parts be put above instead of much below. You should know, too, the bindings send the blood forth in phlebotomy, but tight ones hold it back.

15. I think that the sanguine and bilious tend to acid belching, and perhaps for those patients it ends in black bile.[23]

16. For women shivering commences from the flanks, through the back to the head. For men, too, it is more in the back. They shiver in the interior more than the exterior of the body, such as the lower arms and legs: the skin is porous, as the hair of animals proves.[24]

17.[25] Women to whom nothing happens in the prescribed time have viable babies. Additional symptoms, in what months they occur: pains come in cycles. What moves in seventy (days), is completed in three times (seventy). The fact that after menstruation there is gaping on the right, the left. Dampness because of what is coming on; dryness of regimen.[26] Pains occur around the third day after fifty (days have gone by), the sixth after one hundred.

ten in slightly different versions and in different contexts, elsewhere in the Corpus, and it often is not clear whether one of the versions is a purposeful rewriting of the other (and which one of which), or whether there have been errors in transmission. In the gynecological aphorisms at the beginning of ch. 17, Galen's commentary often agrees with *Epid.* 6.8.6 against our text here, and Galen quotes at one point *Epid.* 6's "better version," but what he quotes is not in our *Epid.* 6. This tells us only that different versions were in circulation. I have tried to follow our Hippocratic manuscripts here, but have occasionally chosen better versions from elsewhere when sense required.

[26] Cf. *Epid.* 6.2.25.

118 μηνιαῖοι δευ|τεραίῳ[178] καὶ τεταρταίῳ.[179] ἃ δεῖ εἰδέναι
ἐς τὸν ἐπτάμηνον· εἰ[180] ἀπὸ τῶν γυναικείων ἀριθμητέοι
οἱ ἐννέα μῆνες ἢ ἀπὸ τῆς ξυλλήψιος, καὶ εἰ ἑβδομήκοντα καὶ διακοσίῃσιν[181] οἱ ἑλληνικοὶ μῆνες γίνονται καὶ εἴ τι προσέτι τούτοισι, καὶ ἤτοι τοῖσιν ἄρσε-
σιν ἢ τῆσι θηλείῃσι ταὐτὰ ποιεῖται ἢ τἀναντία. τῶν
βρωμάτων[182] καὶ πομάτων τὰ ὠμὰ[183] ἐμφυσῶνται· καὶ
τῶν ἐν τῇ κεφαλῇ αἱ ἀκρησίαι καὶ τὰ ἐμφυσήματα
ποιέουσιν. αὔξησις ἔστ' ἂν τὰ ὀστέα στερεωθῇ. τῶν
ἐπιμηνίων περίοδος, τὰ πρὸ τούτων βάρεα ἀδελφὰ
τῶν ὀκταμήνων πόνων. πρὸ τόκων[184] τὰ γάλακτα τῆς
μὲν τροφῆς ὑπερβαλλούσης τῆς δὲ ὀκταμήνου ἀπαρ-
τιζούσης. τρωμάτων ἢν ἰσχυρῶν ἐόντων οἴδημα μὴ
120 φαίνηται μέγα | κακόν· τὰ χαῦνα χρηστόν· τὰ ἄνω
μενόμενα κάκιον,[185] διὸ τὰ ἐπιμήνια. ἀδελφὰ τῶν
ὀκταμήνων πρὸς δεκάμηνον τείνοντα[186] κακόν.[187]

18. Οἷσιν οἰδήματα ἐφ' ἕλκεσιν οὐ μάλα σπῶνται
οὐδὲ μαίνονται· τούτων δὲ ἀφανισθέντων ἐξαίφνης
οἷσι μὲν ἐς τὸ ὄπισθεν σπασμοὶ μετὰ πόνων, οἷσι δὲ

---

[178] δευτέρῳ mss.: corr. Asul.

[179] "Die Schmerzen nach Tagen in 50, nach Monaten in drei;
in zwei, vier, sechs, acht" Gal., with many variations

[180] ἢ mss.: "ob" Gal.: corr. Li.

[181] διακοσίοισιν V   [182] τρωμάτων (v.l.) Gal.

[183] τὰ ὠμὰ Smith: "Von den Speisen (und Getränken, var.)
blähen sich die rohen" Gal.: οἱ ὦμοι καὶ οἱ μάστοι ἐμφυσ. mss.

[184] πρὸ τόκων Smith: προτοτόκων mss.: "Vor der Zeit der
Geburt" Gal.   [185] τρωμάτων . . . κάκιον om. Gal.

[186] τείνοντα Smith: τεινόντων γενόμενα mss.

Monthly ones in the second and fourth (months). What should be known for the seven-months child. Whether the nine months should be numbered from the menses or from the conception; whether the Hellenic months contain two hundred seventy days, and whether there is an addition to those; and whether the same (calculations) are made either for males or for females, or the opposite. Harsh food and drink cause flatulence, and gas in the area of the head is caused by unwholesome living.[27] Growth, until the bones become solid. The cycle of menstruation: the heaviness before it is akin to the pains at eight months. Milk appears before the birth when there is a surplus of nutriment and the eight-month period is complete.[28] When wounds are severe, if the swelling does not appear it is very bad. The best swellings are loose.[29] When material remains above it is more harmful, therefore menstruation. Symptoms connected to the eighth month stretching to the tenth month are bad.

18. Those who have swellings at wounds do not have convulsions or delirium. When swellings disappear suddenly, those whose wounds are behind will have spasms

[27] ἀκρησίαι = "Ausschweifung," "loose behavior," in Galen's commentary, where he remarks that it refers not only to sexual activity, but generally to one's manner of living.

[28] Galen omits the following three sentences, "When . . . menstruation."

[29] Cf. *Aph.* 5.66–67.

---

[187] διὸ . . . κακὸν] "Deswegen ist das Monatliche ein Gegenstück zu dem achten (? = τῶν ὀκτώ). Wenn sich die Sache bis zu dem zehnten Monat hinzieht, so ist es schlimm" Gal.

ἐς τοὔμπροθεν ἢ μανίαι ἢ ὀδύναι πλευροῦ ὀξέαι[188] ἢ
δυσεντερίη ἐρυθρή.[189] τὰ οἰδήματα τὰ παραλόγως
ῥηΐζοντα κίβδηλον, οἷον τῷ τοῦ Ἀνδρονίκου παιδίῳ τὸ
ἐρυσίπελας ἐπαλινδρόμησεν, ἢν μὴ ἐς τὸ αὐτὸ ἐλθὸν
χρηστόν τι σημαίνῃ. τοῦτο ἔκ τε γενέσιος περὶ τὸ οὖς,
περὶ ἥβην διεδόθη,[190] ἑτέρῳ τριταίῳ ἐκ γενετῆς γενο-
μένῳ ἀπεπύησεν ἐναταίῳ· γίνεται οὗτος ἑβδομαῖος
ὑγιής. κακοηθέστερα τὰ ἀφανιζόμενα ἐξαίφνης.

ΤΜΗΜΑ ΤΕΤΑΡΤΟΝ

1. Ἡπατῖτις ἐν ὀσφύϊ μέχρι τοῦ μεγάλου σπονδύλου κάτωθεν καὶ σπονδύλοισι προσδιδοῖ, ἐντεῦθεν
μετέωρος δι᾽ ἥπατος καὶ διὰ φρενῶν ἐς καρδίην· καὶ
ᾔει[191] μὲν εὐθεῖα ἐς κληῖδας· ἐντεῦθεν δὲ αἱ μὲν ἐς
τράχηλον αἱ δὲ ἐπ᾽ ὠμοπλάτας αἱ δὲ ἀποκαμφθεῖσαι
κάτω παρὰ σπονδύλους καὶ πλευρὰς ἀποκλίνουσιν ἐξ
ἀριστερῶν μὲν μία | ἐγγὺς κληΐδων, ἐκ δεξιῶν δὲ [ἐπί
τι αὐτὴ χωρίον][192] ἄλλη [ἡ δὲ][193] σμικρὸν κατωτέρω
ἀποκαμφθεῖσα ὅθεν μὲν ἐκείνη ἀπέλιπε προσέδωκε
τῇσι πλευρῇσιν ἔστ᾽ ἂν τῇ[194] ἐπ᾽[195] αὐτῆς τῆς καρδίης
προστύχῃ ἀποκαμπτομένη ἐς τὰ ἀριστερά· ἀποκαμ-

122

188 ὀξέαι ἢ ἐμπύησις (v.l.) Gal. *Aph.*
189 δυσεντερίη ἐρυθρή mss. (V.l.) Gal.: "wenn diese Gesch-
wulst intensiv rot ist" Gal.: ἢν ἐρυθρά μᾶλλον ᾖ *Aph.* 5.65
190 τοῦτο . . . διεδόθη] "Denn sie folgten hintereinander seit
der Geburt unaufhörlich um die Scham herum" Gal.
191 Li.: εἴη V: ἡ HIR

with pain, those whose wounds are in front will have delirium or sharp pains in the side, or bright red dysentery. Swellings that are eased for no apparent reason are a deceitful sign (as in the case of Andronicus' child, where the erysipelas returned) unless by returning to the same state they offer a favorable sign. This affection by the ear which was congenital was dispersed towards the pubis,[30] another's which appeared on the third day after birth festered on the ninth. He was cured in seven days. Things that disappear suddenly are more dangerous.

SECTION 4

1. The liver vein is in the loins as far as the great vertebra below, and it gives forth on to the vertebrae. From there it rises through the liver and through the diaphragm to the heart. It goes straight from the heart to the clavicles, and at that point some veins bend to the neck and some towards the shoulder blades, and some bend down and incline along the spine and ribs. From the left side one near the clavicles, from the right another, bending off slightly lower, gives forth on to the ribs from the place where that one left off, until it meets the one at the heart itself as that one bends off to the left. And turning back on to the verte-

---

[30] Some ancient commentators interpreted this to mean "was dispersed as he neared adolescence." Galen read a quite different text: "From birth they came continuously one after the other . . ."

---

192 om. Smith following Gal. comm.
193 om. Smith   194 τοι V
195 ἐξ HIR

φθεῖσα[196] δὲ κάτω ἐπὶ σπονδύλους καταβαίνει ἔστ' ἂν ἀφίκηται ὅθεν[197] ἤρξατο μετωρίζεθσαι, ἀποδιδοῦσα τῇσι πλευρῇσι[198] τῇσιν ἐπιλοίποισιν ἁπάσῃσιν, καὶ ἔνθεν καὶ ἔνθεν ἀποσχίδας παρ' ἑκάστην διδοῦσα[199] μία ἐοῦσα.

Ἀπὸ μὲν τῆς καρδίης ἐπί τι χωρίον ἐν τοῖσιν ἀριστεροῖσι μᾶλλον ἐοῦσα, ἔπειτα[200] ὑποκάτω τῆς ἀρτηρίης ἔστ' ἂν καταναλωθῇ καὶ ἔλθῃ ὅθεν ἡ ἡπατῖτις ἐμετεωρίσθη. πρότερον δὲ πρὶν ἢ ἐνταῦθα ἐλθεῖν, παρὰ τὰς ἐσχάτας δύο πλευρὰς ἐδικραιώθη·[201] καὶ ἡ μὲν ἔνθα ἡ δὲ ἔνθα τῶν σπονδύλων ἐλθοῦσα κατηναλώθη.

Εὐθεῖα δὲ ἀπὸ καρδίης πρὸς κληῖδας τείνουσα ἄνωθεν τῆς ἀρτηρίης ἐστί, καὶ ἀπὸ ταύτης ὥσπερ καὶ παρ' ὀσφῦν κάτωθεν τῆς ἀρτηρίης.[202] ἀίσσει ἐς τὸ ἧπαρ ἡ μὲν ἐπὶ πύλας καὶ λοβὸν ἡ δὲ ἐς τὸ ἄλλο ἐξ ἧς[203] ἀφώρμηκε σμικρὸν κάτωθεν φρενῶν. φρένες δὲ προσπεφύκασι τῷ ἥπατι ἃς οὐ ῥηΐδιον χωρίσαι. δισσαὶ δὲ ἀπὸ | κληΐδων αἱ μὲν ἔνθεν αἱ δὲ ἔνθεν ὑπὸ 124 στῆθος ἐς ἦτρον· ὅπη δὲ ἐντεῦθεν οὔπω οἶδα. φρένες δὲ [κάτω][204] κατὰ τὸν σπόνδυλον τὸν κάτω τῶν πλευρέων, ᾗ νεφρὸς ἐξ ἀρτηρίης, ταύτῃ ἀμφιβεβηκυῖαι.[205] ἀρτηρίαι δὲ ἐκ τούτου[206] πεφύκασιν[207] ἔνθεν καὶ ἔνθεν ἀρτηρίη τόνον ἔχουσα.[208] ταύτῃ δέ πη παλινδρομή-

---

[196] καὶ ἀποκαμ. V   [197] ἔνθεν V
[198] ἔστ' ἂν . . . πλευρῇσι om. HIR (add. in marg. H)
[199] δίδου HIR   [200] ἐπὶ τὰ mss.

brae it descends until it arrives at where it began to ascend, giving out on to all the remaining ribs, and on this side and that it gives branches to each, though it is single.

From the heart to an area it is more on the left; then it is below the artery until it is used up and arrives at the place from which the liver vein ascended. But before it arrives there, it divides by the last two ribs. One part goes on one side of the vertebrae, the other on the other side, and they are used up.

The straight vein stretching from the heart to the collar bone is more above the artery, and away from it, just as, in the loins, it is below the artery. It hurries to the liver, one part to the gates and the lobe, the other heads off to the other area from which it started out, a little below the diaphragm. The diaphragm is attached to the liver and cannot be easily separated. There are pairs of veins that go, one on one side, the other on the other, from the clavicles under the chest to the lower belly. Where they go from there I do not yet know. The diaphragm is at the vertebra below the ribs where the kidney separates from an artery, and the diaphragm bestrides the artery. The arteries grow out from it, on one side and the other, an artery having a nerve. At this point the liver vein, having run back down from the

---

201 Erotian: ἐδιαιρέθη HIR: ἐδιώχθη V: ἐδιχώθη Hipp. *Oss.* Gal. *Plac.*: "es entstehen ihr zwei Ausläufer" (v.l.) Gal. comm.

202 κάτωθεν τῆς ἀρτηρίης om. Gal. comm.

203 ἐξῆς HIR   204 del. Smith

205 Gal. comm. ("reitend") Gal. *Plac.*: ἀμφιβεβήκει αἱ V (αἷς HI: οἷς R)   206 Galen: τοῦ τείνοντος mss.

207 ἐξεπεφύκασιν HIR   208 Gal. *Plac.* (and probably Gal. comm.): ἀρτηρίης τόνον ἔχουσαι mss.: -ῆσι . . . ἔχουσαι *Oss.*

65

σασα ἀπὸ καρδίης ἡ ἡπατῖτις ἔληγεν. ἀπὸ δὲ τῆς
ἡπατίτιδος διὰ τῶν φρενῶν[209] αἱ μέγισται δύο ἡ μὲν
ἔνθεν ἡ δὲ ἔνθεν φέρονται μετέωροι. πολυσχιδεῖς δὲ[210]
διὰ τῶν φρενῶν εἰσιν, ἀμφὶ ταύτας[211] καὶ πεφύκασιν·
ἄνωθεν δὲ[212] φρενῶν αὗται [δὲ][213] μᾶλλόν τι ἐκφανεῖς.[214]

2. Δύο δὲ τόνοι ἀπ᾽ ἐγκεφάλου ὑπὸ[215] τὸ ὀστέον τοῦ
μεγάλου σπονδύλου ἄνωθεν καὶ πρὸς[216] τοῦ στομάχου
μᾶλλον. ἑκατέρωθεν τῆς ἀρτηρίης παρελθὼν ἑκάτερος
ἐς ἑωυτὸν ἦλθεν ἴκελος ἑνί· ἔπειτα ᾗ[217] σπόνδυλοι καὶ
φρένες πεφύκασιν ἐνταῦθα ἐτελεύτων, καί τινες ἐν-
δοιαστοὶ πρὸς ἧπαρ καὶ σπλῆνα ἀπὸ τούτου τοῦ
κοινωνήματος ἐδόκεον τείνειν. ἄλλος τόνος ἐκ τῶν
ἑκατέρωθεν σπονδύλων[218] παρὰ ῥάχιν παρέτεινεν ἐκ
πλαγίων[219] σπονδύλων, καὶ τῇσι πλευρῇσιν ἀπένε-
μεν.[220] ὥσπερ αἱ φλέβες, οὗτοι διὰ φρενῶν ἐς μεσεντέ-
ριόν μοι δοκέουσι τείνειν, ἐν δὲ τούτοισιν ἐξέλιπον.
αὖτις δ᾽ ὅθεν φρένες[221] ἐξεπεφύκεσαν, ἀπὸ τούτου
126 ξυνεχεῖς ἐόντες κατὰ μέσον | κάτωθεν ἀρτηρίης τὸ
ἐπίλοιπον παρὰ σπονδύλους ἀπεδίδουν ὥσπερ αἱ φλέβες μέχρι κατηναλώθησαν διελθοῦσαι ἐς[222] τὸ ἱερὸν
ὀστέον.

[209] νεφρῶν HIR   [210] τε HIR Gal. *Plac.*
[211] ταύταις HIR   [212] δὴ HIR
[213] del. Smith   [214] ἐμφανέες HIR
[215] mss. (v.l.) Gal. comm.: παρὰ Gal. comm. Gal. comm. *De
artic.*   [216] mss. (v.l.) Gal. comm.: πρὸ Gal. comm.
[217] οἱ HIR

heart, leaves off. And from thc liver vein through the diaphragm the two largest veins are carried above, one on this side, one on that. And there are many branching veins through the diaphragm, and they grow around it, but above the diaphragm they are more visible.

2.[31] Two nerves go from the interior of the head under the bone of the great vertebra above and more in front of the esophagus; proceeding on both sides of the trachea individually they come to themselves so as to be as one. Then where the vertebrae and the diaphragm join they end and some obscure ones seem to go towards liver and spleen from that juncture. Out of the vertebrae on each side a nerve stretches along the backbone out of the oblique parts of the vertebrae and distributes to the ribs. Like the blood vessels these seem to me to stretch through the diaphragm to the mesenterion, and they stop at that point. But again from where the diaphragm is attached they are continuous down the center below the artery for the rest of the way and give forth along the vertebrae like the blood vessels until they are used up as they arrive at the sacrum.

[31] This unique early description of the nerves is somewhat obscure. Though the author calls them *tonoi* and uses the verb *teinein*, "stretch," words appropriate for a bowstring and suggesting that the nerves are for pulling, he seems also to speak of them as channels like the blood vessels.

---

218 ἑκατέρωθεν ἐκ τῶν κατὰ κληῖδα σπονδύλων Oss.
219 πλαγίου HIR
220 Oss.: ἀπένεμον VIR: ἀπείνεμον H
221 om. V
222 διελθοῦσαι ἐς mss.: πᾶν διελθόντες Oss. Li.

3. Ἐν Αἴνῳ ὀσπριοφαγεῦντες ἐν λιμῷ[223] ξυνεχεῖς, θήλειαι, ἄρσενες, σκελέων ἀκρατεῖς ἐγένοντο καὶ διετέλεον· ἀτὰρ καὶ ὀροβοφαγέοντες γονναλγεῖς.

4. Ἐπιτηδεύειν ὀξυθυμίην ἐμποιεῖν καὶ χρώματος ἀναλήψιος[224] ἕνεκα καὶ[225] ἐγχυμώσιος, καὶ εὐθυμίας[226] καὶ φόβους καὶ τὰ τοιαῦτα· καὶ ἦν μὲν τὸ ἄλλο σῶμα ξυννοσέῃ ξυνιᾶσθαι, εἰ δὲ μή, τοῦτο.

5. Ἡ Στυμάργεω οἰκέτις ᾗ οὐδὲ αἷμα[227] ἐγένετο ὡς ἔτεκε θυγατέρα, ἀπέστραπτο[228] τὸ στόμα,[229] καὶ ἐς ἰσχίον καὶ σκέλος ὀδύνη. παρὰ σφυρὸν τμηθεῖσα[230] ἐρρήϊσεν·[231] καίτοι καὶ τρόμοι τὸ σῶμα πᾶν[232] κατεῖχον· ἀλλ᾽ ἐπὶ τὴν πρόφασιν δεῖ ἐλθεῖν,[233] καὶ τῆς προφάσιος τὴν ἀρχήν.[234] |

ΤΜΗΜΑ ΠΕΜΠΤΟΝ

128

1. Ὁκόσοι πυρροί, ὀξύρρινες, ὀφθαλμοὶ σμικροί, πονηροί. ὁκόσοι πυρροί, σιμοί, ὀφθαλμοὶ μεγάλοι,[235] ἐσθλοί. ὑδρωπιώδεις χαροποί, πυρροί, ὀξύρρινες, ἢν

[223] Gal.: ἐν Αἴνῳ HIR: ἐναίμῳ V
[224] ἀναλήψιας HIR   [225] om. HIR
[226] Gal.: εὐθυμίης mss.
[227] ᾗ οὐδὲ αἷμα Gal. Trem. (7.603 K): ἡ ἰουδουμαία V: ἡ ἰδουμαῖα HIR: οὐδὲ αἷμα Gal. Venes. adv. Eras. (11.162 K): "der nicht einmal Blut" Gal. comm.: ᾗ οἰδήματα conj. Langholf, cf. Epid. 2.2
[228] ἐπέστραπτο VHI: ἐπέστραψε R
[229] Gal. Venes.: τὸ στόμα τοῦ αἰδοίου HIR: τὸ σῶμα Gal. comm.: στόμα τῆς μήτρας Gal. Trem.

3. In Aenus those who continually ate beans during a famine, both men and women, became weak in the legs and continued so. And those who ate vetch had knee ailments.[32]

4. It is appropriate to induce anger for the sake of restoring color and humors, also to induce happiness, fear, and the like. If the rest of the body is ill also, it should be treated at the same time. Otherwise just this.

5. The house servant of Stymarges, who did not even bleed when she bore a daughter, had the mouth of her womb retroverted. She had pain in her hip joint and leg. Phlebotomy at the ankle relieved her. But trembling seized her whole body. One must approach the cause, and of the cause, the source.

SECTION 5

1. Those with a ruddy complexion, sharp nose, small eyes, are bad (sickly). Those with ruddy complexion, flat nose, large eyes, are good. Hydropics are gray-eyed, ruddy,

[32] Cf. *Epid.* 6.4.11.

---

230 mss. Gal. *Venes.*: παρὰ σφυρὸν φλεβοτομηθεῖσα Gal. *Trem.*: "die Gebärmutterader angeschnitten" Gal. comm.

231 Gal. *Trem.*: ἐρράϊσε Gal. *Venes.*: ἐρρύησε mss.

232 Gal. *Trem.*: κατὰ σῶμα πάντα HIR: τὸ σῶμα πάντα V: τὸ σῶμα περικατεῖχον Gal. *Venes.*

233 Gal. *Trem.*: διελθεῖν mss.: χρὴ ἐλθεῖν Gal. *Venes.*

234 mss. (v.l.) Gal. comm.: ἀφόρμην Gal. comm. Gal. *Trem.*: τρόφην Gal. *Venes.*

235 om. HIR

μὴ[236] φαλακροὶ ἔωσιν. ἰσχνοφωνίην[237] κιρσὸς λύει ἐς
τὸν ἀριστερὸν[238] καὶ τὸν δεξιὸν ὄρχιν· ἄνευ τούτων[239]
τοῦ ἑτέρου οὐχ οἷόν τε λύεσθαι. μεγάλοι, φαλακροί,
τραυλοὶ ἰσχνόφωνοι, ἐσθλοί. νοσήματα δὲ ἔχουσι
τραυλὸς ἢ φαλακρὸς ἢ ἰσχνόφωνος ἢ δασὺς ἰσχυρῶς
μελαγχολικά. [νοσήματα δ' ἔχουσιν][240]

2. Ὅσοι τῇ γλώσσῃ παφλάζουσι χειλῶν[241] μὴ
ἐγκρατεῖς ἐόντες, ἀνάγκη λυομένων ἐμπύους γίνεσθαι.
ὀδύνη ἐν τοῖσι κάτω χωρίοισιν ἰσχυρὴ ἦ[242] κωφότης
λύει, καὶ αἷμα πολλὸν ἐκ τῶν ῥινῶν, ἢ μανίη.

3. Ἢν λεχοῖ[243] σπασμὸς ἐπιγένηται, πῦρ ποιεῖν[244]
καὶ ἐς κύστιν κηρωτὴν ἐγχέας πολλὴν χλιαρὴν κλύζειν.

4. Ἢν τῆς κεφαλῆς τὸ ὀστέον καταγῇ, διδόναι
γάλα καὶ οἶνον, ἴσον ἴσῳ· ἢν δὲ ἕλκος ἦ, φλεβοτομεῖν
τὰς ἔσω ἢν μὴ πυρεταίνη· ἢν δὲ παραφρονέη, τὴν
κεφαλὴν καταβρέχειν ἢν μὴ τὰ ὑποχόνδρια ἐπηρμένα
ἦ· ἢν τὴν κεφαλὴν ἀλγέη, ἐς στῆθος ἔρχεται, | ἔπειτα
130 ἐς τὸ ὑποχόνδριον, ἔπειτα ἐς τὸ ἰσχίον. πάντα δὲ οὐχ
οἷόν τε ἀλγεῖν.

5. Ἀνεμίην φλεβοτομίη.

6. Τῷ φαρμάκῳ τὸν ῥόον ἴσχειν ἐπαλείφων, ὅδε
γὰρ ὁ ῥόος ἐκ τῆς μεγάλης φλεβός· ἢν δὲ αὐτόματον
ῥέῃ πολλόν, νηστευέτω ἄλλο[245] ἢ γάλα, δύο ὕδατος,

---

[236] πύρροι . . . μὴ om. HIR   [237] ἰσχοφωνίην (v.l.) Gal.
[238] τὴν ἀριστερὴν V: "linken Schenkel" Gal.
[239] τουτέου HIR. After this sentence Galen's commentary has
a hiatus until section 6.1

sharp-nosed, unless they be bald. Varicosity resolves weakness of voice on to the left and the right testicle. Without one or the other it cannot be broken up. Large, bald, lisping, weak-voiced, good. However, those who lisp, who are bald, weak-voiced, or shaggy, have melancholic affections.

2. Those who stammer with the tongue and cannot control the lips must, when that is resolved, become empyemic. That is resolved by severe pain in the lower area. or deafness, also by much blood from the nose, or delirium.[33]

3. If spasms come on one who has given birth, build a fire and, pouring a large quantity of warm, wax-based salve into a bladder, use as a clyster, warm.

4. If the bone of the head be broken, give milk and wine, half and half. If there be a wound, phlebotomize the inner veins unless there be fever. If there is delirium, soak the head unless the hypochondria are swollen. If there is headache, it proceeds to the chest, then to the hypochondrium, then to the hip joint—they cannot all be in pain at once.

5. For flatulence, phlebotomy.

6. Restrain the flow by anointing with the drug. For this flow is from the great vein. If there is much flow spontaneously, let the patient fast except for milk, two parts of wa-

---

33 Cf. *Epid.* 2.5.9, 2.6.5 below.

---

240 del. Li.

241 χειρῶν mss.: corr. Asul. (cf. Hipp. *De crisibus* 43)

242 ὀδύνη . . . ἰσχυρὴ ἢ Li., from *De crisibus* 43: ὀδύνην . . . ἰσχυρὴ (om. ἢ) mss.

243 ms. D (Li.): λέχοις mss.

244 ποιεῖ mss.: corr. Lind.   245 om. HIR

τέσσαρας γάλακτος. τὰς ἀγόνους πυριᾶν καὶ φαρμακεύειν.

7. Ὅσοι ἐξαπίνης ἄφωνοι, ἢν ἀπύρετοι ἔωσι, φλεβοτομεῖν.

8. Φλέγματος κατάρροοι· ἐκ τῶν μαζῶν ἕλκουσιν οἱ ὀφθαλμοί, καὶ ἐξερεύγεται κατὰ τὰς ῥίνας ἐς τὸν πνεύμονα.

9. Οἷσι βὴξ ξηρή, οὐ λύεται ἢν μὴ ὀδύνη ἰσχυρὴ ἐς τὰ ἰσχία ἢ ἐς τὰ σκέλεα ἢ ἐς τὸν ὄρχιν.

10. Ἢν ὑδρωπιῶντα βὴξ ἔχῃ, ἢν μὲν αὐτίκα λειποθυμέῃ, θερμοῖσι πᾶσι διαχρήσθω· ἢν δὲ μή, θωρῆξαι, καὶ σιτίων ἐμπλῆσαι, τάμνειν δὲ τὰς ἔσω.

11. Τοῦ νοσήματος τοῦ μεγάλου ἐν τάσει[246] γενομένου λύσις ἰσχίων ὀδύνη, ὀφθαλμῶν διαστροφαί, τύφλωσις, ὀρχίων οἴδησις, μαζῶν ἄρσις.

12. Ἢν πυρετοὺς ἔχοντος τὰ περὶ τὸ πρόσωπον ἰσχνὰ ᾖ ἐν ἡμέρῃ γονίμῃ, τὴν ἐπιοῦσαν λύσις.

13. Ὕδρωψ,[247] ἢν οἴδημα ἔχων ἐν τοῖσι σκέλεσι βήσσῃ.

14. Ἢν τὸ οὖς ἀλγέῃ, τῷ γάλακτι διαχρήσθω.

15. Ἢν μὴ ἐν τῇ γονίμῃ μεθῇ ὁ πυρετός, ὑποτροπιάζειν[248] ἀνάγκη.

---

[246] ἔθει De crisibus 44
[247] ὑδρὼψ Diosc. ap. Gal. Glos.
[248] ὑποτροπάζειν V

ter, four of milk. To women who are barren give a vapor bath and apply drugs.

7. For those who have sudden loss of voice, if they are without fever, phlebotomize.

8. Fluxes of phlegm: the eyes draw from the breasts and it regurgitates through the nose to the lungs.

9. For those with a dry cough: it is not resolved save with a sharp pain in the hips, or the legs, or the testicle.

10. If a cough possess one with dropsy, if he faint suddenly, use all heating treatments. If not, fortify him with wine, fill him with food, and phlebotomize the inner veins.

11. When the great malady (epilepsy) occurs with rigidity,[34] the resolution comes with pain in the hips, turning about of the eyes, blindness, swelling of testicles, swelling in the breasts.

12. If, in one who has a fever, the area of the face is thin on a productive day, there will be a resolution on the following day.[35]

13. It indicates dropsy if one with a swelling in the legs develops a cough.

14. If the ear be pained, let the patient use milk.

15. If a fever does not remit on a productive day, it will necessarily recur.

[34] The parallel passage at *De Crisibus* 44 says "When the great malady has become habitual," and may be correct.

[35] γόνιμος is a frequent word in sections 5 and 6 of *Epid.* 2, but not elsewhere in ancient medical texts in the same connections. I translate it as "productive." *Aph.* 4.61 has a version of this same aphorism in which it substitutes the word περισσός, "uneven."

16. Οὗ ἂν ἡ φλὲψ ἡ ἐν τῷ ἀγκῶνι σφύζῃ, μανικὸς καὶ ὀξύθυμος· ᾧ δ᾽ ἂν ἀτρεμέῃ, τυφώδης.

17. Τρῶμα ἢν αἱμορραγήσῃ, μὴ βρέχειν τὸ ἕλκος, τὴν κεφαλὴν δὲ βρέχειν θερμῷ.

132   18. Ἢν καρδιώσσῃ, θερ|μὸν ἄρτον μετ᾽ οἴνου ἀκρή-
του διδόναι.

19. Ἐμέτου λύσις· ὕδωρ θερμὸν διδόναι πίνειν, καὶ ἐμείτω.

20. Ὅσα σφακελίζει, ἀπολαβόντα τὴν φλέβα ἑλ-
κῶσαι καὶ ὑγιῶσαι.[249]

21. Σπασμοῦ χειρὸς δακτύλων ἄνευ πυρετοῦ, σχά-
σαι ἢν μὴ τὴν κεφαλὴν ἀλγέῃ· εἰ δὲ μή, ὕδωρ θερμὸν
καταχεῖν.

22. Ὀφθαλμῶν, σποδίου δωδέκατον, κρόκου πέμπτον, πυρῆνος ἕν, ψιμυθίου ἕν, σμύρνης ἕν·[250] τὸ ὕδωρ
κατὰ τῆς κεφαλῆς ψυχρὸν καταχεῖν, καὶ διδόναι σκόροδα σὺν μάζῃ.

23. Κιρσοὶ δὲ φαλακρῶν, ἢν μὴ μεγάλοι ἔωσι,
μανιώδεις.

24. Ἀλφοῦ καὶ λέπρης, τίτανος ἐν ὕδατι, ὡς μὴ
ἑλκώσῃς.

25. Χορίων κάθαρσις ἢν ὑπερέχῃ· ἐλλέβορον[251]
πρὸς τὰς ῥῖνας προστιθέναι ὥστε πτάρνυσθαι, καὶ
ἐπιλαμβάνειν τὰς ῥῖνας τῆς πταρνυμένης καὶ τὸ[252]
στόμα.

[249] ὑγιὴς V   [250] ἕν add. Lind.
[251] om. HIR   [252] om. V

16. One whose blood vessel at the arm-bend throbs is wild and passionate. He whose vessel is still is stuprous.

17. If a wound bleeds, do not soak the wound, but soak the head in warm water.

18. For heartburn give warm bread and neat wine.

19. To cure vomiting, give warm water to drink, and let him vomit that.

20. For ulceration, separate the vein, wound it, and induce healing.

21. For spasm of the fingers without fever, open a vein unless there be pain in the head. Otherwise pour on warm water.

22. For eyes, a twelfth part of spodium, a fifth part of saffron, one part of "pit," one of white lead, one of myrrh. Pour cold water over the head and give garlic with barley cake.[36]

23. Varicosities of bald people, unless they be large, indicate mania.

24. For white or scaly leprosy, gypsum in water; be careful not to cause ulceration.

25. Purging of the afterbirth if it remains: hellebore in the nostrils to cause sneezing, and close the mouth and nose of the woman as she sneezes.[37]

---

[36] Some ingredients here are obscure. Spodium is probably an oxide, e.g., of copper, "pit" is probably the stone of a fruit or a lump of frankincense.

[37] Cf. *Aph.* 5.49.

ΕΠΙΔΗΜΙΑΙ

ΤΜΗΜΑ ΕΚΤΟΝ

1. *Ἦν ἡ κεφαλὴ μεγάλη καὶ οἱ ὀφθαλμοὶ σμικροί, τραυλοί, ὀξύθυμοι. οἱ μακρόβιοι πλείους ὀδόντας ἔχουσιν. οἱ τραυλοί, ταχύγλωσσοι, μελαγχολικοὶ κατακορεῖς. ἀσκαρδαμύκται[253] ὀξύθυμοι. μεγάλη κεφαλὴ ὀφθαλμοὶ μέλανες καὶ μεγάλοι, ῥὶς παχείη καὶ σιμή,[254] ἐσθλοί. χαροποί, μεγάλοι < . . . >[255] κεφαλὴ σμικρή,[256] αὐχὴν λεπτός, στήθεα στενά, εὐάρμοστοι. κεφαλὴ σμικρή,[257] οὐκ[258] ἂν εἴη τραυλὸς οὐδὲ φαλακρὸς ἦν μὴ γλαυκὸς ᾖ.

2. Σπασμῶν, φωνὴ ἐν γονίμῳ λύεται, ἀπήλλακται τοῦ μεγάλου νοσήματος.

134   3. Λεχοῖ[259] δὲ πυρεταινούσῃ καὶ ἀλγεούσῃ, ὕδωρ καταχεῖν καὶ πτισσάνην παχείην διδόναι τρὶς τῆς ἡμέρης, θερμήν.

4. Παιδία τρέφεται ἑβδόμῳ μηνὶ ἢ ἐνάτῳ ἢ δεκάτῳ, καὶ ἵσταται τῇ φωνῇ, καὶ ἰσχὺς ἕπεται καὶ τῶν χειρῶν κρατεῖ. τῆς φωνῆς λυομένης πάντα λύεται, ἡ γὰρ φύσις τῇ φθέγξει ὁμοίη, λύεται δὲ ἐν γονίμῃ.

---

253 ἀσκαρδαμύκται: "vielzwinkernd" Gal. comm.
254 ῥίνα παχείην καὶ σιμὴν mss.: corr. Smith
255 I conjecture a lacuna. "Wenn das Auge in reines Graublau sticht, der Kopf gross, der Halz kurz, das Brustbein weit ist, festzornig" Gal. comm.
256 Gal.: μεγάλη mss. (v.l.) Gal.
257 κεφαλὴ σμικρὴ om. (v.l.) Gal. comm.
258 οὐδ᾽ HIR
259 R: λέχοι VHI

SECTION 6

1. If the head is large and the eyes small, if they are stammerers, they are quick to anger. People who are long-lived have more teeth. Stammerers and rapid talkers are severely melancholic. People who do not blink are quick to anger. Those with large head, large dark eyes, thick, blunt nose, are good. Blue eyed, large. . . .[38] Those with small head, thin neck, narrow chest, are equable. If one's head is large he will not stammer or be bald unless his eyes be gray.

2. Spasms: the voice is released on a productive day. He is freed from the great disease (epilepsy).

3. If a woman who has given birth has fever and pain, pour water over her, give her thick barley gruel, warm, thrice daily.

4. Children are nurtured in the seventh month, or the ninth, or tenth. They are established in speech, strength follows, they control their hands. When the voice is released, everything is released, for one's nature is like speech, and it is set free on a productive day.

[38] Galen's commentary, as read and translated by Hunain Ibn Ishaq, presents this sentence thus: "Those with bright blue-gray eyes, large head, short neck and broad breastbones are prone to anger." Galen's commentary reports much discussion of this physiognomic material among ancient commentators, medical people, and others. What *Epid.* 2 has to say is not out of line with ancient physiognomic lore.

ΕΠΙΔΗΜΙΑΙ

5. Ἢν αἱ φλέβες σφύζωσιν ἐν τῇσι χερσὶ καὶ τὸ πρόσωπον ἐρρωμένον καὶ ὑποχόνδρια μὴ λαπαρὰ[260] ᾖ, χρονίη ἡ νοῦσος γίνεται· ἄνευ σπασμοῦ οὐ λύεται, ἢ αἵματος πολλοῦ ἐκ τῶν ῥινῶν, ἢ ὀδύνης ἐς τὰ ἰσχία.

6. Τοῦ λαιμοῦ ὕδωρ θερμὸν[261] κατὰ τῆς κεφαλῆς καταχεῖν ἢν μὴ ψῦχος ᾖ· ἢν δὲ μή, ἄλητον ὡς θερμότατον διδόναι καὶ οἶνον ἄκρητον.

7. Ταραχῆς[262] γαστρός, κυάμους ἐφθοὺς διδόναι ἢν μὴ τὰ ἄνω κατακορέα[263] ᾖ, ἢ κύμινον διδόναι τρώγειν μετὰ τῶν κυάμων.

8. Ἀπόληψις δὲ[264] τοῦ νοσήματος οὐκ ἂν γένοιτο εἰ μὴ ἐν γονίμῃ ἡμέρῃ, οὐδὲ ἂν ἀρχὴ γένοιτο ἢν μὴ ἀγόνῳ ἡμέρῃ καὶ μηνὶ ἔτει δὲ γονίμῳ.

9. Λίτρον αἰγύπτιον καὶ κορίανον καὶ κύμινον τρίβοντα σὺν ἀλείφατι ξυναλείφειν.

10. Ὅσα θνήσκει,[265] ἀνάγκη γονίμῳ ἡμέρῃ καὶ γονίμῳ μηνὶ καὶ γονίμῳ ἔτει. προλέγειν δὲ ὀρθῶς ἂν ἔχοι θάνατον ἢ ὀδύνας ἰσχυράς, οἷον ἢν τὰ ὄμματα μὴ ἔρρωται ὁ θάνατος ἐν τάχει. ἢν[266] ἐν γονίμῳ ἔτει γίνηται ἀπ᾽ ἀμφοτέρων γονίμους[267] ἀνάγκη γενέσθαι·

260 λίπαρα V
261 ψυχρὸν (v.l.) Gal. comm.
262 ταραχῇ (i.e. -ῇ) V
263 κατακορέα: "rein" Gal. comm.
264 om. V
265 θνήσκειν V
266 ὧν V
267 γόνιμον Gal ("muss es gedeihen")

78

5. If the veins in the arms throb and the face is strong and the hypochondria be not sunken, the disease becomes chronic. It is not resolved without spasm, or much blood from the nose, or pain in the hips.

6. For the throat, pour warm water over the head unless the weather be cold. Otherwise give meal as hot as possible, and undiluted wine.

7. For upset intestines, give boiled beans unless the upper intestine be saturated (bilious), or give cumin to nibble with the beans.

8. Removal of the illness could not occur save on a productive day, nor could it begin except on a barren day and month, but in a productive year.[39]

9. Egyptian nitre (sodium carbonate) and coriander and cumin: grind them, add to anointing oil, and apply.

10. All deaths must occur on a productive day, in a productive month and a productive year. And one might be able correctly to predict death or severe pain. For example, if the eyes do not function, death will come quickly. If it be in a productive year they (day and month) are necessarily productive from the two of them, but if it is in an un-

[39] One is tempted by Erotian's interpretation, "odd" and "even" for "productive" and "barren," but there is force in Galen's argument here in favor of interpreting "critical" and the opposite. Galen tries to connect these to the critical numbers in other Hippocratic works, especially *Prognostic*, which contain both odd and even numbers, 4, 7, 11, etc. (see Loeb Hipp. vol. I). I do not feel that I have made out just what the author or assembler of these aphorisms had in mind. Galen relates this and the following aphorism to epilepsy.

ἢν δὲ ἀγόνῳ ἔτει καὶ ἀγόνῳ ἡμέρῃ, θνήσκειν ἀνάγκη
γονίμῳ ἡμέρῃ.[268]

11. Τοῦ ἀριθμοῦ τρίτη[269] ἰσχυροτάτη.[270]

12. Κυνάγχην καὶ ὀφθαλμίην φλεβοτομεῖν.[271]

13. Τρωθέντος ἐντέρου, ἡ ἀναπνοὴ ἔρχεται | κάτω
136 ἀφανὴς κατὰ τὸ τρῶμα καὶ κενοῦται τὰ στήθεα. διδόναι οὖν γάλα[272] καὶ οἶνον ἴσον ἴσῳ.

14. Ὧν κατακορέα τὰ στήθεα, ψελλοί, μανιώδεις,
καὶ φαλακροί· τούτων ὅσοι ἐκ γενεῆς καὶ στρεβλοί,
ἀσύνετοι, ἠλιθιῶντες[273] ἢ μαινόμενοι· οἷσι δὲ μὴ, ἑτέρου κακοῦ λύσις.

15. Περὶ φύσιος δύναμιν πλείστην ἔχει τιτθός,
ὀφθαλμὸς δεξιός. ταὐτὰ τῶν κάτω καὶ ὅτι ἐμπέφυκε
τοῖσι δεξιοῖσι τὰ ἄρσενα.

16. Γυναιξὶν ἐπιμήνια ὥστε ἴσχειν, σικύην μεγί-
στην[274] παρὰ τὸν τιτθὸν προσβάλλειν.

17. Τρίμηνον παιδίον πάντα δηλοῖ καὶ γάλα[275] τότε
ἔχει.

---

268 θνήσκειν . . . ἡμέρῃ om. HIR (add. in marg. H)
269 τρίται (v.l.) Gal.
270 ἰσχυροτάτην (om. τοῦ . . . τρίτη) Rufus apud Gal.
271 φλεβοτομίη HIR
272 γάλα ἢ V
273 ἢ λιθιῶντες mss. Gal.: corr. Erm.
274 μεγάλην Gal.
275 μέγα HIR

productive year and on an unproductive day, the death will necessarily come on a productive day.[40]

11. In numbering, the third day is the most powerful.

12. For sore throat and ophthalmia, phlebotomy.

13. When the intestine is wounded, breath comes below by the wound invisibly, and the chest is emptied. Give milk and wine in equal proportions.

14. Those whose chest is saturated[41] stammer, tend to mania, are bald. Those of them that are congenitally cock-eyed are witless, silly, or maniacal. If not, there is a cure for the other evil.

15. The right breast and the right eye have the greatest force with regard to nature. The same with the lower parts, also because the male is engendered on the right.

16. To restrain menstruation in women, apply a very large cupping instrument to the breast.[42]

17. The three-month child exhibits everything, and the mother has milk then.

[40] I find the Greek here translatable, though not entirely coherent. Two of the three productive times (day, month, year) must coincide. It is still not clear what produces productiveness. Galen reads this section as three distinct aphorisms, of which he considers the first true (interpreting productive as critical). The second, on prediction, he considers potentially true but badly put. The third he takes to be a statement about birth and death in which productive and its opposite mean even and odd. That he considers un-Hippocratic nonsense.

[41] Galen interprets "hairy"; "bilious" is the common interpretation.

[42] *Aph.* 5.50.

18. Ἢν πολλὸν ῥέῃ γάλα, ἀνάγκη ἀσθενεῖν τὸ ἐν γαστρί.[276] ἢν στερεώτεροι ἔωσιν οἱ τιτθοὶ ὑγιηρότερον τὸ ἔμβρυον.

19. Φλὲψ ἔχει παχείη ἐν ἑκατέρῳ τιτθῷ· ταῦτα μέγιστον ἔχει μόριον ξυνέσιος.

20. Στραγγουρίην λύει φλεβοτομίη.

21. Ἢν τὰ ἄνω χωρία σπαργᾷ τὰ περὶ τὴν κεφαλήν, ἑλκέων κάθαρσις, ἔμετος.

22. Ὕδρωψ, ἀπὸ γαστρὸς ταραχῆς ἢ ἀπὸ βηχός.

22b. Καρκίνου γινομένου, τὸ στόμα πικραίνεται· διδόναι δὲ πίνειν ἐλατήριον δὶς ἢ τρὶς ἢν μὴ ψέλλος ᾖ· ἐπιδεῖν δὲ[277] χαλκοῦ ἄνθος, καύσας ἔστ' ἂν[278] πυρρὸν ᾖ, καὶ σπογγίην, ἢν μὴ ψέλλος ᾖ.

23. Ἀλύκης, φρίκης, χάσμης, οἶνος ἴσος ἴσῳ ἢ γάλα.

24. Ὠτὸς περιωδυνίην,[279] σικύην προσβάλλειν.

25. Ὅτι ἂν τῶν ἄνω πονέῃ, ὀδύνη ἐς τὰ ἰσχία ἢ ἐς τὰ γούνατα, καὶ ἆσθμα λύει πάντα τούτων γινομένων.[280]

138   26. Εἰλεοῦ λαπαροῦ, | ψυχρὸν οἶνον πολλὸν ἄκρητον κατ' ὀλίγον[281] διδόναι, ἔστ' ἂν ὕπνος ἢ σκελέων ὀδύνη γένηται· λύει δὲ καὶ πυρετὸς καὶ δυσεντερίη ἄνευ ὀδύνης. ἢν ὑποχόνδριον τεταμένον ᾖ, πιέζειν τῇ χειρὶ καὶ λούειν.[282]

---

[276] τὸ παιδίον V
[277] δεῖ HIR
[278] ἕως ἂν HIR

18. If much milk flows, the fetus must be weak. If the breasts are more solid the embryo will be healthier.

19. There is a thick vein in each breast. These things have the largest part in understanding.[43]

20. Phlebotomy relieves strangury.

21. If the upper parts around the head are swollen, use cleansing of wounds, vomiting.

22. Dropsy, from intestinal upset or cough.

22b. When a cancer has developed, the mouth becomes bitter. Give a purgative to drink twice or thrice, unless the patient be a stammerer. One must bind on rust of bronze, having heated it red hot, and a sponge, unless the patient be a stammerer.

23. For anxiety, shivering, yawning: wine with equal parts of water, or milk.

24. For painful ear, apply a cupping implement.

25. If there is distress in any upper area, pain to the hips or the knees or asthma resolves all these things when they occur.

26. For mild ileus,[44] give much chilled wine undiluted in small quantities until sleep or leg pain occurs. Fever and dysentery without pain also relieve it. If the hypochondrium is stretched, knead it with the hand and bathe the patient.

[43] Cf. *Epid.* 2.6.32, below.
[44] Intestinal obstruction.

---

279 περιωδυνίη V   280 γενόμενον HIR
281 Hipp. *De crisibus* 56: κατὰ λόγον mss.: "allmählich" Gal. comm.
282 HIR (v.l.) Gal.: λύειν Gal.: καὶ λούειν om. V

27. Παρωνυχίης, κηκὶς μέλαινα ἐν μέλιτι.

28. Ὕδατος ἀφιεμένου, γάλακτος ὀκτὼ κοτύλας δοῦναι πιεῖν· ἢν δὲ ἐμέῃ καὶ μὴ πίνῃ, μυττωτὸν δριμύν.

29. Ὥστε ἔχειν γυναῖκα ἐν γαστρί· πωλύπια ὑπὲρ φλογὸς ὀπτῶντα, ὡς θερμότατα καὶ πλεῖστα, ἡμίφλεκτα διδόναι τρώγειν,[283] καὶ τρίψαντα λίτρον αἰγύπτιον καὶ κορίανον καὶ κύμινον, κόλλικας ποιεῦντα, προστιθέναι τῷ αἰδοίῳ.[284]

30. Ἢν ἐκ κραιπάλης κεφαλὴν ἀλγέῃ, οἴνου ἀκρήτου κοτύλην πιεῖν·[285] ἢν δὲ ἄλλως κεφαλὴν ἀλγέῃ, ἄρτον ὡς θερμότατον ἐξ οἴνου ἀκρήτου[286] ἐσθίειν.

31. Ἢν ἄνθρωπον θέρμη ἔχῃ μὴ ἀπὸ χολῆς μηδὲ ἀπὸ φλέγματος ἀλλ᾽ ἢ ἀπὸ κόπου ἢ ἄλλως πυρεταίνῃ, ὕδωρ θερμῆναι πολλόν, ἔπειτα ὑπερχέων τὴν κεφαλὴν βρέχειν μέχρι τοὺς πόδας ἱδρώσει·[287] καὶ ἄλητον ἕψεσθαι ὡς παχύτατον. ἐπὴν δὲ ἱδρώσῃ τοὺς πόδας ὡς πλεῖστον, ἄλητον[288] καὶ θερμότατον ἐσθίων καὶ οἶνον ἄκρητον ἐπιπίνων, περιστειλάμενος ἱματίοισιν ἀναπαυέσθω. 31b. Εὐσκόπως[289] ἐμεῖν,[290] ναρκίσσου δύο ἢ τρεῖς κεφαλὰς ἐπὶ τῷ δείπνῳ ἐσθιέτω.

32. Τῷ μέλλοντι μαίνεσθαι τόδε προσημαίνει·[291] αἷμα συλλέγεται αὐτῷ ἐς[292] τοὺς τιτθούς.

---

[283] τράγειν V   [284] πρὸς τὸ αἰδοῖον V   [285] πίνειν V
[286] σὺν οἴνῳ ἀκρήτῳ HIR   [287] μέχρις ἂν . . . ἱδρώσῃ
HIR   [288] ἄλητον ὡς πλεῖστον mss.: I correct this from
Galen's comm.   [289] εὐκόπως HIR
[290] Smith: ἡμῖν mss.: "Erbrechen" Gal.

27. Suppuration under the fingernail: black oak gall in honey.

28. When water is taken away, give eight cotyls[45] of milk to drink. If there is vomiting and the patient will not drink, give astringent *muttatos*.[46]

29. To produce conception, give, as hot as possible, inkfish roasted over a flame, very hot and half-cooked, to nibble. Grind Egyptian nitre, coriander, and cumin, make small balls and insert in the vagina.[47]

30. If the patient suffer in the head from a hangover, give him a half pint of neat wine to drink. For other pains in the head have him eat very warm bread dipped in neat wine.[48]

31. If heat seize a man, not from bile or phlegm, but from fatigue or other feverish condition, heat much water, pour it over his head to soak it until the feet sweat. Boil thick gruel and, after the feet sweat generously, have him eat the gruel hot, drink undiluted wine, cover himself with cloaks, and rest. 31b. To vomit prudently, eat two or three heads of narcissus after dinner.

32. For one who is going to go mad, this sign foretells it: blood gathering in the breasts.[49]

[45] About four pints.

[46] A paste of honey, cheese, and garlic, though Galen says in his commentary here that it is a soup of bread and onion or garlic.

[47] Cf. the recipe in *Epid.* 2.6.9 above.

[48] Cf. above, *Epid.* 2.5.18.

[49] Cf. *Aph.* 5.40.

[291] προσημαίνει τὸ σημεῖον HIR
[292] ἐπὶ HIR

ΕΠΙΔΗΜΙΩΝ ΤΟ ΤΕΤΑΡΤΟΝ

V 144
Littré

1. Μετ᾽ ἰσημερίην καὶ μετὰ Πληϊάδα οἷα τὰ ἀνεσθιόμενα καὶ βλεννώδεα· ᾧ[1] τὴν κεφαλὴν ᾦξα ἀπῆλθεν ὑπὲρ τοῦ ὠτός. τῷ παρὰ Λεωκύδεος ἐν ποδί. Φανοδίκου οἱ δάκτυλοι οἱ ἐν τῷ ποδὶ ἐπί τοῦ στήθεος. ὁ τμηθεὶς τὴν κνήμην, ταύτῃ μὲν καὶ ἐμελάνθη ᾗ τὸ μέγα ἕλκος ἐν τῷ ἔξω τῆς κνήμης, καὶ ἐκ τοὔπισθεν ἤει· ἐπεὶ καθαρὸν ἐγένετο,[2] πλευροῦ ὀδύνη καὶ στήθεος κατ᾽ ἴξιν, ἀριστεροῦ, καὶ πυρετοί. ἀπέθανεν ἀπὸ τοῦ πυρετοῦ.

2. Τὸ χολῶδες τῷ σχοινοπλόκῳ κατακορές, καὶ τὰ καυστικά. καταφερομένῳ περὶ ἰσημερίην κάτω αἷμα πολὺ διῆλθε. γέροντι πάνυ σφόδρα ἀπεγένετο οὐ πρόσω τεσσαρεσκαίδεκα ἡμερέων. τῷ δὲ στιγματίῃ[3] παρὰ τἀντιφίλου[4] καυστικῷ κριθέντων ἑβδόμῃ χολώδει τυφώδει,[5] τρίτῃ μετὰ κρίσιν ἤει[6] οὕτως. αἷμα ἔπτυε.[7] περιεγένετο, καὶ ὑποστροφὴ ὕστερον ἐγένετο. ἐκρίθη ὡς εἰκὸς περὶ Πληϊάδων δύσιν τὸ πρῶτον, μετὰ δὲ Πληϊάδων δύσιν χολώδης ἐς μανίην. κρίσις[8] περὶ ἐνάτην ἄνευ ἱδρῶτος.[9]

---

[1] ᾇ HIR   [2] ἐγίνετο HIR
[3] Li.: στίγματι ᾗ V: στήγματι ᾗ HI: στήγματι ᾗ R

EPIDEMICS 4

1. After the equinox and the Pleiades, the affections were like consuming ulcers and mucous sores. The man whose head I opened had drainage behind the ear, the one in Leocydes' house on the foot. Phanodicus' toes were towards the ball of the foot. The man whose calf was cut developed a blackness on the outside of the calf where there was a large ulcer, which spread from the rear. After it became clean he had pain in the ribs and chest on the corresponding side, the left, and fever. He died from the fever.

2. The biliousness of the ropemaker was extreme, and the caustic fever. He was prostrated by it around the equinox and passed much blood below. An old man lost much, not before fourteen days. The branded slave near Antiphilus', who had caustic fever with crisis on the seventh day, biliousness and delirum, had the same evacuation on the third day after the crisis. He spat up blood. He survived and had a later relapse. It is likely that the first crisis was around the setting of the Pleiades. And after the setting of the Pleiades he was bilious to the extent of madness. A crisis about the ninth day, without sweating.

---

⁴ παρ' ἀντιφίλου HIR   ⁵ τυφώδει ἦ I: τ. ἢ HR
⁶ ἢ I: ἡ HR   ⁷ ἔπτυσε HIR
⁸ κρίσις περιεγένετο HIR   ⁹ ἰδρώτων HIR

3. Περὶ ἰσημερίην ὁ Καλχηδόνιος[10] ἀπὸ πυλέων μετακομισθεὶς παρ' ἀγορήν, ῥήγματος, περὶ μαζὸν δεξιὸν ὀδυνώμενος, ἔπτυεν | ἄλλοτε καὶ ἄλλοτε ὑπόχλωρον· γαστὴρ χαριέντως.[11] ἱδρὼς ἀρξάμενος ἑβδόμῃ εἶχε τὰ πολλὰ ἄχρις[12] ὀγδόης. ἐκρίθη τεσσαρεσκαιδεκάτῃ. περὶ τεσσαρακοστὴν ἤρθη περὶ τὰ ὦτά οἱ ἀμφότερα οὐκ ἀπεοικός. ἐδόκει ἔμπυος ἔσεσθαι, οὐκ ἐγένετο.

4. Στῆθος Ἀριστοδήμῳ ἐκαύθη τῷ Φίλιδος, ὁμοίως[13] ἀπέβη ἐκ πτώματος καὶ τούτῳ. προυπῆρχε δὲ ὀδύνη τις ἀνωτέρω.

5. Μετ' ἰσημερίας φθινοπωρινὰς ὑποστροφαί, καὶ ἄλλως ἄχρι[14] τροπέων χειμερινῶν.

6. Μεθ' ἡλίου τοῦ θερινοῦ ἡ Ἀχελῴου ἑκταίη ἀπέφθειρεν ἐμετώδης[15] ἐοῦσα καὶ φρικώδης· καὶ ἱδρῶτες· κρίσις τεσσαρεσκαιδεκάτῃ· ὁποσάμηνον οὐκ οἶδα· ἄρσεν δὲ καὶ ἄλλο πρὸς τὰς εἴκοσιν ἔφη, εἰ ἀληθέα οὐκ οἶδα.

7. Περὶ ἡλίου τροπὰς χειμερινὰς βόρεια ἦν. ἰκτεριώδεις ἐγένοντο κατακορέως, καὶ οἱ μὲν φρικώδεις, οἱ δὲ καὶ[16] οὔ. γλῶσσαι ξυγκεκαυμέναι τρίτῃ, καὶ ὄχλοι περὶ ἕκτην καὶ ἑβδόμην, οὗτοι[17] μακρὰν ἀποτείνοντες ἐς[18] τεσσαρεσκαιδεκάτην. γαστέρες ἀντεχόμεναι καὶ ἐν τῇσι φαρμακείῃσιν οὐχ ὑπακούουσαι κατὰ λόγον τῶν πυρετῶν, καὶ ἀνίδρωτες· σπλῆνες ἔστιν οἷσι σμι-

---

[10] καρχηδόνιος V   [11] χαριέντως om. HIR
[12] μέχρι V

88

3. Near the equinox, the Chalcedonian carried from the gates to the agora, with a fracture, severe pain by the right breast, spat greenish matter periodically. His belly did well. Sweat, beginning on the seventh day, was practically continuous until the eighth. Crisis on the fourteenth. About the fortieth he had mild swellings by both ears. It appeared he would become purulent. He did not.

4. The chest of Philis' son Aristodemus was cauterized. He, too, developed his affection from a fall. A pain in the upper area was the beginning for him.

5. Relapses after the fall equinox, and generally until the winter solstice.

6. At the summer sol(stice), Achelous' wife aborted on the sixth day, with vomiting and shivering. There were sweats, crisis on the fourteenth day. How many months pregnant I do not know. She said that (she had lost) another, a male, towards the twentieth day. I do not know whether that was true.

7. There was northerly weather toward the winter solstice. Patients became thoroughly jaundiced; some shivered, some did not. Their tongues were burnt on the third day. Upsets toward the sixth and seventh, which lasted a long time, to the fourteenth day. Intestines held back, not obedient to purgatives as they are in fevers. No sweating. Some had small, hard spleens. They had tension toward

---

13 Li.: οἵως mss.
14 μέχρι HIR
15 αἱματώδης HIR
16 om. HIR
17 καὶ οὗτοι HIR
18 om. HIR

κροὶ σκληροί·[19] πρὸς ὑποχόνδρια δεξιὰ ἐντεταμένοι[20] καὶ πρὸς χεῖρα βορβορύζοντες,[21] αἱμόρροοι, καὶ οὔροισιν ἡ κάθαρσις καὶ ἡ κρίσις. πολλῷ δὲ μᾶλλον εἴχοντο κάτω, καὶ γὰρ τοῦτο ἀπελαμβάνετο[22] ὑπὸ τὸν χρόνον τοῦτον οἷσι μὴ οὕτως ἦν, σπλῆνες δὲ[23] ἐπηρμένοι, αἱμορραγίη ἐξ ἀριστεροῦ. ἥλιος | ἐτράπετο, τὰ χειμερινὰ χειμερίως ἐν βορείοισι, μετὰ δὲ ὀλίγον νότια ἦν ἐφ᾽ ἡμέρας πεντεκαίδεκα, μετὰ δὲ ταῦτα νιφετὸς τεσσαρεσκαίδεκα ἡμέρησιν· ἀμφὶ ταῦτα τοῦ ἔτεος ἰκτεριώδεις κατακορεῖς οὐ κρινόμενοι εἰλικρινέως, φιλυποστροφώδεα. μετὰ[24] χιόνας νότια ὑπεγένετο καὶ ὑέτια. κόρυζαι κατερράγησαν[25] καὶ ξὺν πυρετοῖσι καὶ ἄνευ,[26] ἑνὶ δέ τινι καὶ ἐς ὀδόντας ἐκ τοῦ μέσου[27] προηληγηκότι ἐπὶ δεξιὰ καὶ ὀφρῦν καὶ ὄμμα. ἦσαν δὲ καὶ βραγχώδεις, καὶ φάρυγγες φλεγμαίνουσαι, καὶ οἱ σπόγγοι καλεόμενοι ἀνεῖχον καὶ τὰ παρὰ τὰ ὦτα ἐπάρματα μαλακὰ καὶ γνάθον ξὺν[28] πυρετῷ καθίστατο. ἀρχομένοισι πυρεταίνειν ἐγίνετο ἐπάνω καὶ ἐπὶ θάτερα τὰ πολλὰ τούτων, καὶ οἱ σπόγγοι εἰσὶν οἷσιν ὑπὸ τὸ μετόπωρον καὶ χειμῶνα, ἀτὰρ καὶ τὰ πιτυρώδεα· καὶ ἀπέφθειραν πολλαὶ παντοίως καὶ ἐδυστόκεον. ἕκτη τῇ παρθένῳ κριθέντα, ἕκτη ὑπετροπίασεν, ἐκρίθη δὲ δι᾽ ἕκτης. πάντα ἐν τούτοισι τοῖσι χρόνοισιν ἑκταῖα ὀγδοαῖα ἐκρίνετο.

8. Περὶ Πληϊάδων δύσιας ἡ Μαιανδρίου τοῦ τυφλοῦ αὐτίκα χλωρὸν καὶ αὐτίκα πυῶδες ἔπτυσε· περὶ

---

[19] om. HIR (H adds it above the line)

the right hypochondrium, rumbling when felt by the hand. Hemorrhages, purging by urine, and crises followed. But those whose affections did not follow this course (and they were far more numerous in that period) were bound up below. Their spleens were swollen; they had hemorrhage from the left nostril. Then the sun turned, winter was stormy and northerly and a little later southerly for fifteen days. Then it was snowy for fourteen days. In that part of the year jaundices, deep-colored ones, without clear crises; tendencies to relapse. After the snow, southerly weather and rains came on. Flows of phlegm broke through, both with and without fever. It went to the teeth of one who had previously had pain from the center to the right around the eyebrow and eye. There were hoarseness and inflamed pharynxes. The glands called sponges retained it, and soft swellings developed beside the ears and by the jaw with fever. At the beginning the fever occurred in the upper regions and on both sides for the most part, and the sponges for some towards fall and winter, and also the scrofulous eruptions. Many women aborted in many fashions, and had difficult births. For the maiden, reaching crisis on the sixth day, it returned on the sixth and reached crisis on the sixth. All affections reached crisis on the sixth or eighth day in that period.

8. Near the setting of the Pleiades the wife of blind Maeandrias suddenly started spitting greenish purulent

---

20 ἐκτεταμένοι HIR

21 ὑποβορβορύζοντες HIR

22 ἐπελαμβάνετο HIR

23 δὲ καὶ HIR

ἕκτην καὶ ἥπατος ζύμωσις καὶ κάτω ὑποχώρησις
ὀλίγη. σαρκοπυώδεα ἄνω· ὀλίγα λευκὰ πλατέα ἀν-
έπτυσεν. ἀπόσιτος. ἀπέθανεν ἐγγὺς εἰκοσταίη.

9. Ἡ[29] ἐκ τῶν γειτόνων Θέστορος οἰκέτις· ἐκ καυ-
στικῶν ὑποχωρήσιος,[30] χολώδεος συχνῆς ὑποχονδρίου ἐντεταμένου, τῇ ἕκτῃ ἐξ ἐπισχέσιος ἡ κοιλίη λεπτὰ
συχνὰ ἐς ἅπαξ διῆλθε, καὶ εὐθέως ἵδρωσε καὶ ἐκρίθη,
καὶ ἡ κοιλίη ἔστη· ἐς δὲ τὴν αὐτὴν ὥρην ῥιγώσασα
ἐπυρέτηνε[31] καὶ ἐς τὴν αὐτὴν πάλιν ὥρην.

10. Ἡ Θερσάνδρου, λευκοφλεγματώδης οὐ πάνυ
150 ἐοῦσα, θηλάζουσα | ἐπυρέτηνεν ἐν ὀξεῖ.[32] ταύτῃ γλῶσ-
σα[33] ξυνεκαύθη τῶν ἄλλων ξυγκαιομένων ὑπὸ τὸν
χρόνον τοῦτον. γλῶσσα δὲ[34] ἐτρηχύνετο ὥσπερ χαλα-
ζώδει πυκνῷ, καὶ ἑλμίνθια κατὰ στόμα. περὶ τὰς
εἴκοσιν[35] οὐ τελέως ἐκρίθη.

11. Περὶ δὲ Πληϊάδων δύσιας ὁ ἐκ[36] Μητροφάντου
τὴν κεφαλὴν πληγεὶς ὑπὸ ἑτέρου παιδὸς ὀστράκῳ, καὶ
ἀπογενόμενος δωδεκαταῖος ἐπυρέτηνε·[37] προφάσιος
δέ, σμήχουσα[38] ἔτριψε τὰ περὶ τὸ ἕλκος τις,[39] καὶ
μετεψύχθη. ἐξήρθη[40] τὰ χείλεα αὐτίκα, διελεπτύνθη τὸ
δέρμα πανταχῇ πρόσω ἀπὸ τοῦ ἕλκεος.[41] πρισθέντι δὲ
οὐ βραδέως οὔτε πύον ἐρρύη οὔτε ἐκουφίσθη, παρὰ δὲ

[24] μετὰ δὲ HIR   [25] κατεάγησαν V
[26] ἄνευ πυρετῶν HIR
[27] μέσου] ἀπομέσου HI: ἀπὸ μέσου R
[28] παρ᾽ ὦτα καὶ γνάθον ἐπάρματα μαλθακὰ καὶ ξὺν HIR
[29] om. IR   [30] ὑπὸ χλωρίσηος V
[31] ἐπυρέταινε IR   [32] ἐπυρέταινεν ὀξύ HIR

matter. Near the sixth day infection (lit.: fermentation) of the liver, small bowel movement. Purulence of flesh above. She spat up a few white flat fragments, could not eat. Died near the twentieth day.

9. The servant of Thestor's neighbors: after caustic fever, continuous bilious excrement, hypochondrium tight; on the sixth day, after being bound up, her intestine passed, once, much thin matter. She began straightway to sweat, and reached a crisis. Her bowels settled. Towards the same hour her fever and shivering recurred, and again towards the same hour.

10. Thersander's wife, slightly leucophlegmatic, was nursing; she developed acute fever. Her tongue was burned, and she was parched generally at that time. Her tongue grew hard like thick hailstones, and there were worms in her mouth. Incomplete crisis about the twentieth day.

11. At the setting of the Pleiades, the boy from Metrophantus' house, wounded in the head with a potsherd by another child, became feverish after twelve days had passed. The explanation: the woman who washed the wound rubbed the area around it and it took a chill. The lips of the wound puckered and the skin all around it grew thin. He was trephined without delay, but no pus ran off, nor was he eased. It was expected that he would fester be-

---

33 ἡ γλῶσσα HIR   34 τε HIR
35 περὶ δὲ τὴν εἰκοστὴν HR: περὶ δὲ τὴν η′ I
36 ἐν HIR   37 ἐπύρεσσε HIR
38 ὅτι σμήχων HIR   39 om. HIR
40 καὶ ἐξήρθη HIR
41 πολλαχῆ ἀπὸ τοῦ ἕλκεος πρόσω HIR

τὸ οὖς ἐδόκει παραπυήσειν⁴² ἐπὶ τῇ γένυϊ ἐπὶ τὰ
ἀριστερά, ταύτῃ γὰρ καὶ τὸ ἕλκος· ἔπειτα τοῦτό τε οὐκ
ἀπεπύει καὶ ὁ ὦμος ὁ δεξιὸς ἐνεπύησε ταχέως. ἀπέθανε περὶ τέσσαρας καὶ εἴκοσιν.

12. Μετὰ Πληϊάδων δύσιας ὁ τὸ οὖς⁴³ περὶ εἴκοσιν
ὕστερον ἄφωνος ἄφωνος τὰ ἐπὶ δεξιά τε ἀκρατής. ἀπύρετος
ἵδρωσε. δεξιὸν οὖς, δεξιὸς ὀφθαλμὸς ἑστήκει οὐ κάρτα, καὶ ἐφείλκετό⁴⁴ τι ἐκ τοῦ κάτω μέρεος. ἀριστερὰ δὲ
ἴλλαινεν ὡς ὀδυνώμενος.⁴⁵ τράχηλος σκληρὸς ἀπεγένετο τὴν αὐτὴν⁴⁶ ὥρην ἴσως. ὠδυνήθη ὕστερον.

13. Μετὰ Πληϊάδων δύσιν ὁ θεράπων ὁ τοῦ Ἀττικοῦ
ὑπὸ τεταρταίου ἁλισκόμενος, τυφώδης⁴⁷ ἱδρύθη. ἕτερος τὴν αὐτὴν ὥρην ἀληθεῖ τυφομανίῃ.⁴⁸ ἐς ἰσχία καὶ
σκέλεα ἦλθεν ὀδύνη. ἐπαύετο, ποσταῖος οὐ γινώσκω.
ταύτην τὴν ὥρην φρικώδεις ἐμετώδεις καὶ μετὰ κρίσιν
152 ἀπόσιτοι καὶ χολώδεις, καὶ σπλῆνες μεγάλοι | σκλη-
ροί, ὀδυνώδεις, καὶ αἱμορραγικοί, τοῖσι δὲ τὴν αὐτὴν
ὥρην μετὰ Πληϊάδων δύσιας ἐκ ῥινῶν αἷμα χολῶδες⁴⁹
ἐπισπλήνοισιν.

14. Ἐν Κραννῶνι τῇ Νικοστράτου ληφθείσῃ
τεσσαρεσκαιδεκάτῃ ἔφθασεν⁵⁰ αὐτίκα ἀκράτεια⁵¹ τραχήλου καὶ τῶν ἄλλων, καὶ σῖτος ἐγκατεκλείσθη μέχρι

⁴² παραπυΐσκειν HIR
⁴³ οὖς ἀλγήσας HIR   ⁴⁴ ἐφήλκωτό V
⁴⁵ ἴλλαινεν ὡς ὀδυν. Smith: ἴλλαεν (-αινεν R) αἰνῶς ὁ
ὀφθαλμὸς ὀδυνώμενος HIR: ἴλλαεν αἰνῶς ὀδυν. V
⁴⁶ τὴν αὐτὴν Smith: τρίτην mss. et edd.
⁴⁷ τυφλώδης VHIR (τῦφ ωδης H)

94

side the ear, at the jaw, on the left side (that was the side of the wound). As it happened, that failed to fester, and the right shoulder quickly developed an abscess. He died around the twenty-fourth day.

12. After the setting of the Pleiades, around twenty days later the man with ear trouble lost his voice and had loss of faculty on the right side. Sweat without fever. The right ear and right eye would not stabilize, and there was drawing from the part below. On the left he squinted as from pain. His neck became hard perhaps about the same time. Afterward his pain recurred.

13. After the setting of the Pleiades the servant of Atticus, taken by a quartan, settled into a coma. Another at the same season, genuine *typhomania*.[1] Pain came into the hips and legs, stopped, I don't know which day. At that time they had shivering, vomiting, no appetite after the crisis, were bilious. Spleens large, hard, painful and hemorrhagic. Some in the same season after the setting of the Pleiades had bilious blood from the nose in affections of the spleen.

14. In Crannon, Nicostratus' wife, who had had a seizure, on the fourteenth day suddenly experienced weakness in the neck and other parts, and food was shut in (con-

---

[1] Typhomania: the word "comatose" in the previous sentence is *typhodes*; *typhomania*, as Galen says in his glossary, should mean lethargy with delirium, although elsewhere he indicates that there was doubt about the meaning in antiquity.

---

48 τυφλομανίη HIR (-φ o- H)   49 χλοώδες HIR
50 ἔφθασαν V   51 Foës: ἀκράτεα mss.
52 om. V

δεκάτης. πνεῦμα πυκνὸν σμικρόν·[52] ἀκρασίη· ψηλα-
φῶσα δακτύλους, παραλέγουσα,[53] ἱδρῶτες. εἱλκύσθη
ἐπὶ δεξιὰ[54] τράχηλος, στόμα, ὄμμα, ῥίς. οὔρων ὑπόστασις λευκὴ ὀροβώδης, ἑτέρη λευκὴ ξυσματώδης,
ἄλλη[55] ὑπόχλωρος λεκιθώδης· ταύτῃ ἔστιν ὅτε ὡς
πιμελῶδες ἐφίστατο,[56] τοῦτο ἀθρόον, οὐκ ἐπὶ πολὺ
διεσκεδασμένον, οἷον τὸ ἐναιώρημα διεστηκός, οἷον
ἔξω ἐξ οἵου[57] τὸ ὑφιστάμενον ἔπειτα οὐρεῖται. καὶ τὸ
μέν τι τοιοῦτον· τὸ δὲ ἱδρυμένον ἄλλο τοιοῦτον ὀλίγον
ἐπὶ πλατὺ διεσκεδασμένον, ἄλλο τεταραγμένον. ἕτερον τοιοῦτο ἐναιώρημα νεφελίου ὑπομέλανος δοκέοντος πάχος ἔχειν, χαύνου δέ. ἄλλο λεπτόν. ἄλλο
ἐναιώρημα λεπτὸν τοιοῦτον. ἄλλο οἷον ἵππου. ἄλλο
οἷον τὰ ζοφώδεα.[58]

15. Ὁ πρῶτος παρενεχθείς, μειράκιον· τούτῳ οὖρον
καθαρὸν λεπτόν, πάντων διαχώρησις λεπτὴ πολλὴ
ἄχολος, γλῶσσα τρηχέη πάνυ, πυρετὸς περικαής,
ἄγρυπνος, κοιλίη κυρτή. οὗτος παρέκρουσεν, οἶμαι
ὀγδόη, τρόπον τὸν ἀκόλαστον, ἀνίστασθαι, μάχε-
σθαι, αἰσχρομυθεῖν ἰσχυρῶς, οὐ τοιοῦτος ἐών. τούτῳ,
154   ἀθρόον | οὔρων πολλῶν ἐλθόντων λεπτῶν ἐξ ἐπισχέ-
σιος. ὕπνος ἐγένετο ξυνεχής, καὶ ἱδρὼς κρίσιμος δοκέων ἐξ οὐ τοιούτου,[59] ἴσως περὶ δεκάτην. ἔπειτα
ἐξεμάνη τε αὖτις καὶ ἀπέθανε ταχέως ἑνδεκαταῖος,

---

[53] παραλγέουσα mss.: παραλέγειν Gal. 7.950 K
[54] τὰ δεξιὰ HIR
[55] ὀρροβ. . . . ἄλλη HIR: ξυσμ. ἄλλη ὀρροβ. ἐτ. λευκὴ V

stipation?) till the tenth. Fast, shallow breathing, loss of control, groping around with the fingers, delirious talk, sweat. Neck drawn to the right, also her mouth, eye, nostril. There was one sediment in her urine: white, like vetch seed; another, white, like scrapings; another, yellowish, like egg yolk. With that one sometimes a scum like fat; it was thick, not much dispersed, like the separated, suspended matter that occurs after urine has been withheld. Some was of that sort, some quite stable. In another instance a small amount of that kind scattered over a flat plane. In another instance it was turbid. Another: suspended matter of blackish clouds, that appeared to have thickness but were unsubstantial. In another instance it was thin; another of that kind but thin suspended matter. Another was like that of a horse, and another opaque.

15. The first to have delirium, a young man. His urine pure and thin. Much feces of all sorts, thin, not bilious. Tongue very hard. Burning fever, sleepless, belly bulging. He was delirious on the eighth day, I think, in the irrepressible way: leapt up, fought, used very foul language. (He wasn't that type.) He passed great quantities of thin urine after retention. He developed continuous sleep, and sweat which seemed critical, after not being so, perhaps towards the tenth day. Then his madness resumed and he died suddenly on the eleventh day. The cause, I think: he

---

56 πιμελῶς δὲ ἐφίστατο I: πιμελὴ διεφίστατο H
57 οἵου οὐ HIR
58 V adds πόσα ("How many there were!")
59 τοιοῦτος HIR

προφάσιος οἶμαι πιεῖν[60] ἄκρητον συχνὸν πρὶν ἐκμα-
νῆναι ὀλίγῳ. ἔτεα[61] αὐτῷ εἴκοσιν ἐγγύς.

16. Τοῦ φθινοπώρου ἤμεσε χολὴν μέλαιναν ἡ Εὐ-
μένεος, καὶ ὀδμαὶ[62] δὲ πρόδηλοι καὶ οἱ φρικώδεις
πυρετοὶ καὶ αἱ καρδιαλγίαι χολώδεις, βραχέα ἀνεμοῦ-
σα καὶ τὸ ἑλμίνθιον· διαχωρήματα λεπτὰ πάντα τὸν
χρόνον.

16b. Πρὸ Πληϊάδων δύσιος ὀλίγον τε[63] περὶ αὐτὰς
αἵ τε αἱμορραγίαι, καὶ βραχύτεροι οἱ πυρετοὶ καὶ
ὑποστροφώδεις αὐτίκα βραχείῃσιν ὑποστροφῇσι, καὶ
ἀπόσιτοι καὶ ἑφθοὶ καὶ ἀσώδεις καὶ καρδιαλγεῖς καὶ
θηριώδεις ἐν τῇσι κρίσεσι καὶ ῥιγώδεις καὶ χολώδεις.

17. Μειράκιον ξεῖνον τρίτῃ αἷμα ἐκ ῥινῶν πολύ, καὶ
τετάρτῃ καὶ πέμπτῃ· ἕκτῃ ἵστατο. κοσμίως παρέκρουσεν ἑβδομαῖος· γαστὴρ ἑστήκει, κωματώδης ἦν. ὑποστροφὴ τριταίῳ, ἐξέλιπε γαστὴρ ἀντεχομένη. οὖρον
οὐκ οἶδα· περὶ κρίσιν οἷον ἔδει.

18. Περὶ δὲ Πληϊάδων δύσιας νότια καὶ ὑέτια ἦν.
μειράκιον,[64] μυξώδεα ὑπόχολα πέπονα γλίσχρα δια-
χωρήματα συχνά·[65] πῦρ | ξυνεχές,[66] γλῶσσα ξηρή.
ἐκρίθη ἑκταῖος, ἑβδομαῖον αὖτις ἔλαβεν· ἀφῆκεν αὐ-
θημερὸν τρόμῳ. ὠτὸς ῥεῦμα κατ᾽ ἀριστερὸν γλίσχρον
παχὺ ἕκτῃ.

19. Τῷ παιδίῳ τῷ φαγεδαινωθέντι ὀδόντες οἱ ὑποκάτω καὶ τῶν ἄνω καὶ τῶν κάτω οἱ ἐμπρόσθιοι ἀν-

[60] τοῦ πιεῖν HIR
[61] ὀλίγῳ ἔτεα H: ὀλιγοετέα V: ὀλίγα ἔτεα IR

drank much undiluted wine just before going mad. He was near twenty years old.

16. In the fall Eumenes' wife vomited black bile. There was an obvious odor; fever with shivering; bilious heartburn; she vomited small quantities, including worms. She had thin bowel movements throughout.

16b. Before the setting of the Pleiades and for a time near them, hemorrhages, briefer fevers which tended to relapse immediately with brief relapses. They were unable to eat, languid, nauseous, cardalgic, had worms at the crisis, shivering, biliousness.

17. The young foreigner, on the third day, much blood from the nostrils, also on the fourth and fifth. It stopped on the sixth. He was delirious in a decorous way on the seventh. His bowel was stopped. He was comatose. Downturn on the third day, bowel let loose from being stopped. I do not know how the urine was. At the crisis it was as it should be.

18. Around the setting of the Pleiades weather southerly and rainy. A young man, feces mucous, bilious, ripe, slimy, frequent. Continous fever. Dry tongue. Crisis on the sixth day. It seized him again on the seventh, but departed the same day with trembling. Thick, sticky flow from the left ear on the sixth day.

19. The child with the phagedaenic ulcer (eroding sore): his lower teeth and the upper ones in front floated

---

62 αἱ ὀδμαὶ HIR
63 Li.: ὀλίγον τὸ V: ὀλίγωντο I: ὠλιγώντο H: ἐλέγοντο R
64 κατὰ τότε μειράκιον HIR
65 συχνὰ διαχωρήματα HIR
66 συνεχέες HIR

ἔπλεον· ἔγκοιλον εἶχον. ὀστέον ὧν μὲν ἐκ τῆς ὑπερῴης
ἀπέρχεται, μέση ἵζει ἡ ῥίς, ὧν δὲ οἱ ἄνω ὀδόντες οἱ
ἔμπροσθεν, πλατεῖα ἄκρη. ἀριθμούμενος[67] ὁ πέμπται-
ος ἀπὸ τῶν ἔμπροσθεν, τέσσαρας ῥίζας κατὰ δύο
ξυνεζευγμένας ὡς πρὸς ἑκάτερον τῶν γειτόνων ὀδόντων, ἄκρας ἀποκεκαμμένας ἐς τὸ ἔσω μέρος πάσας.
παρὰ τὸν τρίτον ὀδόντα ἀποπυήματα πλείω ἢ περὶ
τοὺς ἄλλους πάντας, καὶ τὰ ἐκ ῥινῶν παχέα ῥεύματα
καὶ ἀπὸ κροτάφων ὀδύναι ἐκ τούτου μάλιστα γίνονται.
ἐσθίεται οὗτος μάλιστα. ὁ πέμπτος ἐκ μέσου μὲν
κόνδυλον εἶχε, δύο δ' ἔμπροσθεν· ὁ σμικρὸς πρῶτος
ἔνδοθεν κατὰ τοὺς δύο ἐβέβρωτο. ῥίζαν μίαν παχέην
ὀξέην εἶχεν ὁ ἕβδομος. Τῷ Ἀθηνάδεω[68] παιδίῳ ἄρσενι
ὀδὼν ὁ[69] ἐπ' ἀριστερὰ κάτω, ἄνω δὲ ὁ[70] ἐπὶ δεξιά.
τούτου οὖς δεξιὸν ἐνεπύησεν οὐκ ἔτι ἀλγέοντος.

20. Μετὰ[71] Πληϊάδα εὐδίαι ἐπινέφελοι καὶ ὁμί-
χλαι.[72] κρίσιες πεμπταῖαι καὶ ἑκταῖαι καὶ ἑβδομαῖαι,
ἔτι δὲ καὶ μακρότεραι. ὑποστροφώδεις οἱ πυρετοὶ καὶ
ἔς τι[73] πλανώδεις καὶ ἀπόσιτοι καὶ χολώδεις, καὶ
δυσεντερίαι ἀπόσιτοι πυρώδεις. περὶ Πληϊάδων δύσιας νότια ἰσχυρῶς ἦν καὶ[74] αἱμορραγίαι καὶ τριταιοφυεῖς[75] καὶ ἠπιαλώδεις. ὁ ἐν τῷ σκυτείῳ ἡμορράγη-
σε, κατακορὴς διαχώρησις | ὀλίγη, ἐκρίθη ἑβδομαῖος
ῥίγει. ὁ παῖς ὁ παρὰ τὸ ἔσχατον καπηλεῖον ἡμορράγησε τεταρταῖος πολλόν. αὐτίκα ἐφλυήρει· γαστὴρ
ἀντίσχετο, ὑποχόνδριον ὀδυνῶδες,[76] σκληρόν. πρὸς

158

---

out, leaving a hollow. In those whose bone goes away from the palate, the nose sits down in the middle; in people whose upper front teeth are missing, it has a flattened tip. The number five tooth was separated from the first four, whose roots joined into two—each of the neighboring teeth joined—all of the tips were bent inward. There was more festering around the third tooth than around any other, and thick flows from the nostrils, and pains in the temples came especially from this one. This tooth was consumed especially. The fifth had a "knuckle" in the middle and two in front. The small first tooth was corroded inside by the two adjoining. The seventh had a single, sharp, thick root. The male child of Athenades: lower left tooth, upper right. He had festering in the right ear after the pain had gone.

20. After the Pleiad, pleasant weather with clouds and mist. Crises on fifth, sixth, seventh, and even later days. Fever tended to relapses, to produce delirium, loss of appetite, biliousness, and there were dysenteries with loss of appetite, high fever. About the setting of the Pleiades, much southerly weather. Hemorrhages, tertian-like fevers, nightmare fevers (*epialoi*). The man at the shoe shop had hemorrhage, a few bilious feces. Crisis on the seventh day with shivering. The boy (slave?) in the last shop hemorrhaged much on the fourth day. Straightway talked nonsense. Bowels stopped up. Painful, hard hypochondrium.

---

69 ὀδόντες οἱ V (cf. *Epid.* 5.44)
70 οἱ V   71 μετὰ δὲ HIR
72 ὀμίχλιαι HIR   73 Lind.: τὸ mss.
74 om. HIR   75 τριταιοφυέες HIR
76 ὀδυνώδεες VHIR: corr. Asul.

βάλανον ἑκταίῳ ὑπῆλθε κακὰ χλωρά. ἑβδόμῃ πρωὶ
ῥιπτασμὸς πολύς, βοή, φλεβῶν σφυγμοὶ παρ' ὀμφαλόν.

20b. Ἐν τοῖσιν ὀξυτάτοισι τῶν πυρετῶν οἱ σφυγμοὶ
πυκνότατοι καὶ μέγιστοι. οἷον τὸ ἐς δείλην παροξύνεσθαι, τοιοῦτον ἐν πάσῃ τῇ νούσῳ. πρὸς τὰς ἀρχὰς δὲ
καὶ οἱ παροξυσμοί, καὶ τὸ πρωιαίτερον σκεπτέον καὶ
τὸ ξυνεχέων καὶ τὸ τοῦ ἐνιαυτοῦ.

20c. Μετὰ Πληϊάδων δύσιας νότια.[77] πέμπτῃ κρι-
νόμενα, διαλείποντα ‹μίαν›,[78] μίαν λαμβάνει. τὰ δὲ
φολλικώδεα ἐπιφλυκταινούμενα, οἷα τῷ Ἀκανθίῳ λα-
τύπῳ.[79]

20d. Περὶ δὲ Πληϊάδων δύσιας ὑποψωρώδεα καὶ τὰ
τρηχέα, οὐκ ἐπιδακρύοντα μὲν μᾶλλον[80] ταύτην τὴν
ὥρην, ἀτὰρ καὶ τὰ λειχηνώδεα,[81] οἷα[82] τῇ Πυθοδώρου
καὶ τῷ καπήλῳ[83] ξὺν πυρετῷ ἅμα ἀρχομένῳ σχεδόν.
τῇ Πυθοδώρου ἰσχία ἀκρατέα.

20e. Μετὰ Πληϊάδων δύσιας φρικώδεις, αἱμορραγικοὶ ἐκ ῥινῶν. ὁ μέν γ' λάβρως, ὁ[84] σκυτεύς, ἐκρίθη
ἑβδομαῖος, μίαν διαλιπών, μίαν ἐλάμβανεν αὖτις,
τετάρτῃ ἐκρίθη. ἄλλος τῶν παρὰ Λεωκύδεος ἐκρίθη
ἑβδόμῃ, ἄλλος τετάρτῃ. Μόσχος, ἐνάτῃ λάβρον ἐξ
ἀριστεροῦ, βραχὺ δ' ἐκ δεξιοῦ·[85] πρὸς τὰς τεσσαρεσκαίδεκα ἐς κρίσιν ἢ ἔδει· ἀρξάμενα παρωξύνετο ἁμαρ-

---

[77] νότια ἦν V   [78] add. Deichgräber
[79] λαπύτῳ HIR   [80] τρηχέα τὰ κνησμώδεα οὐκ ἐπιδακρύοντα μᾶλλον μὲν HIR

102

After a suppository on the sixth day he passed greenish, bad material. On the seventh, early, tossing about, loud crying out, throbbing of blood vessels by the navel.

20b. In the most severe fevers the pulse is strongest and most frequent. When exacerbations are towards afternoon, the same occur through the whole disease. One must examine the exacerbations at the beginning, and what occurs earlier, the character of the continuous fevers and that of the year.

20c. After the setting of the Pleiades, southerly weather. Crises on the fifth day, remissions one day, accessions one. Scaliness and blisters, as with the mason Acanthius.

20d. Around the setting of the Pleiades, there appeared slightly itchy, hard areas, without weeping to a greater extent in that season. Also the tumified pustules as with Pythodorus' wife and the shopkeeper, virtually at the commencement of fever. Pythodorus' wife had weakness in the hips.

20e. After the setting of the Pleiades, fevers with shivering, hemorrhages from the nose. The shoemaker, who bled vigorously, had a crisis on the seventh; it remitted one day and again came for one; crisis on the fourth day. Another, one of Leocydes' people, had a crisis on the seventh day, another on the fourth. Moschus bled vigorously on the ninth from the left nostril, and briefly from the right: towards the fourteenth he reached the appropriate crisis. It

---

81 The mss. add ἐξαιρόμενα ταῦτα here; I exclude it as a gloss
82 οἷα V: ἦν οἷ I: ἦν οἱ H: ἦν οἱ R
83 κατηλίῳ HIR   84 ὁ μέντοι HIR
85 δεξιοῦ μυκτῆρος HIR

ΕΠΙΔΗΜΙΑΙ

τάδες βρωμάτων ιζ΄ ἐγένοντο.⁸⁶ παρὰ τὸ οὖς ἐπὶ δεξιὰ
σμικρὸν ἔσωθεν σκληρόν, ἔξωθεν σμικρὸν χαῦνον
ὀδυνῶδες, οὐδὲν ἐπιδιδόν· ἐννεακαιδεκάτῃ ἀπεγέ|νετο
160   νυκτός.

20f. Τοῖσι πάνυ χολώδεσιν, ἐμπύοισι μάλιστα, ὀλῷ
ἰκέλη⁸⁷ ἡ κάθαρσις· οἷον ὁ τὴν σικύην προσβαλλό-
μενος ἐπεὶ ἐς τὸ ἰσχίον ὀδύνη, τούτῳ ἐς σκέλος⁸⁸ κάτω
ἦλθε, καὶ ἐρρήϊσεν. ὁ ἀπὸ τοῦ κεραμέου ἰπνοῦ⁸⁹ κατα-
πεσών, ᾧ οὐ προσεβλήθη αὐτίκα σικύη, ἐκάμφθη⁹⁰
ἔσω, καὶ εἰκοστῇ ἐπαλιγκότησεν, αἱμορραγικὰ⁹¹ καὶ
τρυγώδεα καὶ ἐσθιόμενα.

20g. Ἡ Τενεδίη τεταρταίη ἀπέφθειρεν, ὡς ἔφη,
τριηκοσταῖον ἄρσεν. γαστὴρ ὑγρὰ λεπτά, ξυνεκαύθη
γλῶσσα· κρίσις τεταρταίη.

20h. Μετὰ Πληϊάδων δύσιας σπληνώδεα, καὶ μέχρι
πέμπτης ἔρρει. ἐκρίθη ἑβδομαίῳ⁹² ἐόντι, οὖρον οἷον
ὀρόβων πλύμα, ὅμοιον αὐτὸ ἑωυτῷ, ἔπειτα καθαρόν·
ὑποστροφή. διέλειπέ τε καὶ τῷ Μεγαρέος, πλὴν αἷμα⁹³
οὐκ ἐρρύη. οὖρον οἷον τὸ Ἀντιγένεος·⁹⁴ λευκόν, παχύ,
ὅμοιον.

21. Ἡλίου χειμερινῶν τροπέων ἄστρον οὐ σμικρόν,
πέμπτῃ δ᾽ ὕστερον καὶ ἕκτῃ⁹⁵ σεισμός. ὅτ᾽⁹⁶ ἐν Περίν-

⁸⁶ ἐπεγένοντο V   ⁸⁷ ὅλως ἐπὶ σκέλεα V (cf. Epid.
2.3.1, Gal. Gloss. s.v. ὅλον)   ⁸⁸ σκέλεος V
⁸⁹ ἵππου mss.: corr. Struve   ⁹⁰ ἐκαύθη HIR
⁹¹ αἱμορραγικαὶ V   ⁹² ἑβδ. δὲ HIR
⁹³ ὑποστροφὴ τίς διέλειπε ε΄ καὶ τὸ μεγαλόσπληνον αἷμα
V: Langholf conjectures ὑποστροφὴ τῇ ε΄ διέλειπε ε΄

104

started again, grew worse. There were errors in diet at the seventeenth day. By the ear on the right, a small hardness inside, a small spongy area outside, painful; it did not increase; on the nineteenth day, it disappeared in the night.

20f. Very bilious people, especially when there are abscesses, have purgings like ink of cuttlefish;[2] for example, the one who had the cupping glass applied, when he had pain at the hip; it went below to the leg, and he was eased. The one who fell from the kiln in the ceramic factory: he did not have the cupping glass immediately; it turned inward, and on the twentieth day it became nasty: hemorrhages, thick expectoration which was corrosive.

20g. The woman from Tenedos aborted on the fourth day, so she said, a thirty-day-old male fetus. Intestines watery, thin. Tongue parched. Crisis, fourth day.

20h. After the setting of the Pleiades one had splenic problems, hemorrhage until the fifth day. Crisis on the seventh day, urine like wash-water of vetch seeds, uniform throughout, later pure. Relapse. It remitted also for the son of Megareus, but he had no hemorrhage. Urine, like that of Antigenes, white, thick, uniform.

21. During the winter solstice, a large star. On the fifth and sixth following, earthquake. When we were in Perin-

---

[2] "Like ink of cuttlefish" should perhaps be amended to read "towards the legs," the reading of ms. V and of *Epid.* 2.3.1, a parallel passage.

---

94 Ἀρτιγένεος HIR
95 καὶ ἕκτῃ] ἑκάστων HIR
96 ἤ τ᾽ HIR

θῳ ἦμεν, ἡ ἀσθματώδης ἡ Ἀντιγένεος, ἢ οὐκ ᾔδει[97]
εἰ ἔχει· ἐρυθρὰ ἐπιφαινόμενα ἄλλοτε· | καὶ ἄλλοτε
162 γαστὴρ σμικρή, ἄλλοτε μεγάλη, οἴομαι. βήσσουσα
γὰρ ἐτύγχανεν ὁδοιοπορήσασα[98] θᾶσσον. μεὶς[99] ἦν
ὄγδοος. ἱδρύθη προπυρεταίνουσα.

22. Ἡ δὲ τοῦ Ἀπημάντου ἀδελφεοῦ γυνὴ ἀπέφθειρεν ἑξηκονθήμερον ὡς ἔφη ἑβδομαίη θῆλυ. περὶ δ᾿
ἐνάτην ἐθορυβήθη. μετὰ δὲ κρίσιν[100] τὰ δεξιὰ ἤλγει
ὡς ἂν ὑπὸ ἀποστροφῆς. αὕτη εἶχε ταχέως καὶ ἀπέφθειρεν ἑτέρην[101] λευκοῖσιν. θυγατέρα ἔτεκεν ἐρυθροῖσιν[102] ὡς ἔδει.

23. Φρικώδεις ἀσώδεις ἀπόσιτοι ὑποστροφώδεις
χολώδεις αἱμορραγικοὶ ὑπόσπληνοι ὀδυνώδεα τρόπον
ἐκ τῶν ἀριστερῶν οἱ πλεῖστοι. τῇ Ἀπημάντου, ᾗ
ἔστραπτο, ταύτῃ τὸν ὀφθαλμὸν τὸν δεξιόν, τὸν δὲ
κενεῶνα ἐπὶ τὰ ἕτερα. ἡ Ἀριστοφῶντος θυγάτηρ τῇ
τρίτῃ καὶ τῇ πέμπτῃ ἐπυρέτηνε. ξηρὴ διετέλει τὰ
πλεῖστα, γαστὴρ μέντοι ταραχώδης. ταύτῃ δύσκριτα
ὑπὲρ τριήκοντα ἐπαύσατο. φλύκταιναι μὲν ἐκ κόπων
οὐ κάρτα ἰσχυρῶν ἀφικνεῦνται, ἐς ἑβδόμην ὑποπέλιοι.
ῥῖγος τῇ ὄπισθεν τοῦ Ἡρώου παιδίσκῃ ἐγένετο. αἱ δὲ
λευκαὶ μεγάλαι οὐδ᾿ αὐταὶ πάνυ χρησταί. τῶν κατ-
164 οιχέων καὶ ὑπνωδέων καὶ μὴ ξυμπιπτόντων καὶ χολῆς
ἀντεχομένης, καὶ ἢν ὑγρὴ ᾖ ἢ σκληρή, οὐ ξυμπί-
ποντες.

[97] εἰδυῖα HIR
[98] ὁδοιπορῆσαι mss.: corr. Erm.

thus the asthmatic woman, Antigenes' wife, who did not know whether she was pregnant. She had red patches on her skin sometimes. Sometimes her belly was small, sometimes large, I think. She had come down with a cough after a hurried journey. It was the eighth month. It had become established after a spell of fever.

22. The wife of Apemantus' brother on the seventh day aborted a female fetus of sixty days, she said. Around the ninth she was upset. After the crisis she had pain in the right side as though it had turned back up. She conceived again immediately, and aborted another girl with white fluxes. She gave birth to a girl with red fluxes, as is proper.

23. Patients were shivering, nauseous, without appetite, relapsing, bilious, hemorrhagic, splenic, mostly of a painful sort on the left. In Apemantus' wife it turned about: she had a painful eye on the right and painful flank on the left. Aristophon's daughter was feverish on the third and fifth day. She was dry throughout for the most part, but her belly was upset. It stopped around the thirtieth day without a clear crisis. Blisters came up from work that was not severe; on the seventh day they were somewhat livid. The young woman behind the Heroon had shivering, and large white ones (blisters?), but they were not very productive. In catalepsy and drowsiness, if there is no collapse, even if the bile (i.e. bilious feces?) be withheld, whether it be wet or dry they do not suffer collapse.[3]

---

[3] I give the general sense of this apparently corrupt sentence.

---

[99] μὴν HIR   [100] Asul.: κράσιν mss.
[101] Smith: ἑτέρη mss.
[102] ἐν ἐρυθροῖσιν HIR

Ζωίλου τοῦ τέκτονος τρομώδεις σφυγμοὶ νωθροί. οὔρησις καὶ κοιλίη μετρίως ἀχρόως. ἤτρου ἔντασις ἑκατέρωθεν ἐς ἰθὺ μέχρι ὀμφαλοῦ σὺν ὀξεῖ. ἀπόσιτος, διψώδης.

24. Ἡ παρὰ τῇ Κόμεω[103] ἀγορηνόμου θυγάτηρ· ἐν γαστρὶ ἀσήμως ἀρξαμένη διμήνου ἔμετος φλεγματώδης, ὁτὲ δὲ χολώδης, ἔρρηξε. χαλεπῶς ἔτεκε, τελέως ἐκαθάρθη. ἔμετος ὁμοίως μέχρι τριηκοστῆς, ἔπειτα γαστὴρ ἐταράχθη, καὶ ὁ ἔμετος ἐπαύσατο. λειεντερίη. γυναικεῖα οὐκ ἐγένετο δύο ἐτέων.[104] χειμῶνος ἔσχεν αἱμορροΐδας.

25. Οἱ δύο ἀδελφεοὶ οἱ τοῦ Κέκροπος οἰκεῖοι, οἷσι τὰ μέλανα κατ᾽ ἀρχὰς διῄει, ὑπότρυγα καρυκοειδέα,[105] ἐκ κατακορέων ἀφρωδέων χλοώδεις[106] ἦσαν. ὃς τὸν οἶνον[107] ἐκ ξυνθήκης ἦρεν, ἐπυρέτηνεν[108] αὐτίκα. τριταίῳ ἡμορράγησεν, τετάρτῃ καὶ πέμπτῃ καὶ ἑβδόμῃ καὶ ὀγδόῃ· ἐκρίθη· κοιλίη ὑγρή. ὁ ἐκ μετάλλων,[109] 166 ὑποχόνδριον δεξιὸν | ἐκτεταμένον,[110] σπλὴν μέγας, κοιλίη[111] ἐντεταμένη, ὑπόσκληρος,[112] πνευματώδης, ἄχροος· τούτῳ ἐς γόνυ ἀριστερόν· ὑποστροφή· δι᾽ ὅλου ἐκρίθη. ὁ Τιμένεω εἶχέ τι πνεύματος, ὡς χλοώδης εἶναι· ὁτὲ[113] ἐς χεῖρας χλοώδης. ὁ τῆς λεχοῦς ἀνὴρ ὁ παρὰ τὰ σιτοδόκα[114] ὁ ἰκτεριώδης, πρὸς ὃν ἑβδομαῖον

103 Τηκομαίω mss.: corr. Meineke   104 Foës: ἑτέρων mss.
105 βαρυκοείδεα HIR: καὶ ῥυκκοειδέα V bis (i.e. also in the repetition of this section after *Epid.* 4.60) (cf. Erot. 49.6N) ·
106 χολώδεες V bis   107 ὄνον HIR
108 ἐπυρέταινεν HIR

Zoilus the carpenter had a sluggish trembling pulse. Urine and intestines slightly off color. Stretching of lower abdomen on both sides in a straight line up to the navel with a point. No appetite. Thirsty.

24. The daughter who lived with the wife of Comes, the market supervisor, conceived with no definite signs; at two months there broke out phlegmatic vomit, sometimes bilious. She had a difficult delivery. Complete purgation. Similar vomiting to the thirtieth day. Then her belly was upset and the vomiting ceased. Lientery. For two years her menses did not occur. She had hemorrhoids in the winter.

25. The two brothers who lived with Cecrops had black stools to begin with, like wine lees, like blood sauce: then after they had very bilious, foamy bowel movements, they were chlorotic. The one who lifted the wine[4] from the depository became feverish immediately. He hemorrhaged on the third and again on the fourth, fifth, and seventh and eighth days. Crisis. Moist intestines. The one from the mines: right hypochondrium stretched, spleen large, belly tight, rather hard; flatulent. He was pale. His went to the left knee. Relapse. General crisis. Timenes' son had breathing symptoms and became chlorotic. Paleness sometimes extended to his arms. The husband of the recently delivered woman, the one who lived by the granary, the one with jaundice: I came to him on the seventh day, he

4 Or, following mss. HIR, "lifted the donkey."

---

109 τῶν μετάλλων Gal. *Diff. resp.* 7.951
110 ἐντεταμένον Gal.   111 καὶ κοιλίη HIR
112 ὑπόσκληρα HIR   113 εἶναι ὁτὲ] εἰνέω τὲ vel sim.
HIR   114 σιτοδόκου HIR

ἐσῆλθον,[115] ὀγδόῃ ἀπώλετο οὔτε οὐρῶν οὔτε διαχω-
ρέων· ὑποχόνδρια μεγάλα καὶ σκληρά, καὶ πνεῦμα
πυκνόν. οὗτος[116] ἀπολλύμενος οὐδ᾽ ὑπὸ τοῦ[117] πόνου
ἐνότισε τὸ μέτωπον. ἡ τούτου γυνὴ ἐξέβαλε θῆλυ
ἑβδόμῃ ἐν[118] ἑβδόμῳ μηνί, ἐφάνη δὲ τετάρτῃ· ταύτην
μὲν ἐν ἀρχῇ ποδῶν ὀδύνη εἶχε, λήγοντος[119] δὲ τοῦ
πυρετοῦ πνεῦμα οὐκ ἐλύθη, ἀλλ᾽ ἐγκατελείφθη·[120] ἐς
χεῖρας ὀδύνη καὶ βραχίονα· ὑποστροφὴ διὰ μηκίστου
ἔλαβεν· ῥῖγος·[121] ἐπέσχητο οὖρα πρὸ κρίσιος. τῇ
παρὰ Τιμένεω ἀδελφεῇ[122] ξὺν ῥίγει ἐκρίθη· τοῦ χλο-
168   ώδεος δὲ[123] λήγοντος ταχέως χεῖρες καὶ ὦμοι, τού|των
δὲ ληγόντων κεφαλῆς, ὀφθαλμῶν· τὰ ἄνω βλέφαρα
ἐπήρθη, καὶ ἐδακρυρρόει· οὐκ οἶδα τὰ λοιπά· ἐκρίθη
περὶ ἑβδόμην τοῦ πρώτου. ὁ[124] Μενάνδρου ἀμπελουρ-
γὸς ὡσαύτως, πλὴν ἡ[125] γαστὴρ κατ᾽ ἀρχὰς λεπτὰ
ἐδίδου, ἔπειτα ἔστη, καὶ τὸ οὖρον. ἐκρίθη· οὐκ ἐρρίγωσεν ἑβδόμῃ, ἦρα ὅτι γαστὴρ προεταράχθη;

Ὁ Ποτάμωνος, τούτῳ[126] κοιλίη οὐ διῄει ἑβδο-
μαίῳ·[127] πρὸ κρίσιος δύο ἡμέρας οὐκ ἐρρίγωσε· διὰ
τοῦτο οὐδὲ τὸ οὖρον ἔσχετο. ὁ ὀδοὺς τοῦ[128] Ἡγησι-
στρατίου, ᾧ τὸ[129] ἀπόπημα παρ᾽ ὀφθαλμόν, καὶ ἀπ-
επύησε παρὰ τὸν ἔσχατον, καὶ αὖτις ἐξιήθη ὁ ὀφθαλμὸς καὶ κατὰ ῥῖνας πύον ἧκε[130] παχύ. παρὰ τὸ οὖλον
σαρκία σμικρὰ στρογγύλα ἀπῆλθε. τούτῳ παρὰ τὸν

---

115 Gal. *Diff. resp.* 7.951: εἰσῆλθεν mss.
116 οὕτως V   117 om. HIR   118 om. HIR
119 διαίροντος Gal.   120 ἐγκατελήφθη HIR

died on the eighth. No urine, no feces. Hypochondria
large and hard. Rapid breathing. As he died he did not
have a damp brow, not even from the agony. His wife
aborted a female child in her seventh month on the seventh day. There were indications, on the fourth day. A pain
of the feet possessed her at the beginning, and when the
fever left her she did not lose the breathing symptoms,
but they became fixed. The pain went to the hands and
arm. The relapse was very lengthy. Shivering. Urine withheld before the crisis. The sister of Timenes (who lived
with him) had a crisis with shivering. When the paleness
stopped she quickly developed trouble in the hands and
shoulders; when they improved, around the head and eyes.
Her upper lids swelled. Many tears. I do not know anything more. Crisis on the seventh day the first time.
Menander's vinedresser went in the same fashion, except
that at first his belly produced thin matter, then stopped;
also the urine. Crisis. He did not shiver on the seventh day.
Was it because his belly was upset earlier?

Potamon's son: he did not have loose bowels on the
seventh, and for two days before the crisis did not have
shivering. Because of that his urine was not withheld.
Hegesistratius' tooth: he had a suppuration by his eye and
also by his back tooth. The eye later healed and he exuded
thick pus through the nostrils. At the gum, small, round,
fleshy, globs came away. He seemed to be going to suppu-

---

121 Li.: οἶος mss.   122 ἀδελφῆ V   123 om. V
124 ὁ δὲ HIR   125 om. V   126 τούτου V
127 after ἐβδομαίῳ V repeats ὁ Ποτάμωνος
128 τῇ V   129 ᾧ τὸ] ὅτῳ HIR   130 ἢ καὶ V

τρίτον ἐδόκει ἀποπυήσειν,[131] ἔπειτα ἀπετρέπετο. ἐξαίφνης δὲ ᾤδησεν ἡ γνάθος καὶ ὀφθαλμοί.

Οἷσιν ἐς τοὺς ὀφθαλμοὺς ἀποστάσιες ἐν καύσοι-
σιν, ἐξέρυθροι γνάθους καὶ αἱμορραγικοί, ἀτὰρ καὶ
οἷσι παρὰ τὰ ὦτά ἐστιν· ἴσως δὲ καὶ ἀποστάσιες ἐς
ἄρθρα μᾶλλον· οὐ μὴν σάφα οἶδα. τὰ ῥίγεα τρομώ-
δεσιν.

Ὑποχονδρίου ἔντασις, γυναικεῖα ἐπεφάνη δεκάτῃ
καὶ ἑβδόμῃ. ἐκρίθη διὰ τέλεος. ᾗ γε μὴ[132] οὕτως
ἐνέμεινεν, ἐκρίθη τρίτῃ· ἄλλῃ πέμπτῃ· ἄλλῃ ἑβδόμῃ
ἐκρίθη.

Τῷ Ἡγησιστρατίῳ οἱ δύο ἔσχατοι[133] τὰ πρὸς
ἀλλήλους ἐβέβρωντο· ὁ ἔσχατος εἶχεν ἄνωθεν τοῦ
οὔλου[134] δύο κονδύλους, ἕνα μὲν κατὰ βρῶμα, ἕνα δὲ
ἐπὶ θάτερον· ᾗ δὲ οἱ δύο, ταύτῃ ῥίζῃ[135] πλατείᾳ ἰκέλη
170 ἐκ δυοῖν[136] | ξυνέπιπτεν· ἐπὶ τὰ ἕτερα μία ἡμίσεια
στρογγύλη.[137]

Γυναικὶ ᾗ[138] ἡμορράγησε τετάρτῃ καὶ ἕκτῃ, ἑβδομαίῃ ἐκρίθη· ἐξέρυθρος. γυναικὶ καρηβαρικῇ ἰσχυρῶς, ταύτῃ ἐκρίθη περὶ εἰκοστήν· εἰκοστῇ καυσώδης
ὑποχόνδρια· ἑβδόμῃ οὐ κάρτα ἡμορράγησε· διαχωρήματα λεπτά· ἐς ὀφθαλμὸν δεξιὸν περὶ ὀγδόην. ἀνδρὶ
ταὐτὰ πλὴν ἑβδόμῃ[139] ἐκρίθη. ὑπόσπληνος ἐς τὰ ἀριστερὰ ὀγδοηκοσταίῳ· καὶ χρονιώτερα τούτῳ τὰ τοῦ

---

131 ἀποπυήσειν HIR
132 εἴ γε μὴ V: ἤ (ἤ R) γε μὴν (μὴ IR) HIR: corr. Li.
133 οἱ ἔσχατοι HIR   134 ὅλου V

rate by the third tooth, then it turned back. Suddenly his jaw swelled, and his eyes.

When in burning fevers there are apostases to the eyes, patients get red at the jaws and have hemorrhages. Some have them by the ears. And maybe, also, more frequent apostases to the joints; I do not really know. Shivering is associated with trembling.

One woman: stretching of the hypochondria. Menses appeared on the seventeenth day, complete crisis. In cases where the menses were not held back one had a crisis on the third, another on the fifth, another on the seventh.

In Hegesistratius, his two back teeth were eaten away in the area where they met. The back one had two "knuckles" above the gum, one next to where it was eaten away, the other opposite. Where these two were, there was a single plinth-like root formed from two, and on the other side a single half round root.

For the woman who had the hemorrhage on the fourth and sixth, the crisis came on the seventh; she was ruddy. The woman who was so heavy-headed had her crisis on the twentieth. On the twentieth burning heat in the hypochondria. No hemorrhage on the seventh. Thin feces. It went to her right eye on the eighth day. A man had the same except for the crisis on the seventh day. His spleen swelled on the left, eightieth day, and his eye problem

---

135 ῥίζαι mss.: ῥίζα Langholf
136 Smith: δύο mss.
137 πρὸς στρογγύλη HIR
138 γυναικείη V
139 ἑβδόμης mss.: corr. Asul.

ὀφθαλμοῦ, ἴσως ὅτι ὕστερον τῆς κρίσιος καὶ ὅτι πολλῷ.[140]

26. Ἡ Τιμένεω[141] ἀδελφιδῆ[142] πνευματώδης, ὑποχόνδρια καὶ ἐντεταμένα ἐφάνη διὰ χρόνου· εἰ δὲ καὶ
εἶχέ τι νήπιον, οὐκ οἶδα. γαστὴρ τὰ πρῶτα ἑστηκυῖα
καὶ ἐμετώδης τότε, ἔπειτα οὐκ ἔτι. γαστὴρ διῄει πολλὰ
γλισχρόχολα, ὑποχόνδριον οὐκ ἐκώλυεν. ἑνδεκάτῃ ἐς
τὸν μέγαν τῆς δεξιῆς φλεγμονή· κατάρρηξις, καὶ ἐπεὶ
ἐχώρει ἀνωτέρω. τούτου γενομένου βελτίων ἐγένετο·
καὶ γὰρ καταφορὴ ἧσσον καὶ πυρετός, καί τινι καὶ
εὐπνοωτέρη, ὅτι ἄνω ἤμει ἔμετον κακόν. ἑξκαιδεκάτῃ
ἰσχνὸν ἐγένετο καὶ πνεῦμα πυκνὸν καὶ πυρετός· ἀπέθανεν. αὕτη ἐπύρεξε πρὸ τῆς ἀποστάσιος· ἑβδόμῃ
μετὰ τὴν ἀπόστασιν ἀπώλετο· ἦν καὶ αὕτη τοῦ ἐξ-
ερύθρου τρόπου. |

172   27. Ὁ παῖς ὃς ἦν τῆς γυναικὸς τῆς τοῦ Ἀπημάντου
ἀδελφεῆς, ὑποχόνδρια μεγάλα καὶ σπλήν, πνεῦμα·[143]
διαχώρησις γλισχρόχολος[144] ὑπόκοπρος· κοπιώδης ἐξ
ἔργων· εἰκοσταῖος ἐς πόδας, καὶ κρίσις.[145] ἦρα τοῖσι
κοπιώδεσιν ἐς ἄρθρα καὶ οὐκ ἐς ὀφθαλμόν; τὰ ὑποχόνδρια δὲ ἐντεταμένα[146] ἦν καὶ δὴ καὶ ἦν[147] τι βηχίον
ξηρὸν ἰσχύῃ.[148]

28. Τὰ ἐγκαταλιμπανόμενα μετὰ κρίσιν ὑποστρο-
φώδεα, καὶ τὰ ἐν αὐτῇσι τῇσι νούσοισιν ἀποκρινό-

[140] πολλά HIR   [141] ἠμένεω (vel -έω) mss.: ἡ Τιμαίνεω
Gal. *Diff. resp.* 7.953 K.   [142] ἀδελφὴ HIR: om. Gal.
[143] πνευματώδης HIR   [144] πικρόχολος Gal.

lasted longer, perhaps because it came after the crisis, and perhaps because it came with much suppuration.

26. Timenes' niece had difficulty breathing. Hypochondria appeared tight for a long period. Was she carrying an infant? I do not know. Bowels stopped at first, when she tended to vomit, then free. They passed much slimy bilious matter. Hypochondrium did not inhibit it. On the eleventh day inflammation went into the large toe of the right foot. Diarrhea, even when there was evacuation upward. After this occurred, she improved: she had less delirium and fever, breathed slightly better, because she vomited up vile matter. On the sixteenth day it deflated; breathing rapid, and fever. Death. She was feverish before the apostasis.[5] She died on the seventh day after the apostasis. She, too, was a rather ruddy type.

27. The slave (possibly "child") who belonged to the woman who was Apemantus' sister: enlarged hypochondria and spleen, breathing problems. Slimy bilious feces mixed with normal. Easily tired from labor. On the twentieth day it went to the feet and there was a crisis. In fatigued people does it go to the joints and not the eye? However, the hypochondria were stretched, especially if his dry cough asserted itself.

28. Things left behind after the crisis indicate relapse. Also things separated in the disease itself: sputum that rip-

---

[5] *Apostasis*: the settling of the disease in the toe "which was not able to contain it," as *Epid.* 2.1.7 explains in apparent relation to this passage.

---

145 κρίσεις V   146 ἐκτεταμένα HIR
147 ἦν Gal.   148 ἡσυχῇ Gal.

μενα· πτύαλον προπεπαινόμενον,[149] ἡ γαστήρ. ἀκρι-
σίαι[150] καὶ ταῦτα.

29. Ἀπημάντῳ, ᾧ τὰ ἐν τῇ ἕδρῃ, ἀλγήματα ἐν τῷ
δεξιῷ κενεῶνι, καὶ παρὰ τὸν ὀμφαλὸν κάτωθεν ὀλίγον·
καὶ ἐκ δεξιοῦ πρὸ τοῦ ἀλγήματος οὐρεῖ αἱμαλῶδες.
ἔληξε τρίτῃ.[151] καὶ ὁ τέκτων ἐπὶ τὰ ἕτερα· ἐκ τῆς αὐτῆς
ἴξιος καὶ οὗτος οὐρεῖ αἱμαλῶδες.[152] λήγοντος δὲ ἀμφότεροι ὑποστάσεις εἶχον, καὶ τοῦτο[153] τρίτῃ, ἐπεχλι-
αίνετο δὲ πλεῖστα Ἀπήμαντος· ὁ ἕτερος οὐκ ἐνόει εἰ
μὴ ἐπὶ τὰ ἀριστερά· καὶ Νικοστράτῳ προσεγένετό τι
τὰ ὕστατα ἐκ τῶν δεξιῶν, κατώτερον ἢ οἷσιν ἐν τοῖσιν
ἀριστεροῖσι, πρόμακρα δὲ πρὸ τοῦ κενεῶνος μέχρι
πρὸς ὀμφαλὸν[154] ἀμφοτέροισιν.

30. Ἡ γραίη ἡ Κατωσωσιλέω[155] λευκοφλέγμα-
τος,[156] κνῆμαι σκληροῖσιν | οἰδήμασι λευκοῖσι φολλι-
κώδεσι, καὶ πόδες, ἧσσον δέ. ἦν δὲ καὶ ἐν τοῖσι κάτω
τῶν μηρῶν· τοῖσι δὲ πολλοῖσι δυσέξοδον τοῦτο· ἀτὰρ
καὶ ὀσφύι· καὶ λεπτόγαστρος· ὑποχόνδρια ὑπολάπα-
ρα, πνευματώδης[157] δὲ οὐ κάρτα. ἦν δὲ ληγόντων τῶν
πλείστων καὶ μελέτῃ[158] ἄλλη.[159] ἐγλαυκώθη[160] ἡ ὄψις.
ταύτῃ μὲν τῶν ὀμμάτων ἡσυχώτερα δή τι σμικρὸν[161]
ἦν, ἰσχίου δὲ καὶ σκέλεος ὑστερικὰ ἦν[162] δοκέοντα
ἀλγήματα εἶναι. προσθεμένη εὐῶδες ἐξ ἀλήτου καὶ
μύρου ξυνέβη ταχέως ἀναυδωθῆναι καὶ τελευτῆσαι.
χρόνος τῇ ἐς ὀφθαλμοὺς ἀποστάσει ἐνιαυτὸν πέρι· ὁ

174

[149] προσπεπαινόμενον V   [150] Smith: ἀκρασίαι mss.
Gal.   [151] τῇ τρίτῃ V

ens too early, stomach problems. These things are signs of failure of crisis also.

29. Apemantus, who had problems in his seat: pains in the right flank and beside the navel slightly below. Before the pain began, from the right side he made bloody urine. That stopped on the third day. The carpenter, on the other side: he, too, produced bloody urine from the corresponding side. When it stopped, both had deposits, and that on the third day. Apemantus was very warm; the other man did not feel it save perhaps on the left. Nicostratus had it on the right at the last, lower down than the spot where he had it on the left, and protrusions on both sides in front of the flank up to the navel.

30. The old woman in Lower Sosilis had white phlegm. Her lower limbs had hard, white, scabby swellings. Also her feet, but less so. Also below the thighs. In most this is hard to get rid of. Also the lower back. Thin at stomach, soft hypochondria, little difficulty breathing. When most of the symptoms disappeared she had another problem, glaucoma. Then her eye problems improved for a short time, but pains that appeared to be hysterical appeared in her hip and leg. When she was given a fragrant application of ground meal and myrrh she suddenly became voiceless and died. The time for the apostasis in the eyes: about a

152 Smith (cf. Erot. 64.9 N): προούρει αἱματῶδες mss.

153 τούτω V   154 ὀφθαλμὸν V

155 Gal. *Gloss.*: κατὰ Σωσιλέω HIR: κατασωσίλεω V

156 λευκοῦ φλέγματος V   157 Li.: πνευματώδεες mss.

158 μελέτη καὶ HIR   159 ἄλλη ἢ VH: ἄλλη ἤν IR: I
follow Littré's text.   160 ἐγγλαυκώθη V

161 σμικρῶν HIR   162 ἦν VH

φακώδης[163] ᾧ καὶ τὰ ἐξανθήματα ἐξετάκη οὐ τελέως.
ἔστι δὲ καὶ ἀπὸ τῶν συρμάτων τοιαῦτα.

31. Ἐν τῇ Ἱππολόχου κώμῃ παῖς, ᾧ ἐν τοῖσιν
ὑποχονδρίοισι τὸ περιλαμβανόμενον περὶ ἀμφότερα
ἦν ἐν[164] τὠυτῷ, σκληρὸν δὲ καὶ κάτωθεν ἔχον ὑπόλαμ-
ψιν, ἐμφερῆ τῇ τοῦ χαλκέως τῇ ὑδρωπιώδει ἢ λευκὰ[165]
ἐλθόντα μέρος τι ἐλάπαξεν· ἦν δὲ καὶ αὐτὸ τοῦτο πρὸ
τοῦ δεξιοῦ ὁμαλόν τι ὑπερεξηρμένον, τῶν ὑπό τι περι-
φέρειαν ἐχόντων. τούτῳ ὁ ὀμφαλὸς[166] ἐκ γονῆς ἐμελάνθη καὶ ἕλκος βαθὺ ἐγένετο, καὶ ὁ ὀμφαλὸς[166] οὐ
μάλα οὐλὴ ἐγεγόνει, καὶ τὸ αἰδοῖον ἀκρόψιλον ἐγεγόνει, οὐ τοιοῦτον αὐτίκα ἐὸν οὐδὲ ξυγγενικόν. ἐπεί τε
μᾶλλον ἐγίνετο, οὗτος ἀνήμει τὰ[167] πλεῖστα· πυρετός·
ἀποσιτίῃ· ὑγιάνθη. περὶ δὲ ἑβδόμην ἀπὸ κατακλί-
176 σιος,[168] πρόσθεν γὰρ προέκαμεν, | ὕδωρ τε πολὺ
πίνων καὶ ἴσως ἄλλως, ὑποκατεφρόνει, ἐρριπτάζετο,
καί τι ἐσπᾶτο. λήγοντος δὲ τοῦ σπασμοῦ ἔλαθεν
ἀποσβείς. πρὸ δὲ τούτου οὔρησεν ἀθρόον, καὶ φῦσαι
διῆλθον ἐν βρόμῳ, καὶ τὰ ἄνω οὐκ ἐλάπαξεν οὐδέν.
ἀπογενομένου τε εὐθέως, κεῖνά τε ἐλαπάχθη ἰσχυρῶς,
καὶ τἆλλα διεφοινίχθη ὡς μάστιξι πᾶν τὸ σῶμα πλὴν
τούτου ᾗ μάλιστα τὸ ἔξαρμα ἦν, καὶ θερμὸς ἐπὶ

[163] ὁ φακώδης] ὀμφακώδης HIR
[164] om. HIR
[165] λεπτὰ HIR
[166] ὀφθαλμὸς V
[167] ἀνήμει τὰ] ἂν ἢ μὴ μετὰ V
[168] τῆς κατακλίσιος HIR

year. There was freckling where the eruptions were not dissolved completely. Such things come also from chronic scaly ulcers.

31. In the village of Hippolochus, a child: at his hypochondria the surrounding area on both sides was in the same condition, hard and shining underneath, reminiscent of the wife of the bronzesmith, the hydropic one, for whom white matter coming forth emptied out a part. She had the same thing on the front of the right side, similar and protruding, of the sort that have a border below. The boy's navel had become black after birth and a deep ulcer developed. The navel had not become a scar, and the genital became bare on top, though that was not immediate, nor had it been so at birth. But when he grew worse he vomited most (of what he ate). Fever, lack of appetite. He got better. About the seventh day after taking to bed (he had been ill before that) drinking a lot of water, and perhaps from other causes, he was somewhat irrational, tossed about and had some spasms. When the spasms ceased, his fever had disappeared without his noticing.[6] Before that, much urine, much gas passed loudly; the upper parts were not emptied at all. But as soon as there was evacuation, they emptied out powerfully, and on the rest of him he grew red as though he had been whipped over all the body, save in the area where the swelling was greatest, and be-

---

[6] Many interpreters take ἀποσβείς to describe the patient's death rather than the quenching of fever. That would alter the interpretation of other elements in the case history. I take the passage to describe, from the author's point of view, a crisis in which the source of swelling in the hypochondria was finally evacuated by being dispersed to the skin.

συχνὸν ἦν. ᾧ δ᾽ ἐν Ἀβδήροισιν ἐρράγη κάτω, ᾤδει πρόσθεν τὰ ἄνω, διῆε δὲ ἀπυρέτῳ.[169] τὸ δὲ ἔπαρμα ψαυόμενον εἴκελον ἐμπύῳ ἦν.

32. Ἀσθματώδει οἰκέτιδι, ἔνθα τὴν στλεγγίδα,[170] ἠμορράγησεν ἐν[171] τοῖσιν ἐπιμηνίοισιν. ἐγένετο δὲ ἄσθματα, ἐκεῖνα ἐπαύσατο· πυρετὸς ἐγένετο. μαζὸς ἐνεπύησεν ἀριστερὸς ὕπερθεν, καὶ ἀπ᾽ ἀρχῆς καὶ οὖς.

33. Τῇ Ὀλυμπιοδώρου παιδίσκῃ αἷμα ἐκ τοῦ δεξιοῦ, καὶ ἐκρίθη ὡς εἰκοσταίη, οἷα καὶ τοῖσι πυρεταίνουσι, καὶ διαχωρήσιες[172] οἷαι καὶ τοῦ | θέρεος ἐπεδήμησαν· καὶ οἷα ἡ Ἱππώνακτος, καὶ οἷα ἡ[173] οἰκέτις ἡ Ἀριστῆος[174] ὀγδόῃ.

178

34. Ἐν τῇσι τῶν φαρμάκων καθάρσεσι[175] κάτω σημεῖα· οἷα εὐφόρως φέρουσι, καὶ ἃ μὴ[176] παρὰ λόγον γυιοῦνται, καὶ μήτε ἔπαφρα μήτε ὕφαιμα, ἀλλ᾽ οἷα ᾠά ἐστιν· οἷα Ἡρακλείδεω, ἐκαθάρθη πολλὰ καὶ ῥηϊδίως ἔφερε.

35. Τῇ ἐν τῇ Βουλαγόρεω κώμῃ[177] ἀμφιαποφθαρείσῃ[178] ἤρξατο πυρεταίνειν ἐπισπλήνῳ, ἐχούσῃ[179] δὲ[180] καὶ δεξιὸν ὑπό τι μετέωρον μὲν οὔ, ἐντεταμένον δέ. οὗτοι καὶ ἐξέρυθροι γίνονται. μᾶλλόν τι ἐξέρυθρος ἐοῦσα, κοιλίην ἐν ἀρχῆσι τεταραγμένην. προσεδεχό-

---

[169] Smith (exempli gratia): ὠδέε προσθέντα ἄνω δειδιότα ἀπυρέτῳ V: ὧδ᾽ ἔμπροσθεν ἀνώδει ἀπυρέτῳ HIR
[170] τὴν στλεγγίδα] ὅταν ἐτάγγια mss.: στ(λ)εγγίδα Erot. 77.9 N: στρεγγίδα Heraclides (Erot.): corr. Heringa
[171] om. HIR   [172] διαχωρήσῃ V
[173] οἷα Ἱππώνακτος καὶ ὕλη HIR

came exceedingly warm. The male in Abdera who broke forth below had a swelling above initially, but it passed through without fever.[7] But the swelling, touched, was like an abscess.

32. The asthmatic servant, when she was using the strigil, hemorrhaged at the time of her menses. She developed asthma, menses stopped. Fever developed. Left breast festered above; also the ear from the beginning.

33. Olympiodorus' young daughter, blood from the right nostril, crisis on the twentieth day as in those who are fevered, feces as are general in summer. Symptoms like Hipponax' wife and the servant of Aristeus on the eighth day.

34. In purging downward with cathartics, these are the signs: those that they bear easily; those that do not unreasonably prostrate them, and if what is purged is not frothy or bloody, but like eggs: like those of Heraclides who was purged many times but bore it easily.

35. The woman in the village of Boulagoreus, who was wasting generally, began to have fever with an enlarged spleen; her right side, a little below, was not swollen, but a little tight. These patients became quite red. She was rather red and had intestinal upset at the beginning. I ex-

---

[7] The text here is corrupt and unintelligible and I have restored it conjecturally. This section seems to express a parallel with that of the boy in Hippolochus' village.

---

174 Ἀριστείδου HIR   175 καθάρσεις HIR
176 καὶ ἃ μὴ Smith: κἂν μὴ mss.   177 κω V
178 ἀμφιαποκαθαρείσῃ HIR   179 ἐπὶ σπληνὶ ἔχουσι HIR   180 V writes ἐχούσῃ δὲ twice

μην ἐς ὀφθαλμὸν στήριξιν ταύτῃ. ἑβδόμῃ ἁλμῶδες ἐκ
τῶν ὀφθαλμῶν ἦλθε δάκνον δάκρυον, καὶ κατὰ ῥῖνα
καὶ κατὰ φάρυγγα καὶ οὖς ἀριστερόν. πεντεκαιδεκάτῃ
ἵδρωσε σὺν ῥίγει. οὐκ ἐκρίνετο πρὸ τοῦ ῥίγεος. ἐχλω-
ρίασε κάρτα, καὶ προσώπου περίτασις[181] καὶ σύμπτω-
σις. τὸ οὖς κατ᾽ ἴξιν τοῦ σπληνὸς καὶ τοῦ πλευροῦ
ἤλγει.

36. Τοῖσι παιδίοισι γαστέρες ταραχώδεις καὶ βῆχες ξηραί. ἐς ὦμον ἔστιν ὅτε ἀπεπύει[182] ἐν τῇσι
βήξεσι τῇσιν ὑστέρῃσιν. Ὁ κναφεὺς τράχηλον, κεφα-
λήν· ἑβδόμῃ χεὶρ ναρκώδης, ἐν τῇ ἐνάτῃ σκέλος.
νάρκα, βὴξ ἐπαύσατο. Ἡ τὴν γνάθον ἐρυσθεῖσα· ἐν
πέμπτῳ μηνὶ ἐν γαστρὶ εἴλκυστο ἐς τὰ ἀριστερά. |

180   37. Ἐν Κραννῶνι Λυκίνῳ γραμματικῷ[183] ἐκ πυρετοῦ χολώδεος ἐπὶ σπληνὶ καρηβαρίη· αἱμόκερχνα[184]
κατὰ σπλῆνα ἐν χείλεσιν ἀμφοτέροισιν, ἕλκεα στρογγύλα ἔνδοθεν σμικρά. ἔπειτά τι καὶ αἷμα ἐκ τοῦ κατ᾽
ἴξιν σμικρὸν ἐρρύη.

38. Τῇ οἰκέτιδι, ἥν νεώνητον ἐοῦσαν κατεῖδον, ᾗ τὸ
σκλήρυσμα ἐν τοῖσι δεξιοῖσιν ἐνῆν μέγα, οὐ κάρτα
ὀδυνῶδες, καὶ γαστὴρ μεγάλη καὶ περιτεταμένη, οὐκ
εἰκέλη ὑδατώδει, καὶ τἆλλα λαπαρή,[185] καὶ οὐ πάνυ
δύσπνοος, ἄχρως δέ· γυναικεῖα ἑπτὰ ἐτέων οὐκ ἐληλύ-
θει. ἐγένετο δυσεντεριώδης καὶ οὐ τεινεσμώδης. καὶ
ὑπὸ ταῦτα ὀδυνῶδες τὸ ἐν τῷ δεξιῷ ἦν, καὶ πυρετοὶ

---

181 περίστασις VIR
182 Lind.: ἀποπύει (vel -ππ-) mss.

122

pected a determination to the eye for her. On the seventh day a salty biting tear came forth from her eyes, and came down the nose and throat and by the left ear. On the fifteenth day she sweated with shivering. There was no crisis before the shivering. She was very pale, and there was tightness and collapsing of her face. The ear on the same side as the spleen, and the problems in the thorax, was painful.

36. The children: upset stomachs, dry coughs. Sometimes suppuration towards the shoulder in the last stages of the coughing. In the wool carder's case, neck and head. On the seventh day, his hand was numb, on the ninth, his leg. The numbness and the coughing ceased. The woman with the twisted jaw: in her fifth month of pregnancy it had been drawn to the left.

37. In Crannon, Licynus the school teacher after a bilious fever on top of spleen trouble had heaviness of head. Bleeding dry roughness on both lips on the side of the spleen, small round sores inside. Then some blood from the nostril on the corresponding side, a small amount, flowed out.

38. The newly purchased servant girl whom I saw, who had a large hardening on the right (abdomen), not painful, and a large taut belly; not like one with edema, she was slack elsewhere; not affected with difficult breathing, but with bad color. She had not menstruated for seven years. She got dysentery, but without tenesmus. Under these circumstances the swelling on her right became painful.

---

183 γράμματι V
184 Heringa ex Erot. 14.20: αἷμα κέρχα V: ἐκέρχα HIR
185 Smith: λιπαρή mss.

βληχροί,[186] οὐ[187] πρόσω ἑπτὰ ἡμερέων. καὶ κοιλίης ταραχὴ ἠλεκτρώδης[188] ὑπόγλισχρος ὑπέρπολλος ἐφ᾿[189] ἡμέρας τινάς. ὑγιής. καὶ γυναικεῖα μετὰ ταῦτα ἐγίνετο, καὶ τὰ[190] κατὰ τὴν γαστέρα λαπαρά, καὶ χρῶμα εὔχροον, καὶ παχὺ ἦν[191] αὐτῷ.

39. Ἡ Μίνωος, ᾗ ἐκ τομῆς πιεσάντων ἐσφακελίσθη· καθισταμένου ἐς πλεύμονα, ἐπεσήμαινε βραχέως ὅσοισι περιῆν[192] καὶ ἄλλο τι ἐς ἔνδον. |

182 40. Τὰ ἐντεθέντα ἐς τὴν ῥῖνα, ἢν[193] πυρετήνωσιν,[194] ἢν μὲν λυθῇ ἡ ὀδύνη, παχέα κατὰ τὸν μυκτῆρα ῥεῖ· ἢν δὲ μήτε ὀδύνη μήτε πυρετός,[195] λεπτὰ καὶ ἴσως πυρώδεα, οἷον τὸ μὲν λεπτὸν Ἡγησίππῳ ἐς νύκτα προσθεμένῳ, τὸ δὲ παχὺ τῷ[196] ἐν Κορίνθῳ εὐνουχοειδεῖ τῷ Σκελεβρέος.[197] ἦν πέπερι.

41. Αἱ ἐπὶ τῇσι νούσοισιν ἀποστάσιες εἰ κρινοῦσι[198] σημεῖον εἰ[199] πυρώδεα ἐόντα μὴ πυρεταίνουσι,[200] καὶ δύσφορα ἐόντα εὐφόρως φέρουσιν,[201] οἷον τὰ ἐν τῇ ἕδρῃ Χάρωνι. τὰ[202] δὲ Λεαμβίῳ ἑλκώδεος ἐντέρου δοκέοντος εἶναι ἐν δυσεντερίῃ φαρμακευθέντι· ὦμος καὶ ἕδρη ἐφειλκώθη[203] ἀπυρέτου ἐν ἀριστερᾷ. ὁ ἀπὸ τῶν φυσέων μετεωριζόμενος[204] ἐπῆρτο καὶ ὑπήλγει ἐν κενεῶνι. τούτῳ γάλα πλεῖον καὶ ἀκρητέστερον πιόντι

186 γλίσχροι IR   187 καὶ οὐ HIR
188 Li.: κοιλίης ταραχὴ ἠλεκτριώδης Erot. 49.12: κοιλίη ἐταράχη (-χθη HIR) ἠλεκτρώδης mss.
189 ἐς HIR   190 om. V
191 παχείην V   192 ὅσοις ἢ περίη V
193 ῥῖνα, ἢν Foës: τρίτην mss.

There were light fevers, not beyond seven days, and intestinal upset producing amber-like feces, slimy, of considerable quantity for some days. She became healthy. Afterwards her menses occurred, the area of the belly, too, became loose, skin color good, substantial flesh with it.

39. Minos' wife, where the pressing was after an incision (perhaps "a wound"), got gangrene. It went to the lung. She gave brief signs, in the time she survived, of further internal troubles.

40. Things inserted into the nose of fever patients: there is a thick flow from the nostril when the pain is relieved. If the pain and fever are not relieved, it flows thin and perhaps fiery, like the thin flow from Hegesippus for whom it was inserted towards night, and the thick one for the eunuch-like son of Scelebreus in Corinth. Pepper was the substance.

41. Apostases in diseases are a sign whether there will be a crisis, if they do not develop fever in response to feverish conditions, and if they bear difficult conditions comfortably, as in the case of the affection in Charon's seat. And in Leambius' case, after a purge he appeared to have an ulcerated intestine in dysentery. He ulcerated on the left side in the shoulder and seat, but without fever. The man who was inflated with flatulence had swelling and pain in the flank. When he had drunk more milk than usual and

---

194 πυρεταίνουσιν V   195 ὁ πυρετός HIR
196 τῶν V   197 κελευρέος HIR
198 Smith: κρίνουσι mss.   199 ἤν HIR
200 πυρεταίνωσι mss.   201 φέρωσιν HIR
202 τῷ HIR   203 ἐφηλκώθη HIR
204 μεταχειριζόμενος V

καὶ ἐπικοιμηθέντι αὐτίκα ναυσίη καὶ θέρμη ἐγίνετο·[205]
ἔπειτα ὀπτῶντι καὶ ἀντὶ[206] σιτίων ἐσχαροπέπτων[207]
κρίμνων ἀποφαγόντι[208] ξυνέστη ἡ γαστήρ, καὶ πυ-
ῶδές τι ἐπιδιῆγεν. ἕδρης δ' ἐπιφλεγμηνάσης ἀπύρετος
καὶ ἀνώδυνος ἦν ἱκανῶς· τοῦτο λέγω.

42. Ὁ ἐν τοῖσι λιθίνοισι προπύλοισι πρέσβυς
ὀσφῦν ἤλγησε καὶ σκέλεα ἄμφω. καὶ θάτερον κατ-
έβαινεν ἐπὶ μηρούς, καὶ ποτὲ κνήμας, καὶ ποτὲ
γούνατα. ἐγχρονιζούσης μάλιστα πολλαὶ παλινδρο-
184   μίαι[209] | ἐγεγένητο· οἴδημα ἐν ποσίν, ὀσφύϊ, κνήμη·
βραχὺ βουβῶνες καὶ ἔντασις ἤτρου. γαστὴρ σκλη-
ρὴ[210] ἡ πᾶσα καὶ ὀδυνώδης τὰ πλεῖστα· εὑρέθη ἔχων
καὶ κύστιν σκληρὴν καὶ ὀδυνώδεα· καὶ βλαστήματα,
καὶ θέρμαι.

43. Ὅτι τοῖσιν ὄμμασι, τοῖσιν οὔασι, τῇσι ῥισί, τῇ
χειρὶ αἱ κρίσιες, καὶ τἆλλα οἷσι γινώσκομεν. ὁ ἀσθε-
νέων, ἢ ἰδὼν[211] ἢ θιγὼν ἢ ὀσφρανθεὶς ἢ γευσάμενος,
τὰ δ' ἄλλα γνούς· τρίχες, χροιή, δέρματα, φλέβες,
νεῦρα, μύες, σάρκες,[212] ὀστέα, μυελός, ἐγκέφαλος, καὶ
τὰ ἀπὸ τοῦ αἵματος, σπλάγχνα, κοιλίη, χολή, οἱ
ἄλλοι χυμοί, ἄρθρα.

205 ἐγένετο HIR
206 ἀπῶντι HIR (ἁπτῶτι H): ἅπτοντι V: corr. Li.
207 ἐσχαροπέπων HIR   208 φαγόντι V
209 παλινδρομαὶ HIR
210 γαστὴρ σκληρὴ after βουβῶνες mss.: transp. Langholf
211 ἰδὼν Smith: ἱδρῶν VHIR: ὁ δρῶν Li.
212 σάββακες V

less diluted, and had slept, suddenly nausea and hotness developed. Then, when he had roasted and eaten a loaf of coarse barley bread instead of his usual food, his belly became quiet and brought through purulent matter. He was practically without fever and pain, because his seat was inflamed, as I interpret it.

42. The old man at the stone gateway had pains in the loins and both legs. It descended on both sides to his thighs and sometimes to his lower legs, sometimes his knees. Especially as time passed there were many recurrences.[8] He developed swelling on his feet, at his loins, in the lower leg. For a short time the glands swelled and his lower abdomen was stretched. His whole belly was hard and for the most part painful. He proved to have a hard, painful bladder. There were eruptions and areas of heat.

43. Crises, and the other things that give us knowledge, (are known) by the eyes, ears, nose, hand. (The one who knows them is) the ill person, seeing, touching, smelling, or tasting, and knowing in the other ways. (Our sources of knowledge are) hair, complexion, skin, veins, tendons, muscles, flesh, bones, marrow, brain, the things from blood, the intestines, belly, bile, the other humors, joints.[9]

[8] Literally "runnings back." The notion is that the phenomena of the disease move from place to place.

[9] Because ch. 42b, the last part of the case history, comes in the middle of the epistemological aphorisms, Littré transposed it to follow ch. 42. He also adjusted the text of chapter 43 to enhance its parallelism with the (epistemological) opening of *In the Surgery* (Loeb, vol. 3, p. 58). I am not satisfied that he was right, and I have restored the manuscript order.

42b. Καὶ μετὰ ταῦτα ἠλγήκει παρ᾽ οὓς· ταὐτὸν σύστρεμμα ἀδένος οὐκ ἐπιψαύοντος, ὀστέον οὐκ ἀποπυοῦν, καὶ τοῦτο ἐμολύνθη, καὶ τότε εὐθὺς ἐπυρέτηνε.

43b. Σφυγμοί, τρόμοι, σπασμοί, λύγγες, ἀμφὶ πνεῦμα, ἄφοδοι,[213] οἷσι γινώσκομεν.

44. Τοῖσιν ἐμπύοισι τὰ ὄμματα, καὶ ἐκρηγνύμενα μεγάλα ἕλκεα γίνεται, καὶ ταμνόμενα βαθέα. ἀμφοτέρως αἱ ὄψεις ἑλκοῦνται.[214]

45. Ὁ ἀπ᾽ Ἀμφιλόχου κώμης Ἀριστέης τετάρτῃ
186 παρέκρουσε. | κοιλίη ὑπόχλωρα διῄει· καὶ ὕπνοι ὑγιεῖς· λευκόχρως. ἀρχομένων οἷον ὑποτρομώδεις[215] καὶ δακτύλους καὶ χείλεα διαλεγόμενοι, καὶ τἆλλα[216] ταχυγλωσσότεροι προπετέως· ἔρευθος ἐπὶ προσώπου μάλιστα τούτοισιν ἦν. οὗτοι ἐκ θωρήξιος[217] ἢ ἐμέτου χρηστῶς ἐμέσαντες ᾤδεον. ὁ δὲ Κατωμοσαδέω[218] ᾧ λεπτὰ ἄχολα ὑδατώδεα πολλὰ[219] διεχώρει, ὑποχόνδριον ὑπακοῦον καὶ κυρτόν, κωματώδης ἐγένετο περὶ τεσσαρεσκαιδεκάτην. πρὸς κρίσιν μάλιστα ἰόντι ῥῖγος ἄτρομον, διάλυσις, πάρεσις, σύμπτωσις· τὰ τῶν ἀπιόντων· κωματώδης, παραφερόμενος ἐξ ὕπνου, οὐκ ἐξεμάνη· ἐκρίθη[220] περὶ τεσσαρεσκαιδεκάτην, οὐδὲν

213 ἄμφοδοι V   214 ἑλκοῦντο R: ἐκοῦντο I
215 ὑποτρομῶδες VHIR: -ώδεες Paris. ms. K
216 τἆλλα καὶ HIR   217 Li.: ἐκθώρηξαν I: -ήξαντες H: ἐκθόριξαν R: ἐκ θορίξηος V   218 Gal. Gloss. s.v. (cf. Κατωσωσιλέω Epid. 4.30): καὶ Μηδοσάδεω or -έω mss.: Μοδοσαέως (ὄνομα τόπου) Erot. 59.20 N.
219 πολλὰ ὑδατώδεα V   220 ἐκρίθη ἦ V

42b. Afterward he had a pain by the ear. In the same area, gathering, the gland not in contact with it, the bone not purulent; this, too, grew soft and he straightway grew feverish at that time.

43b. The pulse, trembling, spasms, hiccups, things related to breathing, the exits, by which we know.

44. In people with purulence in the eyes, even cracks become large wounds, cut areas grow deep. The iris is wounded on both sides.

45.[10] The man from Amphilochus' village, Aristees, was delirious on the fourth day. Pale, greenish bowel movements, sleep healthy; his complexion was white. At the commencement their fingers trembled somewhat, and their lips when they talked, although otherwise they were rather quick and ready speakers. There was redness of countenance especially in these. After successfully vomiting from drinking heavily of wine, or from a vomitive, they swelled up. The man from Lower Mosades: his feces were thin, not bilious, watery and profuse; his hypochondrium yielded to pressure and was rounded; he became comatose towards the fourteenth day. As he was approaching near the crisis, chill without shivering, loss of faculties, loss of strength, collapse, symptoms of the excrement. Comatose, delirious after sleep, not mad. Crisis around the four-

[10] This confusing section has some thematic threads running through its notes about various cases: complexion in relation to type of disease, delirium and consistency of excrement as predictors of crises, symptoms of the hypochondria in relation to delirium.

τῶν κρινόντων ἐναντίον. ὁμοίως ὁ ἕτερος, ἐν[221] ἀρχῇ-
σιν[222] ὑπόγλισχρα διαχωρήματα, ἐξ οἴων τὸ παχὺ
κρίσιμον, ἀγρύπνῳ· μετὰ ταῦτα ὑπόγλισχρα, ὑπόχολα, πέπονα, χολώδεα, μὴ λεπτά· ἐπὴν δὴ ἄρξηται
ξυνίστασθαι, ταχεῖαι αἱ κρίσιες. ὑποχόνδριον ἐντεταμένον ἐφάνη φλεβονώδεα[223] τρόπον περὶ ἕκτην, ἔπειτα
ἐκοιμήθη ἑβδόμῃ· ἐκρίθη περὶ ἐνάτην. λευκόχροοι,[224]
οὐ πυρροί, ἀμφότεροι. διαχωρημάτων ὑδατωδῶν, ἦν[225]
ἐς αἰθρίην τεθῇ πέλιον,[226] ἄνωθεν λεπτὸν κάρτα εἴκελον ἰσατώδει, κάτωθεν γίνεται ὑπόστασιν ἔχον. οἷσι
κατὰ τὰ δεξιὰ ὑπολάπαρος ἔντασις, φρενιτικοὶ ἢν μὴ
λύηται λήγοντος τοῦ πυρετοῦ. οἷσι δὲ ὑπὸ λαπαρό-
τητος κοιλίης ἐν αὐτῷ τούτῳ οἷον περιλαμβανόμενον
188 ἢ σκλη|ρὸν ὀδυνῶδες γίνεται καὶ πάνυ κακόηθες, οὐκ
ἐθέλει διαχεῖσθαι, ἴσως ἐκ τῶν τοιούτων ἐκπυΐσκον-
ται. ἐν δὲ τοῖσι δεξιοῖσιν αἱ ἐπάρσιες, ὅσαι μὲν ἐπὶ
πολὺ καὶ μαλακαί, μάλιστα[227] πιέζοντι ἢν ὑποβορβο-
ρύζῃ, οὐ πάνυ τι κακόηθεις,[228] οἵη τῷ[229] ἀπ᾽ Ἀμφιλόχου καὶ τῷ Κατωμοσάδεω. οὗτοι κωματώδεις καὶ ἐν
τοῖσιν ὕπνοισι καταφερόμενοι.

46. Αἱ καταστάσιες, καὶ οἷα[230] ἐν ᾗσι μᾶλλον καὶ
ἧσσον γίνεται[231] ὥρῃσι, χώρησιν·[232] τὰς ἀκμὰς ὅτε

---

221 ὁ ἐν HIR   222 ἀρχῇ εἰσὶν V
223 HIR: φλεγμονώδεα V: φλεδονώδεα Li. (cf. Erot. 90.8 N)
224 λεύκοχροι mss.: corr. Foës   225 ἦν om. HIR
226 τεθηπέλειον (-ππ- IR) mss.: corr. Li.   227 καὶ μάλα
καὶ μάλιστα mss.: corr. Langholf   228 κακόηθες V
229 Li.: οἵη τῶν V: οἵηντο τῶ ἐν τοῖσι δεξιοῖσιν HIR

teenth day, with none of the critical signs bad. Similarly the other man; sticky slimy feces at first, the sort which when solid are a critical sign; restlessness. Later somewhat slimy, somewhat bilious, concocted, bilious, not thin movements. When they start to be composed the crisis is coming quickly. Hypochondrium seemed stretched like painful swelling of veins around the sixth day. Then he slept on the seventh. Crisis toward the ninth. Both were of white complexion, not red. Watery movements: if one put them in the open air they developed a thin dark layer on top, like the bluish color of woad, and underneath a sediment. People who have flaccid stretching on the right became phrenitic unless it is reduced when the fever subsides. For those who because of slackness of the intestinal area have in that spot a sense of seizure around it or hardness,[11] it becomes painful and very ill-natured, and will not dissipate. Perhaps from such occurrences they develop a suppuration. Swellings on the right side, as many as are mostly soft, and particularly if they produce rumbling when one presses them, are not ill-natured, like the one in the case of the man from Amphilochus' village and the man from Lower Mosades. They were comatose and were delirious in sleep.

46. Constitutions, and what kinds of things become greater and less in what seasons and places. When (to ex-

---

[11] Perhaps: "those who have a hollowness in the lower abdomen, and a lump or hardness." Liddell and Scott says "loose bowels" for λαπαρότης, a word which occurs only here.

---

[230] οἶαι HIR   [231] γίνονται HIR
[232] χώρησιν, ὥρῃσι V

τῶν νούσων, καὶ πρὸς κρίσιν καὶ τὸ καθ᾽ ἡμέρην· καὶ
τὸ πρώτερον καὶ ὀψίτερον, τρίτῃ, τετάρτῃ περιόδῳ·
καὶ περιόδων, ἐν ᾗσι τὰ κρίνοντα, οἷα τὰ ἐμεύμενα[233]
καὶ μή· αἱ ἀποστάσιες, ᾗσιν ὑποστροφαί. φλέβες
κροτάφων καὶ ξυμπτώσιες καὶ χροιῶν μεταβολαὶ πρὸ
τῶν κρισίων, καὶ οἷσι μὴ κρίνεται, καὶ οἱ ἀλυσμοί,
περιτάσιες, χροιῶν μεταβολαὶ ἐκ τοῦ ἐρυθροῦ ἐς χλωρὸν ἢν μὴ κρίνῃ· καὶ ἢν[234] δέρματος περιτάσιες ἄνευ
κρίσιος· καὶ ξύμπτωσις[235] ὄμματος· τοῦ σώματος
σκληρότης καὶ ξηρή.[236] εἰ[237] μὲν ἑκόντες δακρύουσιν,
οὐ κακόν· οἷσι δὲ ἀκουσίως παραρρεῖ, κακόν.[238] οἷσι
δ᾽ ἐπὶ τῶν ὀδόντων περίγλισχρα γίνεται κακὸν ὑπο-
χάσκειν.

47. Ὃς ἐν τῇ κνήμῃ ἕλκος ἔσχε καὶ τῷ ἀττικῷ
190 ἐχρήσατο, τούτῳ | ἐξανθήματα ἐξαιρόμενα ἐρυθρὰ
μεγάλα· τοῦτ᾽ ἀντὶ τῆς βηχὸς τῆς ὕστερον· οὐ γὰρ
ἔβηξε, τὸ δὲ πρότερον.

48. Ἐν Αἴνῳ ὅσοι φρικώδεις, τρωματίαι κεφαλῆς,
κακοήθεις καὶ ἐμπυητικοὶ ἐκ τεινεσμοῦ, ὁδοιπορήσει
ὀδύνη ποδῶν, καὶ ἐν τῇσι ταραχῇσιν οἱ κόποι· ὧν ἡ
Λινία,[239] ἄπόσιτος, τηκομένη· πυώδης, ἄλλοτε σμικρὸν αἱματῶδες· πόδες[240] ἐποίδεον.

[233] οἷα τὰ ἐμεύμενα] οἰδοῦντα αἱμευμένα HIR
[234] αἱ HIR   [235] ξυμπτώσηος V   [236] ἀκληρωτὴς
καὶ ξηρός V: ἀσκληροτὴς ξηρός HIR: corr. ms. D marg.
[237] οἱ HIR   [238] οἷσι . . . κακόν om. V
[239] οἷον κλινίαι V
[240] ἄλλοτε . . . πόδες om. V: πυώδες for πόδες HIR: corr. Li.

pect) the acmes of diseases, towards the crisis and daily. What happens earlier and what later, in the third, fourth period. Of periods, in which ones the critical things occur, such as vomiting or not vomiting; apostases, which ones are associated with relapses. Blood vessels in the temples; collapses, change of skin color, those that precede crises and those in cases that do not have crises. Tossing about, stretching, change of color from red to pale if there is no crisis. Whether there are stretchings of skin without crisis. Collapse of the eye. Dry hardness of the body. Intentional crying is not bad; an involuntary flow is bad. For those with stickiness around the teeth, gaping of the mouth is bad.[12]

47. The man who had a wound in the calf and who used Atticum:[13] eruptions rose on him, large and red: this instead of a cough later. For he did not have a cough, while previously he had.

48. In Aenus, all who had shivering, who had head wounds, who were malignant and festering after tenesmus, who had pains in the feet when they traveled, all also experienced prostration with intestinal upsets. One was Linia: no appetite, wasting away, purulent (excrement), sometimes a little bloody. Her feet swelled up.

---

[12] This list of things to be taken into account seems to represent peculiar interests of the author of *Epid.* 4, with the exception of the borrowed aphorism about voluntary and involuntary tears.

[13] *Atticum* is unparalleled. Littré suggests κνήκῳ (safflower), Langholf νίτρῳ or λίτρῳ (sodium carbonate). I would as soon think that the text is correct, and that the word means honey from Athens.

ΕΠΙΔΗΜΙΑΙ

49. Ἡ Ἱστιαίου[241] ὑδρωπιώδης ἤδη[242] καὶ ἐπὶ τρία ἔτεα, ἦρος ἀρχομένου ἔβησσεν ἐπὶ πλέον.[243] διεπύησεν ἐς χειμῶνα, ἐξυδατώθη.[244] κείνων δὲ φαρμακευθεῖσα ἐρρήϊσεν. ἀπέθανε, παιδίσκη.[245]

50. Τῶν βησσόντων οἱ μὲν τῆσι χερσὶ ταλαιπωρέοντες, οἷον ὁ παῖς ὁ τὰ κλήματα στρέφων καὶ ὁ Ἀμύντεω, παραλυθέντες αὐτὴν μούνην[246] τὴν δεξιὴν ἀμφότεροι. ἐπαύσαντο. ἔπειτα ἔπαθον τοῦτο βήσσοντες. οἱ δὲ ἢ ἵππευσαν ἢ ὡδοιπόρησαν, ἐς ὀσφῦν, ἐς μηρόν. ξηραὶ δὲ αἱ πλεῖσται, εἰ δὲ μή, βίαιοί γε.[247] |

192
51. Ἡ ἐν Μύριος[248] τρόπῳ οὐ νοσώδει ἐδόκει ἄνευ πυρετοῦ τυφώδης[249] ἐοῦσα. ἔπειτα ἐς τρόμον οἱ ἦλθε παντὸς τοῦ σώματος καὶ τῆξιν καὶ ἀποσιτίην καὶ δίψαν, καὶ ψυχρὴ ἀπεγένετο.[250]

52. Οἱ δὲ νυκτάλωπες οἱ πλεῖστον οὐρήσαντες τὸ ὕστερον βραχύ τι· ἐς δὲ τὰ ὦτα, βήσσοντες καὶ ἐπιπυρεταίνοντες· ἐκρηγνύμενα περὶ ἑβδόμην ἢ ὀγδόην.[251] ἡ ἐν Μύριος παῖς ἐπυρέτηνε, καὶ ἐξ ὠτὸς ἐρρύη πυῶδες περὶ ὀγδόην, οὐκ οἶδα σαφέως. ἔστι δ᾽ οἷσιν ὀδοὺς[252] ὑπῆρχε βεβρωμένος, μάλιστα ὁ τρίτος τῶν ἄνω· ἀντὶ πάντων δ᾽ οὗτος εὑρίσκεται βεβρωμένος· ἐς τοῦτον ὀδύνη, καὶ ἔστιν οἷσι καὶ παραπήμα. οἷσι δ᾽ ἐς ὦτα ἰσχυρῶς, βήσσοντες μᾶλλον ἢ ἐκεῖνοι. οἷσι δὲ

---

[241] Ἱσπέου IR: Ἱππίου H   [242] ἤ HIR
[243] πλέον δὲ HIR   [244] ἐξυδατώθη V
[245] ἡ παιδίσκη HIR   [246] μοῦνον HIR
[247] τε V   [248] ἡ Ἐνμυρίος HI: ὁ ἐνμυρίος R: ἡ Μυρίος V: corr. Meineke   [249] τυφλώδης IR

49. Histiaeus' daughter, already edemic for three years, in early spring her cough increased. She became purulent towards winter. She became very edemic. With purgative drugs, some relief. She died, quite young.[14]

50. Of people with coughs, those who work with their hands, like the slave who tied up the grape vines and the son of Amynteus, they were both paralyzed in the right hand only. It went away and then they developed that affection along with the coughing. People who rode horses or walked got it in the lower back or hams. Most were dry coughs, otherwise they were quite violent.

51. The slave at Myris' house did not seem ill. She was stuprous without fever, then she began to tremble all over the body, wasted away, lost appetite, had thirst and chills.

52. People with night blindness who urinated a great deal, later very little: it went to the ears, they coughed, they became feverish. It broke through on the seventh or eighth day. The young female slave at Myris' house was feverish, pus ran from her ear about the eighth day, I believe. Some had a tooth eroded, especially the third upper: this is most often found eroded. Pain centers there. Some have festering around it. People whose ears were strongly affected coughed more than those (with tooth problems).

[14] The identity of this patient is uncertain. One could translate "the woman who belonged to Histiaeus," and "She died. A maid servant." The word παιδίσκη is ambiguous: from meaning "little girl" it came to be applied to slaves and prostitutes.

---

250 ἦν HIR
251 ἢ καὶ ὀγδόην HIR
252 ὀδῶν I: καὶ ὀδῶν H: ὀδόντων R

καὶ ἐς πύησιν[253] σὺν πυρετοῖσι καὶ ἀπαλλαγῆ[254] ἑβδόμῃ < . . . >[255] ἐπιπαρωξύνθη,[256] ὑποχόνδριον οὐκ
ἐλύθη· μαλαχθείσης ἐνῆν τὰ σμικρὰ ὑπόγλισχρα ξυστρέμματα οὐ χρηστά. οὖρον αἱμοχρῶδες. ἔπτυε δὲ
ἀφρῶδες.[257]

53. Πρὸς ὃν Κυνίσκος[258] εἰσήγαγέ με, ἑβδόμῃ παρωξύνθη, περὶ[259] τεσσαρεσκαιδεκάτην ἐκρίθη μωλύσει.
194   κεκαθαρμένος,[260] φάρυγγα | ὀλίγα ἁλμυρὰ[261] πέπονα
ἀναπτύσας, ἐκ ῥινῶν σμικρὸν ἔσταξεν. ἐκαρηβάρει·
χειρῶν καὶ σκελέων κατάλυσις·[262] κοιλίη λυθεῖσα
ὤνησε. πόδες αἰεὶ θερμοί· ὕπνοι ἦσαν. οἶμαι[263] παρ᾽
οὓς οὐ γενέσθαι ὅτι πέπονα ἔπτυσεν.

54. Ἡ Δημαράτου γυνή, πόδες καὶ ἐν τῆσι φρίκησι
θερμοί. εἴτε ἐς τὸ ἔμπυον ἢ[264] μή, ἀπολεῖται.

55. Οἷον[265] εἶχεν ὁ πρεσβύτης ὁ ἀπογενόμενος· ἅμα
ἠσθένει τῇ ἑωυτοῦ γυναικὶ[266] τῇ κεκριμένῃ·[267] μανικόν
τι ἐνῆν, ἐλθούσης δὲ[268] ἕλμιγγος[269] ὑποπαχὴς καὶ
σίτου ὀλίγου αὐτίκα ἐπαύσατο, καὶ ἐκοιμήθη καὶ ὑγιὴς ἦν. ὁ πρέσβυς οὕτως,[270] καὶ τοῦ σώματος περίτασις τοῦ δέρματος, ἄκραια ψυχρά, λαπαρός, τρομώδης

253 ἐς ἐμπύησιν HIR   254 Smith: ἀπαλλάσσῃ V:
ἀπαλλάσσει HIR   255 see the note on the translation
256 ἐπεὶ παρωξύνθη HIR
257 ἀφώδεες V: ἀφρώδεες HIR: corr. Asul.
258 Κύνικος mss.: corr. Li.   259 περὶ δὲ HIR
260 Smith: κακὰ καθαρῶς V: κακὰ καθαρὸς HIR: "nachdem
er erbrochen hatte" Gal. ad Epid. 6.7.10
261 Langholf: πλατέα mss.: "Salziges" Gal.
262 κατάκλυσις mss.: corr. Li.: "Schlaffheit" Gal.

Those who had festering with fevers got relief in the seventh day.[15] It again grew worse, the hypochondrium remained blocked. When the (belly) softened there were small sticky concretions, not of the best. Urine blood colored. (S)he spat up foamy material.

53. The man to whom Cyniscus took me: he grew worse on the seventh day; crisis on about the fourteenth, with gradual disappearance of the symptoms. After having been purged, and having expectorated through the throat a little salty, concocted matter, he had a small hemorrhage from the nose. His head was heavy, his arms and legs weak. Loosening of the bowels helped. Feet warm continuously. There was sleep. No swelling developed beside his ear, I believe, because he expectorated ripe matter.

54. Demaratus' wife: her feet were warm even during shivering. She was going to die whether or not it developed to suppuration.

55. Like the disease of the old man who died: he was ill at the same time as his wife who reached a crisis. She had some mania, but she passed a rather thick worm, and a small amount of feces. Immediately her trouble stopped, she slept and became healthy. The old man, with similar affection, had in addition stretching of the skin on the body, extremities cold, slackness; at the outset he trembled

---

[15] An apparent lacuna here, perhaps with the patient's name.

263 Langholf: οἱ καὶ mss.: "ich glaube" Gal.
264 εἰ HIR   265 om. HIR   266 om. HIR
267 κεκρυμμένον HIR   268 ἐπεὶ δὲ ἐλθούσης HIR
269 ἕλμινθος R: ἕλμικος HI
270 Smith: οὗτος mss.

ἐν ἀρχῇσι καὶ χείλεα καὶ χεῖρας[271] καὶ φωνήν. παρηνέχθη κοσμίως, ἔχασκεν, οὐ πάνυ δύσπνοος ἦν. ἡμέρῃσιν οὐκ οἶδα πρόσω εἴκοσιν ἀπέθανεν.

56. Οἷς ὑποχόνδρια καὶ κοιλίη ὑποχωρεῖ πιεζεύμενα[272] ἀλέα σὺν[273] βορβορυγμῷ, οἵην ἕλμινθα, καὶ ὁ ἐν Ἀβδήροισι[274] κωλωτοειδέα.

ὅτι ἐστὶ πρὸ τῶν[275] κρισίμων ἡμερέων· τῇ προτεραίῃ καὶ τὰ κακὰ καὶ τἀγαθὰ σημεῖα γίνεται, τὰς ἡμέρας ᾗσιν ἐπιπαροξυνόμενοι χρονίζουσι, καὶ ᾗσι λήγοντες βραχύνουσι.[276] καὶ τὰ ἄπιστα τῶν ῥηϊζόντων, καὶ τὰ διὰ σφῶν αὐτῶν παροξύνοντα.[277] | πτυάλων τοῖσι περιπνευμονικοῖσιν, οἷσι χολώδεα ὅταν μέλλῃ λήγειν, τὰ πάνυ ξανθὰ βραχέα γίνεται· οἷα τὰ ἐν ἀρχῇσιν ἐόντα τοιαῦτα ἐπιφαίνεται, οὐ πάνυ δοκέω ταῦτα ἐκπεπαίνεσθαι, ἀλλὰ κρίνειν, οἷον τῷ[278] παρὰ τῷ διδασκάλῳ καὶ ἄλλοτε οἷον εἶδον.

57. Νίκιππος ἐν πυρετοῖσιν ἐξωνείρασε, καὶ οὐδὲν ἐπέδωκεν ἐπὶ τὸ χεῖρον. καὶ τὸ αὐτό οἱ τοῦτο πλεονάκις ἐγένετο καὶ οὐδὲν ἔβλαψε. προερρέθη ὅτι παύσεται ὅταν οἱ πυρετοὶ κριθῶσι, καὶ ἐγένετο οὕτω. Κριτίας ἐν πυρετοῖσιν ὑπὸ ἐνυπνίων ὠχλεῖτο ὑφ' οἵων οἰδέομεν.[279] ἐπαύσατο καὶ αὐτὸς ἅμα κρίσει.

271 χεῖρες mss.: corr. Lind.

272 πιεζόμενα HIR

273 οὐ HIR

274 οἵην . . . Ἀβδήροισι Smith: βορβολυσμῷ οἷα ἂν ἕλμινθα καὶ ὁ ἐν Ἀβ. V: βορβορυημῷ ὡς ἐν Ἀβ. HIR

275 πρὸ τῶν Li.: πρῶτον mss.

in lips, hands, voice. He was delirious inoffensively, he gaped without much breathing abnormality. He died approaching twenty days; I am uncertain when.

56. People whose hypochondria and belly, under pressure, pass material in a gush with rumbling, pass matter such as a worm, or such as the lizard-like things passed by the man in Abdera.

What precedes the critical days: both the good and bad signs occur the day before. In accord with the days on which patients grow worse, their disease becomes of long duration, with those on which it is slackening, the disease becomes short. There are also improvements which are untrustworthy, and spontaneous exacerbations. Expectoration in peripneumonics, if it is bilious before a remission, becomes very yellow for a brief time. The disease of people in whom that appears at the beginning does not ripen thoroughly, I think, but it does reach a crisis like that of the man at the teacher's house, and such as I have seen at other times.

57. Nicippus had a wet dream in fever, and it made him no worse. The same thing repeatedly occurred and did not harm him. It was predicted that it would end when the fever reached crisis, and so it did. In the same way in a fever, Critias was upset by dreams that cause erections; he, too, stopped at the crisis.

---

276 Li.: βραδύνουσι mss.
277 παροξύνονται V H: παρωξύνοντο IR: corr. Langholf
278 Corn.: τὸ mss.
279 Smith: οἴδαμεν mss.

58. Ἄλκιππος ἔχων αἱμορροΐδας ἐκωλύετο θερα-
πευθῆναι. θεραπευθεὶς ἐμάνη. πυρετοῦ ὀξέος ἐπιγενομένου ἐπαύσατο.

59. Ἐν τοῖσιν ὀξέσι πυρετοῖσι[280] διψώδεις ὑπὸ
ἰητρῶν πεπιεσμένοι τῷ ποτῷ[281] ἢ[282] καὶ ὑπὸ σφέων αὐτῶν δοκέουσι πολὺ ἂν ἐκπιεῖν.[283] ὕδωρ ψυχρὸν δοθὲν
ἵνα ἀπεμέσῃ ὠφελεῖ· χολώδεα γὰρ παρέσται.

60. Ὅτι τὰ νεῦρα αὐτὰ ἐφ' ἑωυτὰ ἕλκει, σημεῖον· ἢν
μὲν τὰ ἐν τοῖσιν ἄνω τῆς χειρὸς τρωθῇ νεῦρα, ἐς τὸ
κάτω νεύσει ἡ χεὶρ ὑπὸ τῶν κάτω νεύρων ἑλκομένη· ἢν
δὲ τἀναντία, ἄλλως.[284]

61. Βήσσουσι ξηρὰ οἱ τοὺς ὄρχιας, καὶ ἐκ τῶν
βηχέων ἐς ὄρχιν. λύεται φλεβοτομηθέντα. καὶ φλεγ-
μαίνοντες βήσσουσιν. οἱ ἐν τοῖσιν ἐπὶ βουβῶσι πυρετοῖσιν ἐπιβήσσουσιν.

---

[280] V adds μᾶλλον καυσώδεες after πυρετοῖσι
[281] Corn.: τόπῳ mss.
[282] om. V
[283] ἐκπίνειν HIR
[284] ἄλλως om. V

58. Alcippus had hemorrhoids, was prevented from being treated. When he was treated he went mad. An acute fever came on and he was cured.[16]

59. In acute fevers, people who are thirsty and are urged to drink by the physician, or even by themselves, think that they could drink a great deal. Cold water given to produce vomiting is beneficial because there will be bilious matter present.

60. That tendons draw towards themselves the following is indication: if the tendons on top of the hand be wounded the hand nods down, drawn by the tendons below. And vice versa.

61. People with testicle problems have a dry cough, and from coughs affections go to the testicle. Treated with phlebotomy they are relieved. People troubled with inflammation have coughs. Those with bubonic fevers cough in addition.

---

[16] It is a standard ancient assumption that hemorrhoids are part of Nature's cure; cf. *Epid.* 6.3.23, *Aphorisms* 6.11–12. Apparently Alcippus tried to cure his hemorrhoids contrary to medical advice.

ΕΠΙΔΗΜΙΩΝ ΤΟ ΠΕΜΠΤΟΝ

1. Ἐν Ἤλιδι, ἡ τοῦ κηπουροῦ γυνή· πυρετὸς εἶχεν αὐτὴν ξυνεχής· καὶ φάρμακα πίνουσα οὐδὲν ὠφελεῖτο· ἐν δὲ τῇ γαστρὶ κάτωθεν τοῦ ὀμφαλοῦ ἦν σκληρόν, καὶ ὑψηλότερον τοῦ ἑτέρου, καὶ ὀδύνας παρεῖχεν ἰσχυράς· τοῦτο ἐβλιμάσθη τῇσι σὺν ἐλαίῳ χερσί,[1] καὶ μετὰ τοῦτο ἐχώρησεν αἷμά οἱ συχνὸν κάτω, καὶ ἐγένετο ὑγιής, καὶ ἐβίω.

2. Ἐν Ἤλιδι, Τιμοκράτης ἔπιε πλέον· μαινόμενος δὲ ὑπὸ χολῆς μελαίνης, ἔπιε τὸ φάρμακον· οὗτος ἐκαθάρθη, τὸ κάθαρμα πολύ, φλέγμα τε καὶ χολὴν μέλαιναν[2] δι᾽ ἡμέρης· πρὸς[3] δείλην ἐπαύσατο τῆς καθάρσιος· καὶ πόνον ἐπόνησεν ἐν τῇ καθάρσει πολύν, καὶ πιὼν ἄλφιτον, ὕπνος ἔλαβεν αὐτὸν καὶ εἶχε τὴν νύκτα μέχρι ἥλιος ἀνεκὰς[4] ἐγένετο· ἐν δὲ τῷ ὕπνῳ οὐκ ἐδόκει τοῖσι παρεοῦσιν ἀναπνεῖν οὐδέν, ἀλλὰ τεθνάναι, οὐδὲ ᾐσθάνετο οὐδενός, οὔτε λόγου οὔτε ἔργου, ἐτάθη δὲ τὸ σῶμα καὶ ἐπάγη· ἐβίω δὲ καὶ ἐξήγρετο.[5]

Chapters 1–14 mss V and HIR. Chapter 14– mss V and M
[1] τῇσι . . . χερσί] ἰσχυρῶς τῇσι χερσὶ σὺν ἐλαίῳ HIR
[2] add διήει HIR   [3] καὶ πρὸς HIR

EPIDEMICS 5

1. In Elis, a continuous fever held the wife of the gardener; when she drank (purgative) drugs she was not benefited. On her belly below the navel was a hardness, higher than on the other side, and it caused strong pains. She manipulated it with her hands, with olive oil, after which much blood was passed below and she became healthy and survived.

2. In Elis, Timocrates drank too much and went insane from black bile. He took the drug[1] and was purged. Much was purged from him: he produced phlegm and black bile throughout the day. Towards afternoon he stopped the purging. He had much pain in the purging. He drank barley broth; sleep took him and held him all night until the sun was high. In sleep he did not seem to those who were present to be breathing, but to have died. He perceived nothing, speech or action, and his body was stretched out and rigid. But he survived and waked up.

[1] Probably hellebore, a "violent gastrointestinal poison, having hydragogue, cathartic and emmenagogue properties" (*Dorland's Illustrated Medical Dictionary*), used by the ancients as the cathartic of choice in mental affections especially.

---

4 om. HIR cf. Erot. s.v.   5 ἐξηγέρετο IR: ἐξήγερτο H

3. Σκόμφος ἐν Οἰνειάδῃσι, πλευρίτιδι ἐχόμενος, ἀπέθανεν ἑβδομαῖος παρακόπτων· φάρμακον δὲ ἔπιε κατωτερικὸν ταύτῃ τῇ ἡμέρῃ, τῇ πρόσθεν κατανοέων, καὶ ἐκαθάρθη οὐ πολλά· καθαιρόμενος δὲ παρέκοψεν.

4. Φοίνικι ἐν Οἰνειάδῃσι καὶ Ἀνδρεῖ, ἀδελφεοῖσιν[6] ἐοῦσιν,[7] ἡ γνάθος ᾤδησεν ἡ ἑτέρη καὶ τὸ χεῖλος τὸ πρὸς τῆς γνάθου καὶ τοῦ | ὀφθαλμοῦ, καὶ οὔτε ἔνδοθεν σκοπέοντι οὐδὲν ἐφαίνετο, οὔτε θύραζε[8] ἀπεπύει, ἀλλ᾿ οἰδέουσα[9] σαπρὴ ἐγένετο ξηρῇ σηπεδόνι, καὶ ἀπέθανεν. καὶ θάτερος τὰ αὐτά· ἀπέθανε δὲ ὁ μὲν ἑβδομαῖος, καὶ ἔπιε φάρμακον, καὶ οὐδὲν ὠφελήθη. τῷ δὲ Φοίνικι ἐξετμήθη κύκλος σαπρός, καὶ τὸ ἕλκος ἐκαθάρθη[10] μὲν τὸ πλέον πρὶν ἀποθανεῖν·[11] ὅμως[12] δὲ ἀπέθανε[13] καὶ οὗτος, πλείονα χρόνον βιούς.

5. Εὐρυδάμας, ἐν Οἰνειάδῃσιν, ἐν περιπλευμονίῃ δεκαταῖος ἤρχετο παρακόπτειν· ἰητρευόμενος κατενόησέ τε καὶ τὰ[14] πτύαλα ἐγένετο καθαρώτερα, καὶ προχωρέουσα ἡ νοῦσος ἐπὶ τὸ βέλτιον, ὕπνος τε αὐτῷ κατεχύθη πολύς, καὶ τὰ ὄμματα ἰκτερώδεα ἐγένετο,[15] καὶ ἀπέθανε πρὸς τὰς εἴκοσιν ἡμέρας.

6. Ἐν Οἰνειάδῃσιν ἀνὴρ νούσῳ εἴχετο· ὁκότε ἄσιτος εἴη ἔμυξεν[16] αὐτοῦ ἐν τῇ γαστρὶ ἰσχυρῶς, καὶ ὠδυνᾶτο· καὶ ὅτε φαγόντι τὰ σιτία τριφθείη,[17] καὶ χρόνος ἐπιγένοιτο μετὰ τὴν βρῶσιν τοῦ σιτίου, μετ᾿

---

[6] Ἄνδρυῖ ἀδελφ. V: ἀνδρυαδελφεοῖς HIR: Ἀνδρέα ἀδελφ. Corn.: corr. Li.   [7] om. HIR   [8] θύραθεν HIR

[9] ἥδε οὖσα V   [10] ἐκαθαρίσθη HIR

3. Scomphus, in Oeniadae, possessed by pleurisy, died on the seventh day, delirious. He drank a drug that purges downward that day, having been mentally all right the previous day. Not much was purged, but in the purgation he became delirious.

4. In Oeniadae, the brothers Phoenix and Andreus: in each, one jaw swelled up, and the lip in front of that jaw, and the area around that eye. Nothing was visible to one looking inside, nor was there a discharge of pus outside. The area swelled up and grew rotten with a dry putrefaction, and he died. And the other also with the same affection. The one died on the seventh day. He drank a (purgative) drug and was not helped. But in Phoenix's case, a round putrid area was cut out and the wound was purged for the most part before he died. But still he too died, after surviving a longer time.

5. Eurydamas, in Oeniadae, in the tenth day of peripneumonia, began to be delirious. He was treated and regained sanity, the sputa became purer and the disease turned for the better. Sleep poured over him and he became jaundiced in the eyes; he died towards the twentieth day.

6. In Oeniadae, a man was held by a disease. When he went without food there was strong grumbling in his stomach, and pain. Whenever he had eaten and the food was used up and time had passed after the eating of the food,

---

11 αὐτὸν ἀποθανεῖν HIR   12 om. HIR

13 ἀπέθανε δὲ HIR   14 om. HIR

15 ἐγένοντο V   16 ἔμυσεν V: ἔμυσσεν HIR (cf. Erotian 36.10.N and *Erotianstud.* 301)

17 Asul.: τρεφθείη VHI: τερφθείη R

οὐ πολὺ ταὐτὸ τοῦτο ἔπασχεν. καὶ ἔφθινε τὸ σῶμα,
καὶ ἐτήκετο, καὶ τροφὴ οὐχὶ ἐγένετό οἱ ἀπὸ τῶν σιτίων
ἐσθίοντι· καὶ ὑπεχώρει ὁ σῖτος πονηρὸς καὶ ξυγκε-
καυμένος. ὁκότε δὲ νεωστὶ βεβρωκὼς εἴη, αὐτὸν τοῦτον τὸν χρόνον ἥκιστα ἔμυζε καὶ τὸ ἄλγος εἶχεν
αὐτόν. οὗτος φάρμακα πίνων παντοδαπὰ καὶ ἄνω καὶ
κάτω οὐδὲν ὠφελεῖτο· φλεβοτομούμενος δὲ ἐν μέρει
ἑκατέρην τὴν χεῖρα ἕως ἔξαιμος ἐγένετο, ἔπειτα ὠφε-
λήθη, καὶ ἀπηλλάγη τοῦ κακοῦ.

7. Εὐπόλεμος ἐν Οἰνειάδῃσιν ὠδυνᾶτο ἰσχυρῶς[18]
ἰσχίον τὸ δεξιόν, καὶ τὸν βουβῶνα, καὶ τὴν πλησίον
ξυμβολὴν πρὸς τοῦ ἰσχίου ἀπὸ τοῦ βουβῶνος, καὶ τοῦ
ἰσχίου τὸ πρόσθεν. τούτῳ αἷμα ἀφῃρέθη ἀπὸ τοῦ
σφυροῦ πολὺ πάνυ, καὶ μέλαν, καὶ παχύ· καὶ φάρμα-
κον ἔπιεν | ἐλατήριον, καὶ ἐκαθάρθη πολλά· καὶ ῥῆων
μέν τι ἐγένετο. αἱ δὲ ὀδύναι οὐκ ἐξέλιπον, ἀλλ᾽ ἔμπυον
ἔσχε τό τε ἰσχίον, καὶ τὴν κοχώνην, καὶ τὸ ἀμφὶ τὸν
βουβῶνα, ἅπερ ὠδυνᾶτο καὶ ἐπὶ πλέον· τὸ δὲ πύον
ἐγένετο πρὸς τὸ ὀστέον μᾶλλον ἢ πρὸς τὸ[19] τῆς
σαρκὸς κατὰ βάθος· καὶ ἐλελήθει χρόνον οὕτως ἔχων
ἕως ἀσθενὴς[20] ἐγένετο. ἔπειτα ἐκαύθη. ἔσχαραι πάνυ
πολλαὶ[21] καὶ μεγάλαι ἐγένοντο καὶ πλησίαι ἀλλήλων,
καὶ πύον ἐρρύη πολὺ καὶ παχύ· καὶ ἔθανεν ὀλίγῃσιν
ἡμέρῃσι μετὰ ταῦτα, καὶ ὑπὸ μεγέθεος τῶν ἑλκέων καὶ
πλήθεος, καὶ ἀσθενείης τοῦ σώματος. οὗτος ἐδόκει ἄν,
εἰ ἐτμήθη εὔροον μίαν τομήν, καὶ πρὸς τόμον ἀφίετο
τὸ πύον, καί, εἰ προσέδει τομῆς ἑτέρης, ταμεῖν εὔροον,
ταῦτα παθὼν ἐν τῇ ὥρῃ ἐδόκει ἂν ὑγιὴς γενέσθαι.

he soon suffered the same thing. He wasted in body and grew thin, he got no nourishment from his food when he ate, and the food passed out useless and burned. But when he had just eaten, there was little grumbling precisely at that time, and the pain was least severe. He drank various drugs to purge upward and downward, and was not benefited. But when he was bled in each arm in turn until he was bloodless, then he was benefited and freed from the trouble.

7. Eupolemus in Oeniadae had severe pains in the right hip and groin and at the connection between hip and groin and in the front of the hip. Very much blood was extracted from his ankle, black and thick. And he drank a purgative drug. He was purged of much material and became easier. But the pains did not leave him: pus possessed the hip, and the perineum, and the groin area, all of which were even more painful. The pus was nearer the depth of the bone than of the flesh, and it was not obvious for a time that he was in that condition until he grew weak. Then he was cauterized. The scars were numerous, large, and close together. Much thick pus ran out. He died a few days after that, from the size and number of the wounds and from weakness of his body. It would appear that, if there had been a single incision adequate for drainage and the pus had been drawn toward the incision and, if another incision had been needed, one adequate for drainage had been cut: if this had been done to him at the right time, it seems that he would have become healthy.

---

[18] om. HIR   [19] om. V

[20] πάνυ ἀσθενὴς HIR

[21] ἐσχάραις πάνυ πολλαῖς HIR

8. Λύκων, ἐν Οἰνειάδῃσι, τὰ μὲν ἄλλα ταὐτὰ ἔπασχεν, αἱ δὲ ὀδύναι καὶ ἐς τὸ σκέλος οὐ πάνυ διεφοίτων, καὶ οὐκ ἐγένετο ἔμπυος· ὑγιὴς δὲ πολλῷ χρόνῳ· φάρμακα δὲ ἔπινε, καὶ σικύας προσεβάλλετο, καὶ ἐφλεβοτομεῖτο, καὶ ἐδόκει ῥήϊον γίνεσθαι ταῦτα πάσχοντι.

9. Ἀθήνησιν, ἄνθρωπος ξυσμῷ εἴχετο πᾶν τὸ σῶμα, μάλιστα δὲ τοὺς ὄρχιας καὶ τὸ μέτωπον· εἴχετο δὲ πάνυ σφόδρα, καὶ τὸ δέρμα παχὺ ἦν καθ᾿ ἅπαν τὸ σῶμα, καὶ οἷόν περ λέπρη τὴν πρόσοψιν· καὶ οὐκ ἂν ἀπέλαβες²² οὐδαμόθεν τοῦ δέρματος ὑπὸ τῆς παχύτητος· τοῦτον οὐδεὶς ἐδύνατο ὠφελῆσαι· διελθὼν δὲ ἐς Μῆλον ᾗ τὰ θερμὰ λοετρά,²³ τοῦ μὲν κνησμοῦ ἐπαύσατο καὶ τῆς παχυδερμίης· ὑδρωπιήσας δὲ ἔθανεν. |

210   10. Ἀθήνησιν ἄνδρα χολέρη ἔλαβεν, ἤμει τε καὶ κάτω διῄει, καὶ ὠδυνᾶτο, καὶ στῆναι οὐκ ἐδύνατο οὔτε ὁ ἔμετος οὔτε ἡ ὑποχώρησις, καὶ ἥ τε φωνὴ ὑπολελοίπει, καὶ κινεῖσθαι οὐκ ἐδύνατο ἐκ τῆς κλίνης,²⁴ καὶ οἱ ὀφθαλμοὶ ἀχλυώδεις καὶ ἔγκοιλοι ἦσαν, καὶ σπασμοὶ εἶχον [ἐκ τῆς κοιλίης]²⁵ ἀπὸ τοῦ ἐντέρου ὁμοίως²⁶ λυγγί.²⁷ ἡ δ᾿ ὑποχώρησις πολλῷ πλείων²⁸ ἦν τοῦ ἐμέτου. οὗτος ἔπιεν ἐλλέβορον ἐπὶ φακῶν χυλῷ, καὶ ἐπέπιε φακῶν χυλὸν ἕτερον ὅσον ἐδύνατο, καὶ ἔπειτα ἐξήμεσε,²⁹ καὶ προσηναγκάσθη, καὶ ἔστη αὐτῷ ἄμφω· ψυχρὸς δὲ ἐγένετο· ἐλούετο δὲ μέχρι τῶν αἰδοίων κάτω πάνυ πολλῷ ἕως καὶ τὰ ἄνω διεθερμάνθη. καὶ ἐβίω· τῇ δ᾿ ὑστεραίῃ ἄλφιτα ἔπιε λεπτὰ ἐφ᾿ ὕδατι.

8. Lycon, in Oeniadae, in other respects had the same affections, but the pains also went into the leg, not far, and he was not purulent. He was healthy for a long time. And he drank (purgative) drugs, and applied the cupping glass, and was bled, and it seemed to get easier when he was so treated.

9. In Athens, a man was taken by itching over his whole body, but especially on his testicles and forehead. It was very severe, and his skin was thick over the whole body, and like lepra (white scale) to the view, and one could not pull out any part of the skin because of its thickness. No one could help him. But he went to Melos where the warm baths are, and was cured of the itching and thick skin. But he became dropsical and died.

10. At Athens, cholera seized a man. He vomited and had diarrhea and pain. Vomiting and diarrhea could not be stopped, and his voice failed, and he could not move from his bed, his eyes were misty and hollowed, he had spasms from the intestine like hiccups. The bowel movements were greater than the vomit. He drank hellebore in lentil broth and drank lentil broth in addition, all he could. Then he vomited and was forced (to eat again). Both (vomiting and diarrhea) stopped, but he grew cold. He bathed the lower parts up to the genitals in much water, until the upper parts too were warm. And he survived. The following day he drank thin barley in water.

---

22 ἀπέλαβεν IR   23 ὑποβὰς γὰρ ἐς Μῆλον ἐς θερμὰ λοετρά Erot. s.v. ξυσμῷ   24 ἐκ τῆς κλίνης οὐκ ἠδύνατο HIR   25 ἐκ (ὑπὸ HIR) τῆς κοιλίης mss.: secl. Smith   26 om. HIR   27 λύγξ HIR   28 πλέον V   29 ἐπήμεσε HIR

11. Ἐν Λαρίσῃ γυναικὶ Γοργίου τὰ ἐπιμήνια τεσσάρων[30] ἐτέων ἴσχετο πλὴν ὀλίγων πάνυ· ἐν δὲ τῇ
μήτρῃ, ἐφ᾽ ὁκότερα ἂν[31] κλιθῇ, σφυγμὸν παρεῖχε καὶ
βάρος. αὕτη ἡ γυνὴ ἐκύησε,[32] καὶ ἐπεκύησε,[33] καὶ
ἀπελύθη τὸ παιδίον ἐνάτῳ μηνί, ζῶον,[34] θῆλυ, ἕλκος
ἔχον ἐν τῷ ἰσχίῳ· καὶ τὰ ὕστερα ἑπόμενα, καὶ αἵματος
ῥεῦμα πολὺ πάνυ ἐπεγένετο καὶ τῇ ὑστεραίῃ καὶ τῇ
τρίτῃ καὶ τῇ τετάρτῃ, καὶ θρόμβοι πεπηγότες, καὶ
πυρετὸς εἶχε μέχρι δέκα ἡμερέων[35] τῶν πρώτων· καὶ
ὑπεχώρει τὸ λοιπὸν αὐτῇ αἷμα ἐρυθρόν· καὶ ᾤδει τὸ
πρόσωπον ἰσχυρῶς, καὶ τὰς κνήμας, καὶ τὼ πόδε, καὶ
ἕτερον[36] μηρόν· καὶ σιτία οὐ προσίετο, δίψος δὲ εἶχεν
ἰσχυρῶς·[37] καὶ τὸ ψυχρότατον ὕδωρ ξυνέφερεν, οἶνος
δὲ οὐδαμῶς· ἡ δὲ γαστὴρ μετὰ τὸ πρῶτον παιδίον[38]
ὀλίγῳ μέν τινι ἐλαπάχθη,[39] πάνυ δὲ οὐ ξυνέπεσεν,
212   ἀλλὰ | σκληροτέρη ἦν. ὀδύνη δὲ οὐ προσῆν. τεσσαρα-
κοστῇ δὲ ἡμέρῃ ἀπὸ τῆς πρώτης ἐξέπεσε τὸ ἐπικύημα,
σάρξ· καὶ ἡ γαστὴρ ξυνέπεσε, καὶ τὰ οἰδήματα πάντα. καὶ τὸ ῥεῦμα τὸ λευκόν, καὶ τὸ αἷμα τὸ ὄζον, καὶ
ὑγιὴς ἐγένετο.

12. Γυνὴ ἐν Φερῇσι περιωδύνει κεφαλὴν πολὺν
χρόνον, καὶ οὐδεὶς οὐδὲν ἐδύνατο ὠφελῆσαι, οὔτε
καθαιρομένη τὴν κεφαλήν· ῥηΐστη δὲ ἐγένετο ὁκότε τὰ
ἐπιμήνια εὐχερῶς οἱ ἦει.[40] ταύτῃ[41] ὁκότε περιωδύνει
τὴν κεφαλήν, προστιθέμενα προσθετὰ εὐώδεα πρὸς
τὴν μήτρην ὠφέλει, καὶ ἀπεκαθάρθη ὀλίγον τι. καὶ
ὁκότε ἐκύησεν ἐξέλιπον αἱ ὀδύναι τὴν κεφαλήν.

11. In Larissa the menses of Gorgias' wife had stopped for four years, save for a very small amount. In her uterus, on whichever side she lay, she had a throbbing and heaviness. She became pregnant and pregnant again (superfetation). The child was delivered in the ninth month; a live girl, with an ulcer on her hip. And the rest came out following, and a very copious flow of blood came on the next and on the third and fourth days, and firm clots. Fever held her for the first ten days. For the remaining time she passed bright red blood. She swelled up greatly in the face and calves and feet and one thigh. She had no interest in food, but was very thirsty. She could take very cold water, but no wine. Her belly after the first baby was a little emptied but did not entirely collapse, but was rather hard. But there was no pain. On the fortieth day after the first the second fetus aborted, simply flesh. And her belly collapsed, and all the swellings. The flow was white and the blood the odorous sort, and she became healthy.

12. In Pherae, a woman had pain in the head for a long time, and no one could help her, not even when she was purged in the head. But she was very much eased whenever her menses flowed freely. Whenever the pains in her head came, pleasant-smelling applications to the uterus helped, and there was some purging. When she became pregnant the pains left her head.

---

30 διὰ τεσσάρων HIR   31 ἦν V
32 ἐκλύισε V: ἐκλύησε HI: corr. R   33 ἀπεκύησε HI
34 ζῶν V   35 ἡμερ. δέκ. R   36 τὸν ἕτερον HIR
37 ἰσχυρόν HIR   38 παιδίον ἤτοι παιδίω HI
39 ἐπαλλάχθη V   40 ἦ V   41 αὕτη HIR

13. Γυνὴ ἐν Λαρίσῃ κύουσα· τῷ δεκάτῳ μηνὶ αἷμα ἐχώρει αὐτῇ πολὺ τεσσαρεσκαίδεκα ἡμέρας, πλεῖστον δὲ τὰς τρεῖς τὰς πρὸ τοῦ παιδίου τῆς ἀπολύσιος. τῇ τεσσαρεσκαιδεκάτῃ ἐξέπεσεν ἐκ τῆς γαστρὸς τὸ παιδίον τεθνεός, ἔχον τὸν δεξιὸν βραχίονα προσπεφυκότα τῇ πλευρῇ· καὶ τό χορίον τρίτῃ ἡμέρῃ τῆς νυκτὸς τὴν αὐτὴν ὥρην ὡς ὅτε τὸ παιδίον· καὶ τὰ λευκὰ ἔπειτα μετὰ ταῦτα ἐχώρει τρεῖς ἡμέρας καὶ νύκτας μετρίως τὰ πολλά·[42] μετὰ δὲ τοῦτο πυρετὸς ἔλαβε δύο ἡμέρας καὶ νύκτας δύο, καὶ ὠδυνᾶτο τὴν γαστέρα πᾶσαν καὶ τὸ ἰσχίον,[43] τὸ δὲ ἦτρον μάλιστα.

14. Ἐν Λαρίσῃ Ἱπποσθένης περιπλευμονίῃ ἐδόκει τοῖσιν ἰητροῖσιν ἔχεσθαι,[44] ἦν δὲ οὐδαμῶς· ἀρχῇ[45] μὲν παλαίων ἔπεσε σκληρῷ χωρίῳ ὕπτιος, καὶ ἐπέπεσεν[46] αὐτῷ, καὶ ἐλούσατο ψυχρῷ, καὶ ἐδείπνησε,[47] καὶ ἐδόκει[48] βαρύτερος γίνεσθαι. τῇ δ' ὑστεραίῃ ἐπύρεξε, καὶ βὴξ ἔσχε ξηροτέρη, καὶ τὸ πνεῦμα πυκνόν. πεμπταῖος δὲ αἱματῶδες ἐχρέμψατο, οὐ πολύ, καὶ παρακόπτειν ἤρχετο· | ὁκότε βήσσοι, τότε ὠδυνᾶτο τὰ στήθεα καὶ τὸν νῶτον.[49] ἑκταίῳ δὲ αἷμα ἐρρύη ἐκ τῶν ῥινῶν πταρέντι, ὅσον τέσσαρες κοτύλαι πρὸς τὴν ἑσπέρην· οὔτε ἐφθέγγετο, οὔτε ᾐσθάνετο οὔτε ἔργου οὔτε λόγου. ἐνδεκαταῖος δὲ ἔθανεν. τὰς δὲ πέντε ἡμέρας τοτὲ μὲν ἔμφρων ἦν, τοτὲ δὲ οὔ· ἐγένετο καὶ ἀπύρετος· σίαλον δὲ οὐδὲν ἀπεχώρει, οὐδὲ ῥέγχος[50] εἶχεν, οὐ γὰρ ἦν σίαλον.

214

---

[42] τὰ πολλὰ Smith: τί πολλὰ VHI: om. R

13. A woman in Larissa, who was pregnant, lost much blood for fourteen days in the tenth month, but mostly in the three before the delivery of the child. On the fourteenth day the child was born dead, with the right arm attached to its side. On the third day the afterbirth, at the same time of night as the child, was born. The white flows came after that for three days and nights, moderately in general. After that fever seized her two days and nights and her whole belly and hips were in pain, the lower abdomen most of all.

14. In Larissa Hipposthenes seemed to the physicians to have peripleumonia. But that was not it. At the beginning, when he was wrestling, he fell on his back on a hard place and his opponent fell on him. He washed in cold water, ate dinner, and seemed to become rather heavy. On the following day he had fever, with a dry cough and rapid breathing. On the fifth day he spat up bloody matter, not much of it, and began to be delirious; when he coughed he had pain in the chest and back. On the sixth day the blood ran from his nose after he sneezed, about four cotyls,[2] toward evening. He did not speak and could not perceive speech or action. On the eleventh day he died. But for five days he was sometimes conscious, sometimes not. He became free of fever. No sputum flowed out, nor did he have wheezing, since there was no sputum.

[2] About one quart.

---

[43] τὰ ἰσχία HIR   [44] συνέχεσθαι Gal. *Diff. resp.* (7.955 K)   [45] ἀρχὴν Gal.   [46] ἐπενέπεσεν HIR   [47] ἔμεινε Gal. Ms. M commences at this point.   [48] ἔδοξε Gal.   [49] τὰ νῶτα Gal.

15. Σκάμανδρος ἐν Λαρίσῃ ἰσχίον ἐσφακέλισε, καὶ ὀστέον ἀφεστηκὸς χρόνιον· ὁ δὲ ἐτμήθη τομὴν μεγάλην καὶ πρὸς τοῦ ὀστέου· ἔπειτα[51] ἐκάη. τότε ἡμέρῃ δωδεκάτῃ ἤρξατο μετὰ τὴν τομὴν σπασμός, καὶ εἶχε μᾶλλον· ἐσπάσατο δὲ τὸ σκέλος τοῦτο μέχρι τῶν πλευρῶν· διεφοίτα δὲ καὶ ἐπὶ θάτερα ὁ σπασμός· ξυνεκάμπτετο δὲ τὸ σκέλος καὶ ἐξετείνετο, καὶ τἄλλα μέλεα ἐκίνει, καὶ αἱ γνάθοι ἐπάγησαν· οὗτος ἔθανε σπώμενος ὀγδόῃ μετὰ τὴν τοῦ σπασμοῦ ἐπίληψιν. ἐθεραπεύετο δὲ χλιάσμασιν ἀσκίοισι καὶ πυρίῃσιν[52] ὀρόβων ὅλον τὸ σῶμα, καὶ ὑπεκλύσθη, καὶ ὑπῆλθε παλαιὴ κόπρος ὀλίγη· καὶ τὸ κατακορὲς φάρμακον ἔπιε καὶ προσκατέπιε, καὶ ὑπῆλθε μέν, οὐδὲν ἀπὸ τοῦ καταπότου ὠφελήθη· καὶ ὕπνος ὀλίγος ἦλθεν· καὶ αὖτις πιὼν τὸ κατακορὲς ἰσχυρὸν ἑσπερινός, ἡλίου ἀνιόντος ἔθανεν. ἐδόκει δ᾽ ἂν πλείονα χρόνον διενεγκεῖν, εἰ μὴ κατὰ τοῦ φαρμάκου τὴν ἰσχύν.

16. Ἱπποκόμος Παλαμήδεος ἐν Λαρίσῃ, ἑνδεκαετής, ἐπλήγη κατὰ τοῦ μετώπου ὑπὲρ τὸν ὀφθαλμὸν τὸν δεξιὸν ὑφ᾽ ἵππου, καὶ ἐδόκει τὸ ὀστέον οὐχ ὑγιὲς εἶναι, καὶ ἐπίδυεν[53] ἐξ αὐτοῦ ὀλίγον αἷμα. οὗτος ἐπρί-
216 σθη μέγα μέχρι τῆς διπλόης· καὶ ἰητρεύετο οὕτως |
ἔχων τὸ ὀστέον, ὃ καὶ πρόσθεν αὐτίκα[54] ἐκπύει.[55] ἐπὶ εἴκοσιν, οἴδημα παρὰ τὸ οὖς ἤρξατο, καὶ πυρετός, καὶ

50 Gal.: ῥῖγος mss.
51 καὶ ἔπειτα M
52 πυρίοισιν MV: corr. mss. recc.

15. Scamandrus in Larissa had mortification in the hip; in time the bone came free. A large incision up to the bone was cut, and then cauterized. Then on the twelfth day after the incision, spasm began, and it increased. That leg was drawn up right to the ribs. And the contraction migrated to the other side. The leg was bent double and very tense and his other limbs trembled and his jaws were fixed. He died drawn up on the eighth day after the spasm came on. He was treated with fomentations made from leather bottles, and with heated vetch seeds over the whole body. He was given an enema and a little old excrement came out. He took the saturated (purgative) drug and repeated it. He did pass excrement. There was no help from what he drank. He slept a little. Having drunk the saturated drug again at evening, he died at sunrise. It seemed that he would have survived longer if not for the strength of the medicine.

16. Hippocomus son of Palamedes in Larissa, eleven years old, was struck on the forehead above the right eye by a horse. The bone did not seem sound and a little blood spurted out of it. He was trephined extensively down to the diploe. And he was cured, despite this condition of the bone, which before was readily festering.[3] On the twentieth day a swelling began by the ear, and fever and shiver-

---

[3] The text is uncertain here and I am translating my conjectural restoration. The author appears to be defending his treatment and arguing that the boy's later problems did not stem from the trephination.

---

53 ἐπήδα mss. (cf. Erotian 36.14 N.)

54 The mss. repeat τὸ ὀστέον after αὐτίκα: corr. Erm.

55 Smith: ἔκνεν mss.: πρισθὲν . . . ἔκηεν Li.

ῥῖγος· καὶ ἡμέρῃ μᾶλλον ᾠδίσκετο καὶ[56] ὠδυνᾶτο τὸ οἴδημα· καὶ ἐπύρεσσεν ἀρχόμενος ἐκ ῥίγεος· καὶ οἱ ὀφθαλμοὶ ᾤδησαν, καὶ τὸ πρόσωπον· ἔπασχε δὲ ταῦτα ἐπὶ δεξιὰ μᾶλλον τῆς κεφαλῆς, παρῆλθε δὲ καὶ ἐς τὰ ἀριστερὰ τὸ οἴδημα· οὐδὲν οὖν τοῦτο ἔβλαπτεν· τελευτῶν δὲ πυρετὸς ξυνεχὴς ἔσχεν ἧσσον· ταῦτα ἦν μέχρι ἡμερέων ὀκτώ. ἐβίω δὲ καυθείς, καὶ καθηράμενος αὐτὸ καταπότου, καὶ περιπλασσόμενος τὸ οἴδημα· τὸ δὲ ἕλκος τῶν κακῶν οὐδὲν αἴτιον ἦν.

17. Ἐν Λαρίσῃ Θεοφόρβου παῖς ἐλέπρα τὴν κύστιν, καὶ διούρει γλίσχρον, καὶ ὠδυνᾶτο καὶ ἀρχόμενος καὶ τελευτῶν τῆς οὐρήσιος, καὶ ἔτριβε τὸ πόσθιον.[57] οὗτος πιὼν τὸ διουρητικὸν δριμὺ ἐς μὲν τὴν κύστιν οὐδὲν ἐχώρησεν, ἐξήμεσε δὲ συχνὸν πυῶδες καὶ χολήν, καὶ κάτω ἕτερα τοιαῦτα διεχώρει, καὶ ὠδυνᾶτο τὴν γαστέρα, καὶ ἐκαίετο ἔνδοθεν, τὸ δὲ ἄλλο σῶμα ψυχρὸν ἐγένετο, καὶ παρελύθη ὅλως,[58] καὶ προσδέχεσθαι οὐδὲν ἤθελεν. τούτῳ ἡλκώθη[59] κοιλίη ἰσχυρῶς ὑπὸ ἰσχύος τοῦ φαρμάκου ἄγαν· ἀποθνῄσκει δὲ μετὰ τὴν πόσιν τριταῖος.

18. Γυνὴ Ἀντιμάχου ἐν Λαρίσῃ ἐκυΐσκετο ἡμέρας ὡς πεντήκοντα, καὶ ἠσιτεῖτο[60] τὸν[61] ἄλλον χρόνον, καὶ ἡμέρας ἑπτὰ τὰς | ὑστέρας καὶ ὠδυνᾶτο τὴν καρδίην, καὶ πυρετὸς ὑπελάμβανεν. οὐχ ὑπεκεχωρήκει[62] τοῦ χρόνου τούτου· ταύτῃ ἐδόθη ἐλατήριον κατάποτον ἰσχυρότερον τοῦ δέοντος, καὶ ἀπήμεσε[63] χολὴν ξυγκεκαυμένην ὑπό τε τῆς ἀσιτίης καὶ τοῦ πυρετοῦ, καὶ γὰρ οὐδὲ ποτῷ ἐχρῆτο οὐδενί, ὀλίγην δέ, καὶ ἀπήμεσε καὶ

ing. And in the daytime the swelling and pain were greater.
He became fevered, beginning with shivering. His eyes
swelled, and his face. He was affected more on the right
than the left of the head, but the swelling spread also to the
left. That did no harm. Finally the fever was less continuous. These things continued until the eighth day. He survived, after being cauterized, purging it with medicine for
drinking, and treated with plasters on the swelling. The
wound was not responsible for his problems.

17. In Larissa, the son of Theophorbus had leprosy in
the bladder. He urinated sticky material and had pain at
the beginning and end of urination, and rubbed his penis.
When he drank the acrid diuretic none of it went to the
bladder, but he vomited a large quantity of pus and bile,
and passed more by the bowels; he had stomach pains and
internal fever though the rest of his body was cold. He was
entirely prostrate, and was not willing to take anything; his
intestines were ulcerated by the excessive strength of the
drug. He died three days after the draught.

18. The wife of Antimachus in Larissa was about fifty
days pregnant. She had no appetite for the rest of the time,
and the last seven days she also had heartburn, and fever
seized her. There were no bowel movements during that
time. She was given a purgative to drink, stronger than
needed, and she vomited bile which was burned by the
fasting and by the fever (for indeed she had taken no
drink). But it was a small amount, and she vomited vio-

---

56 δὲ καὶ M   57 πρόσθιον MV: corr. H   58 ὅλος M
59 εἰλκώθη M   60 εἴσιτε M: ἦσιτε V: ἠσιτέετο Asul.
61 τόν τ᾽ M   62 ὑπεχωρήκει MV: corr. Asul.
63 ἀπήμεσέ τε M

157

βιαίως καὶ θρομβώδεα· κἄπειτα ἠσᾶτο,[64] καὶ ἠφίει
αὐτήν,[65] καὶ ἐδόκει ἀσθενεῖν, καὶ οὐκ ἤθελε πίνουσα
ὕδωρ ἐξεμεῖν. μετὰ δὲ τοῦτο ὀδύνη ἴσχει ἰσχυρὴ τὴν
κάτω κοιλίην, ἥλκωτο γὰρ ὑπὸ τοῦ φαρμάκου, καὶ
ἐχώρει αὐτῇ μετὰ τὴν κόπρον αὐτίκα ὕφαιμον καὶ
ξυσματῶδες· αἰεὶ δὲ πλείων ἐγίνετο καὶ ἡ ἀσθένεια καὶ
ἡ[66] ἄση· καὶ τοῦ καθάρματος ἦσαν πέντε κοτύλαι.
ἔστη δὲ ἡ κοιλίη ὕδατος καταχεομένου πολλοῦ κατὰ
τῆς γαστρός· ἄλλο δὲ οὐδὲν ἐδυνήθη προσδέξασθαι·
ἔθανε περὶ μέσας νύκτας. ἐδόκει δ' ἂν βιῶναι εἰ
ἐδύνατο πίνειν τὸ ὕδωρ καὶ ἐμεῖν αὐτίκα πρὶν ὑπιέναι.

19. Οἰκέτις Αἰνησιδήμου[67] ἐν Λαρίσῃ ἡλκώθη κοιλίην καὶ τὸ ἔντερον ὑπὸ χολῆς αὐτομάτης κινηθείσης,
καὶ ἐξεχώρει καὶ ἄνω καὶ κάτω χολὴ καὶ αἷμα, καὶ
πυρετὸς εἶχεν. ταύτῃ ἐδόθη ἀσθενεούσῃ ἐλατήριον
ἀσθενές, ποτὸν ὑδαρὲς καὶ ὀλίγον, καὶ ἤμεσέ τε ἀπ'
αὐτοῦ πολύ, καὶ κάτω ὑπῆλθε πλέον· καὶ τῆς ἑσπέρης
ἐπανῆλθεν. τῇ δ' ὑστεραίῃ πυρετὸς ἦν, ἀσθενὴς δὲ
ἦν.[68] ἡ δὲ κοιλίη ἥλκωτό τε καὶ ἔτι ὑπεχώρει ταὐτά.
τρίτῃ δὲ ἔθανε δείλης, πυρετοῦ ἐπιλαμβάνοντος πάνυ
ἰσχυροῦ. αὕτη ἐδόκει ἀποθανεῖσθαι πάντως, ἥκιστα δ'
ἂν ὕδωρ πίνουσα ψυχρὸν ἕως ἔμετος εἶχεν· ἐπεὶ δὲ
ἐψύχθη ἡ ἄνω κοιλίη, ἀποκαθαρθεῖσα τῷ ὕδατι, χυλὸν
μεταπιοῦσα ψυχρόν, οὕτω μετεκλύσθη. |

64 ἰήσατο V   65 Smith: αὐτήν mss. and edd.
66 om. M   67 Ἐνεσιδήμου MV: corr. Li.
68 om. M

lently material full of clots. Then she was nauseous, and laid herself down; she seemed weak, and was not willing to drink water and vomit it. Afterward strong pain seized her lower intestine, for it was ulcerated by the drug, and immediately after the feces she passed bloody scrapings. The weakness and nausea kept increasing. There were five cotyls[4] of purged material. Her bowel became stable when much water was poured over her stomach, but she was unable to take anything else. She died in the middle of the night. It looked as though she would have survived if she could have drunk water and vomited immediately, before it went below.

19. The servant of Aenesidemus in Larissa was ulcerated in the stomach and lower intestine by bile which was aroused spontaneously, and she passed bile and blood above and below, and was feverish. She was given a weak purgative, since she was weak, a small amount well diluted, and she vomited much as a result of it, and passed much material by the bowel. And it came on again in the evening. The next day there was fever and she was weak. The bowel was ulcerated and she was still passing the same things. On the third day she died in the afternoon, seized by a very strong fever. It appeared that she would have died in any case, but would have had the best chance if she had drunk cold water until she vomited. And when the upper intestine had been chilled, purged by the water, if she had drunk cold liquid she would have been flushed out in that way.

[4] About 2½ pints.

220   20. Εὔδημος ἐν Λαρίσῃ αἱμορροΐδας ἔχων ἰσχυρὰς πάνυ, καὶ χρονίσας ἔξαιμος ὤν. χολὴ ἐκινήθη, ἀλλ᾽ ἠπίωσε τῷ σώματι, καὶ ἡ κοιλίη ἐταράχθη κάτω· ὑπεχώρει χολώδεα, καὶ αἱμορροΐδες ἐπεῖχον.[69] φάρμακον κατωτερικὸν πιὼν ἀπεκαθάρθη καλῶς, καὶ αὖτις μετέπιε χυλὸν καὶ ἔτι ἐτετάρακτο, καὶ ὀδύνη πρὸς τὰ ὑποχόνδρια προσίστατο. τούτῳ ἐπεχειρήθη τῇσιν[70] αἱμορροῖσι τὴν κοιλίην οὐ καλῶς πως ἔχοντι, ἀλλὰ δεομένῳ θεραπείης ἔτι καὶ ἀπεμέσαι·[71] ἔπειτα δὲ ἐπαλειφθέντος τοῦ καρκίνου πυρετὸς ἐπέβαλε καὶ οὐκ ἀφῆκε πρὶν ἀπέκτεινεν· ὅτε δὲ καὶ ἀφῆκε ῥῖγος ὑπολαβὼν[72] ἧκεν ὁ πυρετός, καὶ ὑπεχώρει αὐτῷ χολὴ καὶ[73] φῦσα· ἡ μὲν διεξῄει, ἡ δὲ ἐνῆν, καὶ ὀδύνη ἐν τῇ κοιλίῃ. αἱ δὲ αἱμορροΐδες ἔξω ἦσαν τοῦ ἀρχοῦ ἀπὸ τῶν ὑποκαθαρσίων ἀρξάμεναι τὸν ἄλλον χρόνον, καὶ ἡ φῦσα διὰ ταύτας ὑπεγίνετο· καὶ πρὸς πταρμὸν ἐπεγίνετο ἡ ἀρχή.

21. Ἐν Λαρίσῃ ἀνὴρ ἐτρώθη ἐκ χειρὸς λόγχῃ πλατείῃ ὄπισθεν, καὶ τὸ ἄκρον διήνεγκε κάτω[74] τοῦ ὀμφαλοῦ, πελιόν, ἀποιδέον, καὶ διῆλθε χωρίον πολύ. ἐπεὶ[75] δὲ ἐτρώθη, ἔπειτα ὀδύνη ἔσχε τὰ πρῶτα ἰσχυρή· καὶ ἐπῳδίσκετο ἡ γαστήρ. τούτῳ ἐδόθη τῇ ὑστεραίῃ κατωτερικόν, καὶ διεχώρησεν ὀλίγον ὕφαιμον, καὶ ἔθανεν. ἐδόκει τούτου τὰ ἔντερα εἶναι οὐχ ὑγιέα, καὶ αἵματος ἡ κοιλίη πλέη[76] εἶναι.

---

[69] αἱμορρόας ἀνεῖχον M
[70] τοῖσιν V

160

20. Eudemus in Larissa had severe hemorrhoids, and as time passed was anemic. Bile was aroused but he was quiet in the body. His lower intestine was upset. He passed bilious material. The hemorrhoids projected. He drank a medicine for purging below and was well purged. Again he drank a barley gruel and was again upset. Pains lodged in the hypochondria. In his case one was treating the hemorrhoids while he was not well in the intestines, but needed therapy and vomiting. Later, after the sore had been anointed, fever came on him and did not leave him before it had killed him. When the shivering left him, the fever would come on and seize him, and he would pass bile and gas: some was passed, but some remained, and he had pain in the belly. The hemorrhoids were outside the anus the whole time, beginning from the purgations; and the gas developed because of them. The beginning followed on a sneeze.

21. In Larissa a man was wounded from behind by a hand-thrown broad spear. The tip passed through below the navel. Livid, swollen over a large area. When he was wounded there was much pain at first, and the belly swelled up. The next day he was given a laxative and passed some bloody matter, and he died. It appeared that his entrails were not healthy and that his lower belly had filled with blood.

---

71 ἀπέμεσε M
72 ὑπολαβὸν V
73 καὶ χολὴ καὶ V
74 μέχρι V
75 ἐπὶ M
76 πλέης M

22. Ἀπελλαῖος Λαρισαῖος εἶχε μὲν ἡλικίην ὡς
ἐτέων τριήκοντα, ἢ ὀλίγον ἀπέλιπεν· εἴχετο δὲ τῇ
νούσῳ· ἐλαμβάνετο δὲ τὰς νύκτας μᾶλλον τῶν ἡμερέ-
ων, ἐν τῷ ὕπνῳ. ἐνόσει δὲ ὡς δύο ἔτεα | πρὸ τοῦ
222 θανάτου· ἤμει δὲ χολὴν πυρρήν. ἐνίοτε ἐπιδιέγροιτο,
ἤμει δὲ καὶ μέλαιναν. οὗτος ἀπὸ τῆς κεφαλῆς καθάρσιος ἰσχυρῆς πάνυ καὶ ἐπὶ πολὺν χρόνον καθαιρόμενος, καὶ φάρμακον δὶς πιών· ἒξ μῆνας αὐτῇ[77] διέσχεν.
ἦν δὲ πολυφάγος· ἔχων δὲ τὸ σῶμα ἐπίχολον, παλαίσας πολλά, μάλα ἐρρίγωσε, καὶ πυρετὸς ἐπέλαβε καὶ
ἡ νοῦσος ἐς νύκτα· τῇ δ' ὑστεραίῃ ἐδόκει ὑγιὴς εἶναι
αὐτῷ, καὶ τῇ ἑτέρῃ· τῇ δὲ ἐπιούσῃ νυκτὶ ἡ νοῦσος
ἐπέλαβε δεδειπνηκότα ἀπὸ πρώτου ὕπνου, καὶ εἶχε
τὴν νύκτα καὶ τὴν ἡμέρην μέχρι δορπιστοῦ· ἔθανε
πρὶν ἐμφρονῆσαι. ἐσπᾶτο περὶ τὰ δεξιὰ πρῶτον τό τε
πρόσωπον καὶ τὸ ἄλλο σῶμα, ἔπειτα ἐπὶ τὰ ἀριστερά·
καὶ ὅτε δοκέοι διαναπαῦσθαι, κῶμα εἶχε, καὶ ἔρ-
ρεγχε, καὶ αὖτις ἐξεδέχετο ἡ νοῦσος.

23. Εὔμηλος Λαρισαῖος ἐπάγη τὰ σκέλεα καὶ χεῖρας καὶ γνάθους, καὶ οὐκ ἐδύνατο οὔτε ἐκτείνειν οὔτε
ξυγκάμπτειν, εἰ μὴ ἕτερος ξυγκάμπτοι καὶ ἐκτείνοι,
οὔτε τὰς γνάθους διαίρειν εἰ μὴ ἕτερος διαίροι· ἄλλο
δὲ οὐδέν· οὔτε ὠδυνᾶτο, οὔτε ἤσθιεν εἰ μὴ μάζαν, καὶ
μελίκρητον ἔπινεν. εἰκοσταῖος ὕπτιος πίπτει καθήμε-
νος καὶ τύπτει τὴν κεφαλὴν πρὸς λίθον σφόδρα, καὶ
αὐτοῦ σκότος κατεχύθη· καὶ ὀλίγον ὕστερον ἀναστὰς
ὑγιὴς ἦν, καὶ ἐλέλυτο πάντα, πλὴν μετὰ τὸν ὕπνον ὅτε

22. Apellaeus of Larissa was about thirty or slightly less. He was taken by the illness.[5] He was taken more at night than in the daytime. He was ill about two years before his death. Sometimes he vomited reddish bile. Sometimes he was wakeful. He also vomited black bile. After a powerful purgation of the head, when he was purged strongly and for a long time and drank the drug twice, it stayed away for six months. He was a copious eater, and, having a bilious body, when he had wrestled a great deal he had severe chills, fever seized him, and the illness seized him towards night. On the next day he felt healthy, and the next. The following night, after he had dined, the sickness seized him after his initial sleep, and continued that night and the next day until supper time. He died before coming to himself. He was drawn up on the right side at first, in the face and the rest of the body, later on the left. And whenever he seemed to get better, coma came on him, he would wheeze, and again the sickness possessed him.

23. Eumelus of Larissa grew rigid in his legs, arms, and jaws. He could not extend them or bend them unless someone else extended or bent them, nor open his jaws unless someone else opened them. But no other symptoms. He had no pain and he did not eat, except barley cakes, and he drank honey water. On the twentieth day he fell backwards while sitting and severely struck his head on a stone, and darkness poured over him. Shortly later he stood up and was better. All was relaxed except that when

[5] As Littré notes, the symptoms as well as the therapy employed here suggest an epileptiform disease.

[77] Smith: αὐτὸν mss.

ἐξέγροιτο ὀλίγον τι ξυνεδέδετο τὰ ἄρθρα· ἔτεα δὲ ἦν
ἤδη δώδεκα ἢ δεκατρία· ἔκαμε δὲ μῆνας τρεῖς ἢ
τέσσαρας. |

224   24. Ἐν Λαρίσῃ, παρθένος αἷμα ἐμέσασα οὐ πολύ,
ἔμπυος γενομένη· πυρετῶν ἐπιλαβόντων οὐκ ἀπηλ-
λάσσετο πρὶν τελευτῶσα ἀπέθανε τρίτῳ μηνί· πρὸ δὲ
τοῦ θανάτου, ἐκωφώθη τὰ οὔατα, καὶ οὐκ ἤκουεν εἰ μή
τις πάνυ μέγα βοῶν. πρὸ δὲ τοῦ ἐμέτου τοῦ αἵματος
προσησθένει.

25. Ἐν Λαρίσῃ ἀμφίπολος Δυσήριδος, νέη ἐοῦσα
ὁκότε λαγνεύοιτο περιωδύνει ἰσχυρῶς, ἄλλως δὲ ἀνώδυνος ἦν. ἐκύησε δὲ οὐδέποτε. ἑξηκονταέτης[78] γενομένη ὠδυνᾶτο ἀπὸ μέσου ἡμέρης, ὡς ὠδίνουσα ἰσχυ-
ρῶς· πρὸ δὲ μέσου ἡμέρης[79] αὕτη πράσα τρώγουσα
πολλά, ἐπειδὴ ὀδύνη αὐτὴν ἔλαβεν ἰσχυροτάτη τῶν
πρόσθεν, ἀναστᾶσα ἐπέψαυσέ τινος τρηχέος ἐν τῷ
στόματι τῆς μήτρης. ἔπειτα, ἤδη λειποψυχούσης αὐτῆς, ἑτέρη γυνὴ καθεῖσα τὴν χεῖρα ἐξεπίεσε λίθον[80]
ὅσον σπόνδυλον ἀτράκτου, τρηχύν· καὶ ὑγιὴς τότε
αὐτίκα[81] καὶ ἔπειτα ἦν.

26. Ὁ Μαλιεύς, ἄμαξα αὐτῷ ἐπῆλθεν ἄχθος ἔχουσα ἐπὶ τὰς πλευράς, καὶ κατῆξε τῶν πλευρέων, καὶ
χρόνον αὐτῷ ὑπέστη πύον κάτωθεν τῶν πλευρέων. ὑπὸ
τὸν σπλῆνα καυθείς, ἔμμοτος ὤν, ἀφίκετο ἐς δέκα
μῆνας. ἀνατμηθὲν τὸ δέρμα, ὀπὴ ἐφάνη ἐς τὸ δέρτρον

[78] ἑξήκοντ. δὲ Μ
[79] ὡς . . . ἡμέρης om. V

he wakened after sleep his joints were slightly bound. He was then twelve or thirteen years old. He was troubled for three or four months.

24. In Larissa, a maiden vomited a small amount of blood when she had become purulent. Fevers seized her, and did not leave her before she finally died in the third month. Before death she grew deaf and could not hear except when one was shouting very loudly. She had been weak before the vomiting of blood.

25. In Larissa, the servant of Dyseris, when she was young, whenever she had sexual intercourse suffered much pain, but otherwise was without distress. And she never conceived. When she was sixty she had pain from midday, like strong labor. Before midday she had eaten many leeks. When pain seized her, the strongest ever, she stood up and felt something rough at the mouth of her womb. Then, when she had already fainted, another woman, inserting her hand, pressed out a stone like a spindle top, rough. She was immediately and thenceforth healthy.

26. The man from Malia: a loaded cart ran over him in the rib area, and broke some of his ribs, and in time pus gathered below the ribs. Cauterized below the spleen and treated with a tampon, he continued for ten months.[6] When the skin was cut, an opening into the peritoneum

---

[6] A lint plug will have been inserted into the wound made by the cautery to keep the wound open and maintain drainage. Even so, a fistula developed.

---

80 λίθῳ M
81 καὶ αὐτίκα V

ἐπὶ θάτερα ἀφίκουσα, καὶ πρὸς τὸν νεφρὸν καὶ πρὸς
τὰ ὀστέα ἐπῆλθε σαπρή. τούτου ἥ τε σχέσις τοῦ
σώματος παρέλαθεν ἐπίχολος ἐοῦσα, καὶ ἐν τῷ σώματι καὶ ἐν τῷ νοσήματι ἦν σηπεδὼν[82] τοῦ δέρτρου
πολλὴ καὶ ἄλλων σαρκῶν, ἃς ἔδει αὐτίκα ἐκβάλλειν,
εἴ τις ἐδύνατο, ξηρῷ φαρμάκῳ, ἕως ἰσχύν τινα εἶχεν ὁ
226   ἄνθρωπος· | ἀπὸ γὰρ τῶν ὑγρῶν οὐδὲν ἐπεδίδου, ἀλλ'
ἐσήπετο. ἀπὸ δὲ τῶν μότων ἰσχομένου τοῦ ὑγροῦ,
ῥῖγός τε ἐλάμβανε καὶ πυρετὸς ἐπελάμβανε, καὶ ἐσήπετο μᾶλλον· ἐπέρρει δὲ αὐτῷ σαπρὸν ὑπόμελαν[83]
δυσῶδες, πρὶν δὲ ἐπιχειρεῖν ἰητρεύεσθαι, οἷον ἑκάστης ἡμέρης συχνὸν διεπέρα ἔξω· ἦν δ' οὐκ εὔροον.
ἐγνώσθη τὸ εἶναι πορρωτέρω τὴν φύσιν τοῦ νοσήματος ἢ ὑπὸ τὸ δέρμα· πάντα ἂν ὀρθῶς πάσχων, ὅμως
οὐκ ἂν ἐδόκει σωθῆναι· καὶ διάρροια[84] ἐπέλαβεν.

27. Αὐτόνομος ἐν Ὁμίλῳ ἐν κεφαλῆς τρώματι ἔθανεν ἑκκαιδεκάτῃ ἡμέρῃ θέρεος μέσου λίθῳ ἐκ χειρὸς
βληθεὶς κατὰ τὰς ῥαφὰς μέσῳ τῷ βρέγματι. τοῦτο
παρέλαθέ με δεόμενον πρισθῆναι· ἔκλεψαν δέ μευ τὴν
γνώμην αἱ ῥαφαὶ ἔχουσαι ἐν σφίσιν ἑωυτῇσι τοῦ
βέλεος τὸ σῖνος· ὕστερον γὰρ καταφανὲς γίνεται.
πρῶτον μὲν ἐς τὴν κληῖδα, ὕστερον δὲ ἐς τὴν πλευρήν,
ὀδύνη ἰσχυρὴ πάνυ, καὶ σπασμὸς ἐς ἄμφω τὼ χεῖρε
ἦλθεν, ἐν μέσῳ γὰρ εἶχε τῆς κεφαλῆς καὶ τοῦ βρέγματος τὸ ἕλκος. ἐπρίσθη δὲ πεντεκαιδεκάτῃ, καὶ πύον
ὑπῆλθεν οὐ πολύ·[85] ἡ δὲ μῆνιγξ ἀσαπὴς ἐφαίνετο.

appeared which led in both directions: a rotten channel ran to the kidney and to the bones. The state of his body was bilious, though one did not notice it. In his body generally and at the site of the disease there was much putridity of the peritoneum and of other flesh, which one needed to remove immediately, if one could, with a dry(ing) drug so long as the man had any strength, since he did not improve from the damp (drugs), but grew purulent. Because the moisture was held by the tampons, shivering and fever seized him and he suppurated more. There flowed forth a black, foul-smelling corruption, and before one took the treatment in hand it flowed out in quantity every day, but it did not flow freely. It was recognized that the nature of the disease was farther off than below the skin. Had he been properly cared for in all respects, it still does not seem that he would have survived. Diarrhea also seized him.

27. Autonomus in Omilus died on the sixteenth day from a head wound in midsummer. The stone, thrown by hand, hit him on the sutures in the middle of the *bregma* (front of head). I was unaware that I should trephine, because I did not notice that the sutures had the injury of the weapon right on them, since it became obvious only later. First he had sharp pain towards the collarbone, later in the sides, and convulsions into both arms, for he had the wound in the middle of the head and the *bregma*. He was trephined the fifteenth day and some pus came out. But the membrane appeared uncorrupted.

---

^{82} καὶ ἐν τῷ νοσήματι ἦν σηπεδὼν] τὴν σηπεδῶνα ξηρὴν M ^{83} ὑπὸ μέλανα M ^{84} διάρροι M ^{85} πολύν M

28. Παιδίσκη,[86] ἐν Ὁμίλῳ, ἐκ τρώματος κεφαλῆς ὡς δωδεκέτις θνήσκει ἐν μέσῳ θέρει τεσσαρεσκαιδεκάτῃ ἡμέρῃ· θύρην τις αὐτῇ ἐνέβαλε καὶ τὸ ὀστέον φλᾷ καὶ ῥήγνυσιν· καὶ ἐν τῷ ἕλκει αἱ ῥαφαὶ[87] ἦσαν. τοῦτο ἐγνώσθη ὀρθῶς πρίσεως δεόμενον· ἐπρίσθη δὲ οὐκ εἰς τὸ δέον, ἀλλ᾽ ὅσον ὑπελείφθη, πύον ἐν αὐτῷ ἐγένετο. ὀγδόῃ, ῥῖγος καὶ πυρετὸς ἐπέλαβεν· εἶχε δὲ οὐκ εἰς τὸ δέον, ἀλλ᾽ ὅσον καὶ τῶν πρόσθεν ἡμερέων, ὅτε πυρετὸς οὐκ εἶχεν. ἐνάτῃ δὲ τὸ λοιπὸν ἐξεπρίσθη, |
228 καὶ ὑπεφάνη ὀλίγον πάνυ πύον ξὺν αἵματι· καὶ ἡ μήνιγξ[88] καθαρὴ ἦν. καὶ ὕπνος μὲν ἐπέλαβεν· ὁ δὲ πυρετὸς αὐτὶς οὐκ ἀφίει· σπασμὸς δὲ χεῖρα τὴν ἀριστερὴν ἐπελάμβανεν· ἐν γὰρ τοῖσι δεξιοῖσι μᾶλλον εἶχε τὸ ἕλκος.

29. Κυρηναῖος[89] ἐν Ὁμίλῳ ἔμπυος γενόμενος τὴν κάτω κοιλίην ἐκαύθη ὕστερον ἡμέρῃσι τριήκοντα τοῦ δέοντος· καὶ ἔσχεν ἐπιεικῶς, καὶ ἐξηράνθη τὸ πύον ἐν τῇ κοιλίῃ. ἐν δὲ τῇ θερμοτάτῃ ὥρῃ ἐσθίων ὀπώρην καὶ ἄλλα σιτία ἀξύμφορα, πυρετὸς ἐπέλαβε, καὶ διάρροια, καὶ ἔθανεν.

30. Ἑκάσων ἐν Ὁμίλῳ ὥσπερ χάτερος ὕστερον ἐκαύθη· ὅμως δὲ ἐξηράνθη πλὴν ὀλίγου ἡ κοιλίη· δυσεντερίη δὲ ἐπέλαβε, καὶ αὐτὴν ἀποφυγὼν ἤσθιε τὸ πᾶν ἕως ὅλος[90] ᾤδησε, καὶ ἐρράγη αὐτῷ πύον κάτω, καὶ διάρροια, καὶ ἔθανεν.

31. Ἑκάσων ἐν Ὁμίλῳ ἀπὸ ἀκαθαρσίης καὶ πονηρῆς καθάρσιος ἐς τὸ ἰσχίον ἐπέστη αὐτῷ ὀδύνη ὀξείη· καὶ αὐτὴ μὲν ἀπηλλάχθη, πυρετοὶ δὲ αὐτὸν ὑπέλαβον·

28. At Omilus, a young girl of about twelve years died in midsummer from a wound in the head, on the fourteenth day. Someone hit her with a door and crushed and shattered her skull. The sutures were in the wound. This was recognized properly as needing trephination. It was trephined, but not sufficiently. As some bone was left, pus developed there. On the eighth day shivering and fever seized her. When free from fever she was not as she should have been, but was as on the previous days. On the ninth day the rest was trephined, and a little pus with blood appeared. The membrane was clean. Sleep seized her, but the fever did not go away again. Spasms seized her left hand, since the wound was on the right.

29. The man from Cyrene at Omilus, when he became purulent in the lower belly, was cauterized later than he should have been by thirty days. He was all right, and the pus dried up in the belly. But in the hottest season, as he was eating fruit and other inappropriate foods, fever seized him and diarrhea, and he died.

30. Hecason, in Omilus, was cauterized late like the other one. Still his belly almost dried up. Dysentery seized him, and when he escaped it he used to eat everything, until he swelled up all over, and pus broke out below, and diarrhea, and he died.

31. Hecason, in Omilus: from lack of purging and poor purging a sharp pain developed in his groin. It went away,

---

86 παιδίσκει M
87 αἱ ῥαφαὶ ἐν τῷ ἕλκει M
88 μῆνιξ M
89 Smith: κυρῖνος V: κυρίνος M: κυρίνιος recc.
90 ὅλως V

καὶ χρόνον πολὺν κλινοπετὴς ὤν, οὔτε πίνων οὐδέν,
οὔτε διψῶν, ἀσθενής τε ὢν καὶ φρικώδης. τούτῳ ἀφη-
ρέθη μὲν τὸ νόσημα· χρηστῶς ὡς ἔδει τὸ σῶμα,
ὠφελεῖτο δὲ ἀπὸ τῶν προσφερομένων· τελευτῶντι δὲ
ἐρράγη τὸ νόσημα κάτω, καὶ ἐχώρει πᾶν ὑπὸ πολλῇ
χολῇ, καὶ παρέκοψε, καὶ ἔθανεν· ἐδόκει δ' ἂν ἐκφυγεῖν
τὸ νόσημα. |

230   32. Ἐν Σαλαμῖνι ὁ περὶ τὴν ἀγκύρην περιπεσὼν
ἐπὶ γαστέρα ἐτρώθη·[91] περιωδύνει δέ· φάρμακον δὲ
ἔπιε, καὶ οὐ διεχώρησε κάτω, οὐδὲ ἀνήμεσεν.

33. Ἡ γυνὴ ἣ ἀπέσφαξεν αὐτὴν ἐπνίγετο· καὶ
ἐδόθη αὐτῇ ὕστερον πολλῷ κατάποτον ἐλατήριον, καὶ
ἐξεχώρησεν αὐτῇ.

34. Ὁ ἐξ Εὐβοίης ἐλθὼν νεηνίσκος, πολλὴν κάτω
κεκαθαρμένος διαλιπών, πεπαυμένος πυρέσσων, εἶτα
δοκέων ἄλλου[92] δεῖσθαι, ἔπιεν ἀσθενῆ, ῥίζην ἐλατήριον. μετὰ τὴν πόσιν τεταρταῖος ἔθανεν ἐκκαθαρθεὶς
οὐδέν· ἀλλ' ὕπνος εἶχε, καὶ οὐκ ἐδύνατο παύσασθαι ἡ
δίψα.

35. Ἡ δούλη ᾗ ἀπὸ καταπότου ἄνω μὲν ἐχώρησεν
ὀλίγη καὶ ἔπνιγε, κάτω δὲ πολλή· τῆς νυκτὸς δὲ ἔθανε·
βάρβαρος δὲ ἦν.

36. Ὁ Εὐβίου ἄνθρωπος πιὼν ἐλατήριον τρεῖς ἡμέρας ἐκαθαίρετο· καὶ ἔθανε· τὴν δὲ χεῖρα ἔμπυον εἶχε
μέχρι τοῦ ἀγκῶνος.

[91] διετρώθη M: διεχώρη V
[92] Francis Clifton: ἄνω mss.

but fever seized him. He was bedridden for a long time; he drank nothing, nor was he thirsty, but was weak and trembling. His illness was cured, his body was in very good condition, and he benefited from the things administered. But at the end, the illness broke out below, and everything was passed, with much bile, and he lost consciousness and died. It had appeared that he would survive the disease.

32. In Salamis, the man who fell on the anchor received a wound in the belly. He had great pain. He drank a (purgative) drug and there was no evacuation below, nor did he vomit.

33. The woman who cut her throat: she choked. She was later given much purgative medicine which produced bowel movements.

34. The young man who came from Euboea: he was purged of much bile below and had a quiet period. Stopping he was feverish. Then, thinking he needed another, he drank a weak medicine, root of *elaterion*,[7] and died four days after drinking it: nothing had been purged, but sleep held him, and his thirst could not be quenched.

35. The slave woman: after a potion she evacuated a little bile above, and choked; passed much below. She died that night. She was a barbarian.

36. The man from Eubius, having drunk *elaterion*, was purged for three days. He, too, died. His arm was suppurating up to the elbow.

[7] The squirting cucumber.

37. Ὁ Συμμάχου παῖς ὑπὸ χολῆς ἀπεπνίγη νύκτωρ καταδαρθών, καὶ πυρετοῦ ἐπέχοντος· φάρμακον δὲ πιών, οὐ κατέσχεν, οὐδὲ ἐκαθήρατο ἡμέρῃσι[93] πρὶν ἀποθανεῖν ἕξ.

38. Ὁ παρὰ τὸν δρόμον οἰκέων τῆς νυκτὸς αἷμα ἐμέσας, τῇ ὑστεραίῃ ἔθανεν, αἷμα ἐμέων πολύ, καὶ πνιγόμενος· ἐς σπλῆνα δέ, καὶ κάτω αἱματῶδες αὐτῷ ἐχώρει πολύ.

39. Παιδίον ὑπὸ οὐρέος[94] πληγὲν[95] τὴν γαστέρα καὶ τὸ ἧπαρ, ἀπέθανε τετάρτῃ,[96] τὸ δὲ[97] πνεῦμα πυκινὸν[98] εἶχε, καὶ οὐ κατενόει, καὶ πυρετὸς εἶχεν. |

232   40. Ἑρμοφίλου υἱὸς ἔκαμεν ἡμέρας ἕνδεκα, πυρετὸς δὲ εἶχε, καὶ ἡσίτει,[99] καὶ οὐχ ὑπήει τὰ σιτία· καὶ πρῶτον μὲν παρενόησε, τῆς δὲ νυκτὸς ἐπαύσατο. τῇ δὲ ἐπιούσῃ ἡμέρῃ ἄφωνος ἔκειτο ῥέγχων, διεστραμμένα ἔχων τὰ ὄμματα, πυρέσσων· πτεροῦ δὲ καθιεμένου ἤμεσε χολὴν μέλαιναν· καὶ κλυσθέντι κόπρος ὑπῆλθε πολλή.

41. Ἀριστίωνος δούλης αὐτόματος ὁ ποὺς ἐσφακέλισε κατὰ μέσον τοῦ ποδὸς ἔνδοθεν ἐκ πλαγίου, καὶ τὰ ὀστέα σαπρὰ γενόμενα ἀπέστη, καὶ ἐξῄει κατὰ σμικρὸν σηραγγώδεα,[100] καὶ διάρροια ἐπεγένετο, καὶ ἔθανεν.

42. Γυνὴ ὑγιαίνουσα, παχεῖα, κυήσιος ἕνεκεν ἀπὸ καταπότου ὀδύνη εἴχετο τὴν γαστέρα, καὶ στρόφος ἐς τὸ ἔντερον, καὶ ᾤδει, πνεῦμα δὲ προΐστατο, καὶ ἀπορίη

93 ἡμέρης καὶ Μ

37. The child of Symmachus choked from bile while asleep at night, while fever, too, held him. Although he drank a drug, it did not hold back the disease, nor was he purged in the six days before his death.

38. The man who lived by the race course: he vomited blood at night, and died the next day, vomiting much blood and choking. And it went to the spleen, and he passed much bloody material below.

39. The child, struck in the belly and liver by a mule, died on the fourth day. His breathing was rapid, he was not conscious, and fever held him.

40. Hermophilus' son was sick eleven days. Fever held him, he could not eat and did not pass the food. At first he was delirious but that stopped during the night. The next day he lay voiceless, wheezing, eyes rolled back, feverish. When a feather was introduced he vomited black bile. With an enema he passed much feces.

41. The foot of Aristion's female slave spontaneously ulcerated in the middle of the foot on the inner side. The bones became corrupted, separated and came off little by little, eroded. Diarrhea developed; she died.

42. A stout woman, who was healthy, after drinking a purgative for the sake of conception was possessed in the belly by pain; she had twisting in the intestines, and swelled up. Breathing (difficulty) became prominent, and

---

94 Erot. 64.16N: συός mss.: ὀρέος Gal. *Diff. resp.* (7.956K)
95 ἐπλήγη Gal.   96 τεταρταῖον Gal.
97 om. Gal.
98 πυκνὸν Gal.
99 ἤσιτε M
100 συριγγώδεα mss.: corr. Erm.

ξὺν ὀδύνῃ· καὶ ἐμημέκει οὐ πολύ· καὶ ἐξέθανε πεντάκις
ὡς τεθνάναι δοκεῖν· καὶ οὔτε ἐμέσασα ἀπὸ ὕδατος
ψυχροῦ ἐχάλα οὔτε τῆς ὀδύνης ἐπιούσης οὔτε τὴν
πνοήν. ὕδατος δὲ κατεχύθησαν ψυχροῦ ἀμφορεῖς ὡς
τριάκοντα κατὰ τοῦ σώματος, καὶ ἐδόκει τοῦτο μόνον
ὠφελεῖν· καὶ ὕστερον κάτω ἐχώρησε χολὴ συχνή· ὅτε
δὲ ἡ ὀδύνη εἶχεν οὐδὲν ἐδύνατο χωρῆσαι· καὶ ἐβίω.

43. Ἄντανδρος ἀπὸ καταπότου, ἐξάντης ἐὼν[101] τἆλλα περὶ δὲ τὴν κύστιν ἐδόκει ἄλγος ἔχειν· ἐκαθήρατο
αὐτίκα ὀξέως πολὺ κάθαρμα· καὶ ἀπὸ μέσου ἡμέρης
ὀδύνη εἶχεν ἰσχυρὴ πάνυ ἐς τὴν γαστέρα· πνῖγμα καὶ
ἀπορίη καὶ ῥιπτασμός· καὶ ἤμει καὶ ἐχώρει οὐδέν, καὶ
τὴν νύκτα ἔπασχε, καὶ ὕπνος οὐκ ἐπῄει. τῇ δ᾽ ὑστε-
ραίῃ, ἐχώρει πολλόν, ὕστερον αἷμα, καὶ ἔθανεν. |

44. Τῷ Ἀθηνάδεω παιδὶ ἄρρενι, τῷ φαγεδαινωθέντι,
ὀδὼν ὁ ἐν ἀριστερᾷ κάτω, ἄνω δὲ ὁ[102] ἐν δεξιᾷ· τὸ οὖς
τὸ δεξιὸν ἐνεπύησεν, οὐκ ἔτι ἀλγέοντος.

45. Ὁ σκυτεύς, κάσσυμα κεντῶν ἐν τῷ ἡπητίῳ[103]
ἐκέντησεν ἑωυτὸν ἐπάνω τοῦ γούνατος ὡς ὁ μηρός,[104]
καὶ ἔβαψεν ὡς δάκτυλον. τούτῳ αἷμα μὲν οὐδὲν ἐρρύη,
τὸ δὲ τρῶμα ταχὺ ἔμυσεν, ὁ δὲ μηρὸς ὅλος ἐμετεωρίζετο,[105] καὶ διέτεινεν ὁ μετεωρισμὸς ἔς τε τὸν βουβῶνα
καὶ τὸν κενεῶνα· οὗτος τῇ τρίτῃ ἔθανεν.

46. Ὁ δὲ[106] παρὰ τὸν βουβῶνα πληγεὶς τοξεύματι,
ὃν ἡμεῖς ἑωράκειμεν,[107] παραδοξότατα ἐσώθη· οὔτε

101 ἐξαντήσεων M: ἐξανθήσεων V   102 Lind.: τὸ mss.
103 ὀπιτίῳ M   104 Smith: ὡς ὅμηρος MV

weakness with pain. She vomited a little. Five times she fainted dead away, so as to appear dead. Having vomited after cold water she got no relief, either when the pain came on or in relation to the breathing. But, when about thirty amphorae of cold water were poured over her body, that alone seemed to benefit her. Afterwards much bile passed below, but when the pain possessed her she could not move her bowels. She lived.

43. Antandrus, after a purgative, seemed otherwise healthy but seemed to have a pain around the bladder. He was purged immediately of much material very quickly. From the middle part of the day strong pain held him in the belly. Choking, weakness, tossing. He vomited nothing, passed nothing by the bowels. He suffered in the night; sleep did not come. The next day he passed much feces, then blood, and died.

44. The male child of Athenades, who had phagedaena[8] in the lower left tooth, and the upper right. And his right ear became purulent when the pain had stopped.

45. The shoemaker, sewing the sole of a shoe, stabbed himself on his needle, above the knee where the thigh is; he pierced it about a finger's depth. No blood flowed out, the wound quickly closed up, the whole thigh became elevated. The swelling extended to the gland at the groin and to the flank. He died on the third day.

46. The man hit by an arrow in the gland at the groin, whom we had seen, was preserved in a most unexpected

---

[8] The eroding disease; cf. *Epid.* 4.19.

---

105 μετεωρίζετο V   106 ὁ δὲ om. M
107 ἑωράκαμεν M

γὰρ ἀκὶς ἐξῃρέθη, ἦν γὰρ ἐν βάθει λίην, οὔτε αἱμορραγίη οὐδεμία ἐγένετο ἀξίη λόγου, οὔτε φλεγμονή,
οὔτε ἐχώλευσεν. τὴν δὲ ἀκίδα ἔστε καὶ ἡμεῖς ἀπηλ-
λασσόμεθα, ἐτέων ἐόντων ἕξ, εἶχεν· ὑπενοεῖτο δὲ τούτῳ μεσηγὺ τῶν νεύρων κεκρύφθαι κάτω[108] τὴν ἀκίδα,
φλέβα τε καὶ ἀρτηρίην οὐδεμίαν διαιρεθῆναι.

47. Ὁ πληγεὶς ὀξεῖ[109] βέλει ἐς τοὔπισθεν σμικρὸν
κάτω τοῦ τραχήλου, τὸ μὲν τρῶμα ἔλαβεν οὐκ ἄξιον
λόγου ἐσιδεῖν· οὐ γὰρ ἐν βάθει ἐγένετο. μετὰ δὲ οὐ
πολλὸν χρόνον, ἐξαιρεθέντος τοῦ βέλεος, ἐτιταίνετο ἐς
τοὔπισθεν ἐρυσθεὶς ὡς οἱ ὀπισθοτονικοί· καὶ αἱ γένυες
ἐδέδεντο· καὶ εἴ τι ὑγρὸν ἐς τὸ στόμα λάβοι, καὶ τοῦτο
ἐγχειροίη καταπίνειν, πάλιν ἀνέκοπτεν ἐς τὰς ῥῖνας,
καὶ τὰ λοιπὰ αὐτίκα ἐκακοῦτο, καὶ δευτέρῃ ἡμέρῃ
ἔθανεν. |

48. Νεηνίσκος ὁδὸν τρηχείην τροχάσας ἤλγει τὴν
πτέρνην, μάλιστα τὸ κάτω μέρος, ἀπόστασιν δὲ ὁ
τόπος οὐκ ἐλάμβανεν οὐδεμίαν ὡς ξυνάγων[110] ὑγρόν.
ἀλλὰ τεταρταίῳ τε ἐόντι αὐτῷ ἐμελαίνετο πᾶς ὁ τόπος
ἄχρι τοῦ ἀστραγάλου καλεομένου καὶ τοῦ κοίλου τοῦ
κατὰ τὸ στῆθος τοῦ ποδός, καὶ τὸ μελανθὲν οὐ περιερ-
ράγη, ἀλλὰ πρότερον ἐτελεύτα· τὰς πάσας δὲ ἐβίου
ἡμέρας εἴκοσιν[111] ἀπὸ τοῦ δρόμου.

236

108 Smith: κατὰ mss.
109 ὀξὺ M
110 συνάγον M
111 καὶ εἴκοσιν M

176

manner. The point was not removed because it was in too deep, nor was there any notable hemorrhage, nor inflammation, nor was he lamed. He had carried the point for six years up to the time of our departure. The suspicion was that the point was buried beneath in the midst of his tendons, and that no vein and artery were lacerated.

47. The man hit by a sharp dart from behind just below the neck: the wound he got was insignificant to look at, since it was not deep. But after a short time, when the dart had been removed, he was stretched and drawn backwards like those with *opisthotonos*.[9] And his jaws were fixed; if he took liquid in his mouth and tried to swallow it, he rejected it again into the nostrils. His condition otherwise deteriorated and he died on the second day.

48. A young man who had sprinted on a rough road had pain in his heel, especially the lower part, but the area did not allow an apostasis because it was collecting moisture.[10] But on the fourth day the whole area became black, up to the so-called *astragalos* (ankle joint) and to the hollow behind the ball of the foot, and the blackness did not break out; rather, he died first. He lived a total of twenty days after his run.

---

[9] The kind of tetanus that draws the patient backwards into a bow shape.

[10] By conjecture ($\sigma\tau\epsilon\nu\nu\gamma\rho\grave{o}\varsigma$ $\mathring{\omega}\nu$ for $\sigma\upsilon\nu\acute{a}\gamma\omega\nu$ $\mathring{\upsilon}\gamma\rho\acute{o}\nu$) Littré produced an interpretation opposite to the one I offer. Littré says, "The place was too dry to receive any abscession." I think, however, that the author's notion is that the area kept absorbing moisture and did not permit the noxious matter to move up the leg and out. This interpretation requires the assumption that he used $\dot{\epsilon}\lambda\acute{a}\mu\beta\alpha\nu\epsilon\nu$, "accepted," to mean "allowed."

49. Ὁ δὲ ἐς τὸν ὀφθαλμὸν πληγεὶς ἐπλήγη μὲν κατὰ τοῦ βλεφάρου, ἔδυ δὲ ἡ ἀκὶς ἱκανῶς· ὁ δὲ ἀθὴρ[112] προσυπερεῖχε. τμηθέντος τοῦ βλεφάρου, ἤρθη πάντα· οὐδὲν φλαῦρον· ὁ γὰρ ὀφθαλμὸς διέμεινε, καὶ ὑγιὴς ἐγένετο ξυντόμως· αἷμα δὲ ἐρρύη λάβρον, ἱκανὸν τῷ πλήθει.

50. Ἡ παρθένος ἡ καλὴ ἡ τοῦ Νερίου ἦν μὲν εἰκοσαέτης, ὑπὸ δὲ γυναίου φίλης παιζούσης πλατέῃ τῇ χειρὶ ἐπλήγη κατὰ τὸ βρέγμα. καὶ τότε μὲν ἐσκοτώθη καὶ ἄπνοος ἐγένετο, καὶ ὅτε ἐς οἶκον ἦλθεν αὐτίκα τὸ πῦρ εἶχε, καὶ ἤλγει τὴν κεφαλήν, καὶ ἔρευθος ἀμφὶ τὸ πρόσωπον ἦν. ἑβδόμῃ ἐούσῃ, ἀμφὶ τὸ οὖς τὸ δεξιὸν πύον ἐχώρησε δυσῶδες, ὑπέρυθρον, πλεῖον κυάθου, καὶ ἔδοξεν ἄμεινον ἔχειν, καὶ ἐκουφίσθη. πάλιν ἐπετείνετο τῷ πυρετῷ, καὶ κατεφέρετο, καὶ ἄναυδος ἦν, καὶ τοῦ προσώπου τὸ δεξιὸν μέρος εἵλκετο,[113] καὶ δύσπνοος ἦν, καὶ σπασμὸς τρομώδης ἦν. καὶ γλῶσσα εἴχετο,[114] ὀφθαλμὸς καταπλήξ· ἐνάτῃ ἔθανεν.

51. Ὁ Κλεομένεος[115] παῖς, χειμῶνος ἀρξάμενος, ἀπόσιτος, ἄνευ πυρετοῦ ἐτρύχετο, καὶ ἤμει τὰ σιτία καὶ φλέγμα· δύο μῆνας ἀσιτίη παρείπετο.

52. Τῷ μαγείρῳ ἐν Ἀκάνθῳ τὸ κύφωμα ἐκ φρενίτιδος ἐγένετο· | τούτῳ φαρμακοποσίη οὐδεμία ξυνήνεγκεν, οἶνος δὲ μέλας καὶ ἀρτοσιτίη· καὶ λουτρῶν ἀπέχεσθαι, καὶ ἀνατρίβεσθαι ⟨μὴ⟩ λίην,[116] θάλπε-

_____

[112] αἰθὴρ mss.: corr. Foës (cf. Gal. Gloss. 19.70K)

49. The man struck in the eye was hit in the eyelid, and the point penetrated some distance, though the barb stuck out. His eyelid was cut and everything removed. Nothing bad; the eye survived and became quickly healthy. There was vigorous bleeding of an adequate amount.

50. The pretty virgin daughter of Nerius was twenty years old. She was struck on the *bregma* (front of the head) by the flat of the hand of a young woman friend in play. At the time she became blind and breathless, and when she went home fever seized her immediately, her head ached, and there was redness about her face. On the seventh day foul-smelling pus came out around the right ear, reddish, more than a cyathus (one-fifth of a cup). She seemed better, and was relieved. Again she was prostrated by the fever; she was depressed, speechless; the right side of her face was drawn up; she had difficulty breathing; there was a spasmodic trembling. Her tongue was paralyzed, her eye stricken. On the ninth day she died.

51. Cleomenes' son, beginning in the winter, had no desire for food, wasted away without fever, and vomited his food and phlegm. Aversion to food lasted two months.[11]

52. The butcher at Acanthus developed a humpback after phrenitis. No drug helped, but red wine and eating bread, refraining from bathing, being massaged with re-

---

[11] Cf. *Epid.* 7.70.

---

[113] ἠλκοῦτο V
[114] ἐπείχετο V
[115] Κλεομένεῳ M
[116] μὴ λίην Smith: λίην MV: λείως recc.

σθαί τε μήτε[117] πολλῷ πυριήματι,[118] ἀλλὰ πρηέως.

53. Τῇ Σίμου τὸ τριηκοσταῖον ἀπόφθαρμα· πιούσῃ τι ἢ αὐτόματον τοῦτο· ξυνέβη πόνος, ἔμετος χολωδέων πολλῶν ὠχρῶν καὶ πρασοειδέων ὅτε πίοι· σπασμὸς εἶχε, γλῶσσαν κατεμασᾶτο. πρὸς τεταρταίην εἰσῆλθον· γλῶσσα μεγάλη, μέλαινα· τοῖν ὀφθαλμοῖν δὲ τὰ λευκὰ ἐρυθρὰ ἦν· ἄγρυπνος· τετάρτῃ δὲ ἔθανεν ἐς νύκτα.

54. Ὀρίγανον[119] ὀφθαλμοῖσι κακὸν πινόμενον, καὶ ὀδοῦσιν.

55. Ἡ ἀπὸ τοῦ κρημνοῦ κόρη πεσοῦσα, ἄφωνος· ῥιπτασμὸς εἶχε, καὶ ἤμεσεν ἐς νύκτα αἷμα[120] πολύ· κατὰ ἀριστερὰ πεσούσης, συχνότερον ἐρρύη· μελίκρητον χαλεπῶς κατέπινεν· ῥέγκος, πνεῦμα πυκνόν, ὡς τῶν θνησκόντων· φλέβες ἀμφὶ τὸ πρόσωπον τεταμέναι· κλίσις ὑπτίη· πόδες χλιηροί·[121] πυρετὸς βληχρός· ἀφωνίη. ἑβδομαίη, φωνὴν ἔρρηξεν· αἱ θέρμαι λεπτότεραι ἔσχον· περιεγένετο.

56. Πυθοκλῆς τοῖσι κάμνουσιν ὕδωρ, γάλα πολλῷ τῷ ὕδατι μιγνὺς ἐδίδου.

57. Χιμέτλων, κατασχᾶν, ἀλεαίνειν τοὺς πόδας, ὡς δὲ μάλιστα ἐκθερμαίνειν πυρὶ καὶ ὕδατι.

117 μὴ V
118 πυρήματι V
119 Smith: ὀρίγανος MV
120 om. V
121 χλιηροὶ πόδες V

straint, and being warmed with not much fomentation, but gently.[12]

53. Simus' wife aborted at thirty days. This was either from drinking something or spontaneously. Pain recurred, vomiting of much bilious material, yellow and green, whenever she drank anything. Spasms possessed her, she kept biting her tongue. Towards the fourth day they went into her tongue:[13] it became black and large. The whites of her eyes were red. Sleepless. She died towards the night of the fourth day.

54. Oregano is a bad thing for eyes when drunk, and for teeth.[14]

55. The girl who fell from the cliff became speechless; restless tossing persisted; she vomited much blood towards night. She had fallen on her left side, and there was a large flow. She had trouble drinking melicrat. Rasping in the throat, rapid breathing as of dying people. Blood vessels around the face tense. She lay on her back; feet warm; mild fever; voicelessness. On the seventh day she broke forth her voice. The fevers were lighter. She survived.[15]

56. Pythocles gave ill people water, and milk which he had mixed with much water.[16]

57. Chilblains: incise them, warm the feet, heat as much as possible with fire and water.[17]

[12] Cf. *Epid.* 7.71.

[13] "They" refers to the symptoms or disease material; my interpretation here is based on the parallel passage, *Epid.* 7.74.

[14] Cf. *Epid.* 7.76d.

[15] Cf. *Epid.* 7.77.

[16] Cf. *Epid.* 7.75.

[17] Cf. *Epid.* 7.76a.

58. Ὀφθαλμοῖσι[122] πονηρόν, φακῆ, ὀπώρη, τὰ γλυκέα, καὶ λάχανα· τοῖσι δὲ περὶ ὀσφῦν καὶ σκέλεα καὶ

240 ἰσχίον ἀλγήμασιν ἐκ | πόνων, θαλάσσῃ, ὄξει, θερ-
μοῖσι καταιονᾶν, καὶ σπόγγους βάπτοντα πυριᾶν,
ἐπικαταδεῖν δὲ εἰρίοισιν οἰσυπηροῖσι καὶ ῥηνικῆσι.[123]

59. Τῶν γναφέων οἱ βουβῶνες ἐφυματοῦντο σκληροὶ καὶ ἀνώδυνοι, καὶ περὶ ἥβην καὶ ἐν τραχήλῳ,
ὅμοια, μεγάλα· πυρετός· πρόσθε δὲ βηχώδεις· τρίτῳ
μηνὶ ἢ[124] τετάρτῳ γαστὴρ ξυνετάκη· θέρμαι ἐπεγένον-
το· γλῶσσα ξηρή· δίψα· ὑποχωρήσιες κάτω χαλεπαί·
ἔθανον.

60. Ὁ τὴν κεφαλὴν ὑπὸ Μακεδόνος καὶ λίθῳ πληγεὶς . . . [125] καὶ ἔπεσεν· τρίτῃ ἄφωνος ἦν· ἀλυσμός·
πυρετὸς οὐ πάνυ· λεπτὸς σφυγμὸς ἐν κροτάφοισιν·
ἤκουεν οὐδέν, οὐδὲ ἐφρόνει, οὐκ ἀτρεμέως. ἀλλὰ τῇ
τετάρτῃ ἐκινεῖτο· νοτὶς περὶ μέτωπόν τε καὶ ὑπὸ ῥῖνα
καὶ ἄχρις ἀνθερεῶνος, καὶ ἔθανεν.

61. Ὁ Αἰνιήτης ἐν Δήλῳ[126] ἄκοντι πληγεὶς ὄπισθε
τοῦ πλευροῦ κατὰ τὸ ἀριστερὸν μέρος, τὸ μὲν ἕλκος
ἄπονος· τρίτῃ δὲ γαστρὸς ὀδύνη·[127] οὐχ ὑπεχώρει·
κλυσθέντι[128] δὲ κόπρος ἐς νύκτα· ὁ[129] κόπος διαλιπών.
ἕδρη ἐς τοὺς ὄρχιας τετάρτῃ, καὶ ἥβην καὶ | κοιλίην

---

122 ὀφθαλμοῖς ὡς εἰ MV (cf. *Epid.* 7.76)
123 ῥηνίκασι V   124 add. Smith: om. MV
125 I indicate a lacuna here in accordance with *Epid.* 7.32
126 ἐν ἰδίῳ MV (cf. *Epid.* 7.33)   127 ὀδύνη δεινή M
128 καυσθέντι mss. (cf. *Epid.* 7.33)
129 οὐ mss. (cf. *Epid.* 7.33)

58. Bad for eyes: lentils, fruits, sweets, garden vegetables. For pains of lower back and legs and hip from exertion, pour over them a mixture of sea water, vinegar, hot water, and, dipping sponges in it, give fomentations and bind them up with wool which still has the oil in it, and with sheepskins.[18]

59. The glands of the fullers swelled up hard, painless, both in the groin and at the neck, similar large swellings. Fever. This had been preceded by coughing. In the third or fourth month, bowels liquified. Fevers came on. Tongue dry; thirst; painful bowel movements. They died.[19]

60. The man hit on the head with a stone by the Macedonian . . . fell down.[20] On the third day he was voiceless. Tossing about; mild fever; light throbbing in the temples. He heard nothing, was not conscious, nor was he still. But on the fourth day, he showed movement; he had moisture around the forehead, under the nose, and down to the chin; and he died.

61. The man from Aenea, struck by a javelin at Delos in the left side of the upper back. He had no pain in the wound, but on the third day he had pain in the stomach. No bowel movements, but feces in the evening with an enema. The prostration was intermittent. In his seat towards the testicles, on the fourth day, and to the pubis and the in-

---

[18] Cf. *Epid.* 7.76b, c.

[19] Cf. *Epid.* 7.81.

[20] I infer from the parallel passage, *Epid.* 7.32, that a sentence has been omitted here because a copyist skipped from the first πληγείς to the second ("he was struck above the left temple, a superficial wound. When hit, he blacked out and fell").

242 πόνος καταιγίζων·[130] ἀτρεμεῖν οὐκ ἐδύνατο· χολώδεα
ἤμεσε κατακορέα· ὀφθαλμοὶ οἱ τῶν λειποθυμεόντων.
μετὰ τὰς πέντε ἐτελεύτησεν· θέρμη λεπτή τις ἐνῆν.

62. Τῷ δὲ καθ᾽ ἧπαρ πληγέντι ἄκοντι εὐθὺς τὸ
χρῶμα κατεχύθη νεκρῶδες· τὰ ὄμματα κοῖλα· ἀλυ-
σμῶν[131] δυσφορίη· ἔθανε πρὶν ἀγορὴν λυθῆναι, ἅμ᾽
ἡμέρῃ πληγείς.

63. Τῇ Πολεμάρχου, χειμῶνος κυναγχικῇ, οἴδημα
ὑπὸ τὸν βρόγχον, πολὺς πυρετός· ἐφλεβοτομήθη·
ἔληξεν ὁ πνιγμὸς ἐκ τῆς φάρυγγος· ὁ πυρετὸς παρείπετο. περὶ πέμπτην γούνατος ἄλγημα, οἴδημα ἀριστε-
ροῦ· καὶ κατὰ τὴν καρδίην ἔφη τι ξυλλέγεσθαι αὐτῇ,
καὶ ἀνέπνει[132] ὡς ἐκ τοῦ βεβαπτίσθαι ἀναπνέουσι, καὶ
ἐκ τοῦ στήθεος ὑπεψόφει·[133] ὥσπερ αἱ ἐγγαστρίμυθοι
λεγόμεναι, τοιοῦτό τι ξυνέβαινεν. περὶ δὲ τὰς ἑπτὰ ἢ
ἐννέα ἐς νύκτα κοιλίη κατερράγη· ὑγρὰ κακά, πολλὰ
νομιζόμενα· ἀφωνίῃ· ἐτελεύτησεν.

64. Ὑποκαθαίρειν τὰς κοιλίας ἐν τοῖσι νοσήμασιν,
ἐπὴν πέπονα ᾖ· τὰς μὲν κάτω, ἐπὴν ἱδρυμένα ἴδῃς·
σημεῖον, ἢν μὴ ἀσώδεις ἢ καρηβαρικοὶ ἔσωι, καὶ ὅταν
αἱ θέρμαι πρηΰνεται, ἢ[134] ὅταν λήγωσι μετὰ τοὺς
παροξυσμούς· τὰς δὲ ἄνω ἐν τοῖσι παροξυσμοῖσι,
τότε γὰρ καὶ αὐτόματα μετεωρίζεται, ἐπὴν ἀσώδεις
καὶ καρηβαρεῖς καὶ ἀλύοντες ἔωσιν.

[130] κατέχει ζῶν MV (cf. *Epid.* 7.33)   [131] ἀλυσμὸν M:
ἀλυσμῦν V: corr. Smith   [132] ἀνέπλεεν V
[133] ὑποψόφεε MV: ὑπεψόφεε Asul.   [134] οἱ V

testines, pain like a thunderclap. He could not keep still. He vomited bilious strongly-colored matter. He had the eyes of those who have fainted. He died after five days. There was a slight fever.[21]

62. The one hit by a javelin at the liver was immediately suffused with a deathly color; hollow eyes; distressful restlessness. He died before the closing of the marketplace (he was struck at daybreak).[22]

63. Polemarchus' wife had quinsy in the winter. Swelling by the windpipe, much fever. She was phlebotomised. The choking in her pharynx was relieved; the fever persisted. About the fifth day, pain and swelling in the left knee. She said she had gathering around the heart, and she breathed like a diver who has surfaced. She made a noise from her chest. It was something like the so-called ventriloquists make. On the seventh or ninth day her intestines broke loose towards night. Much moist excrement of a bad sort, an amount considered to be a lot. Voicelessness. She died.[23]

64. Purge the intestines in diseases when the diseases are ripe (concocted); the lower intestine when you see the diseases are settled. Indication: if the patients are not nauseous or heavy-headed, when the fever is mildest, or when it abates after the exacerbations. The upper intestines during the exacerbations, because then it comes upward spontaneously, when the patients are nauseous, heavy-headed, and fretful.[24]

[21] Cf. *Epid.* 7.33.
[22] Cf. *Epid.* 7.31.
[23] Cf. *Epid.* 7.28.
[24] Cf. *Epid.* 7.60.

ΕΠΙΔΗΜΙΑΙ

65. Ἐκ πτώματος τρωθέντος πήχεως, ἐπὶ σφακε-
λισμῷ πυοῦται πῆχυς· πεπαινομένου δὲ ἤδη γλίσχρος
244 ἰχὼρ ἐκθλίβεται, ὡς καὶ Λεωγε|νίσκῳ[135] καὶ Δημάρχῳ
τῷ Ἀγλευτέλεος· ὁμοίως δὲ καὶ πάνυ ἐκ τῶν αὐτῶν
πύον οὐδέν, οἷον τῷ[136] Αἰσχύλου παιδὶ ξυνέβη· πυου-
μένοισι δὲ τοῖσι πλείστοισι φρίκη καὶ πυρετὸς ἐπεγίνοντο.

66. Τῷ Παρμενίσκου παιδί, κωφότης· ξυνήνεγκε μὴ
κλύζειν, διακαθαίρειν δὲ εἰρίῳ· μοῦνον ἐγχεῖν ἔλαιον ἢ
νέτωπον·[137] περιπατεῖν, ἐγείρεσθαι πρωΐ, οἶνον πίνειν
λευκόν.

67. Τῇ Ἀσπασίου ὀδόντος δεινὸν ἄλγημα· καὶ γνάθοι ἐπήρθησαν· καστόριον καὶ πέπερι διακλυζομένη
ὠφελεῖτο.

68. Τῷ Καλλιμέδοντος ξυνήνεγκε πρὸς τὸ φῦμα τὸ
ἐν τῷ τραχήλῳ, σκληρὸν ἐὸν καὶ μέγα καὶ ἄπεπτον
καὶ ἐπώδυνον, ἀπόσχασις[138] βραχίονος, λίνον καταπλάσσειν πεφυρημένον ἐν οἴνῳ λευκῷ καὶ ἐλαίῳ δεύοντα ἐπιδεῖν . . . ἑφθὸν ἄγαν. ἢ ξὺν μελικρήτῳ ἑψεῖν
καὶ ἀλεύρῳ τήλιος,[139] ἢ κριθῶν, ἢ πυρῶν.

69. Μελισάνδρῳ τοῦ οὔλου ἐπιλαβόντος, καὶ ἐόντος
ἐπωδύνου, καὶ σφόδρα ξυνοιδέοντος, ἀπόσχασις βρα-
χίονος· στυπτηρίη αἰγυπτίη ἐν ἀρχῇ παραστέλλει.

135 M: Λεονίσκῳ V: Λεωγενίσκῳ ms. D
136 om. M
137 ἠβειωπόντεον M: ἠνεπότεον V (cf. *Epid.* 7.63)
138 ἀπόστασις MV (cf. *Epid.* 7.65)
139 τίλληος V

186

65. When his forearm was wounded in a fall, besides mortification, the forearm began to fester. When it had already ripened, a sticky serum was expressed, as in the cases of both Leogeniscus and Demarchus, son of Agleuteles. Similarly, it happens that there is no pus from very much the same condition, as happened to Aeschylus' child. When patients became purulent, most had chills and fever.[25]

66. Parmeniscus' child, deafness. Not to wash it (the ear?) out was helpful, but to clean with wool, and only to pour in olive oil or oil of bitter almonds; walking about, rising early, drinking white wine.[26]

67. Aspasius' wife had a severe pain in the tooth, and her jaws were swollen. She was helped when she rinsed it with castorium and pepper.[27]

68. For Callimedon's son, what helped for the swelling on the neck, which was hard, large, unripe, and painful, was opening the vein of the arm, putting on a plaster of linen saturated in white wine and oil, soaking it and binding it on . . . [28] boiled too much. Or boil it with melicrat and meal of fenugreek, barley, or wheat.

69. In Melisander's case, when his gum took (a disease?),[29] and was painful and very swollen: opening of the vein of the arm. Egyptian alum reduces it at the beginning.

[25] Cf. *Epid.* 7.61.

[26] Cf. *Epid.* 7.63.

[27] Cf. *Epid.* 7.64. Castorium is a potion derived from the beaver's musk gland.

[28] Cf. the parallel passage, *Epid.* 7.65, "not hot or too much boiled."

[29] Cf. *Epid.* 7.66.

70. Ὑδρωπιώδει ταλαιπωρεῖν, ἱδροῦν, ἄρτον ἐσθίειν θερμὸν ἐν ἐλαίῳ, πίνειν μὴ πολύ, λούεσθαι καὶ
κεφαλῆς χλιηρῷ· οἶνος δὲ λευκὸς λεπτὸς[140] καὶ ὕπνος
ἀρήγει.

71. Βίαντι τῷ πύκτῃ, φύσει πολυβόρῳ ἐόντι, ξυνέβη ἐμπεσεῖν ἐς πάθεα χολερικὰ ἐκ κρεηφαγίης, μάλιστα δὲ ἐκ χοιρείων ἐναιμοτέρων καὶ μέθης εὐώδε-
ος[141] καὶ πεμμάτων καὶ μελιτωμάτων καὶ | σικύου
246 πέπονος καὶ γάλακτος καὶ ἀλφίτων νέων· ἐν θέρει τὰ
χολερικὰ καὶ οἱ διαλείποντες πυρετοί.

72. Τιμοχάρει χειμῶνος κατάρρους μάλιστα ἐς τὰς
ῥῖνας· ἀφροδισιάσαντι ἐξηράνθη πάντα· κόπος, θέρμη ἐπεγένετο· κεφαλὴ βαρείη· ἱδρὼς ἀπὸ κεφαλῆς
πολύς, ἦν δὲ καὶ ὑγιαίνων ἱδρώδης· τριταῖος ὑγιής.

73. Μετὰ κύνα, οἱ πυρετοὶ ἐγένοντο ἱδρώδεις, καὶ οὐ
περιεψύχοντο παντάπασι μετὰ τὸν ἱδρῶτα· πάλιν δὲ
ἐπεθερμαίνοντο, καὶ μακροὶ ἐπιεικῶς, ἄκριτοι, καὶ οὐ
πάνυ διψώδεις· ὀλίγοισιν ἐν ἑπτὰ καὶ ἐννέα ἐπαύοντο,
ἑνδεκαταῖοι, καὶ τεσσαρεσκαιδεκαταῖοι, καὶ ἑπτακαι-
δεκαταῖοι, καὶ εἰκοσταῖοι ἐκρίνοντο. Πολυκράτει πυρετὸς καὶ τὰ τοῦ ἱδρῶτος οἷα γέγραπται· ἀπὸ φαρμάκου
κάθαρσις κάτω ἐγένετο· καὶ τὰ τοῦ πυρετοῦ ἤπια·
πάλιν κροτάφοισιν ἱδρώτια, καὶ περὶ τράχηλον ἐς
δείλην, εἶτα ἐς ὅλον·[142] καὶ πάλιν ἐπεθερμάνθη. περὶ
δὲ τὰς δεκαδύο καὶ δεκατέσσαρας[143] ἡμέρας[144] ἐπέτει-

---

[140] λεπτὸς λευκὸς M
[141] δυσώδεος corr. to εὐώδ- V

70. For hydropics: exercise, sweat, eating hot bread dipped in oil, drinking little, washing the head with warm water, light white wine, and sleep are of help.[30]

71. Bias the boxer, a naturally big eater: it happened that he fell into a choleric condition after eating meat; especially after rather bloody pork, and fragrant wine and pastry, honey cakes, ripe cucumber, milk, and young barley. In summer choleric problems and remittent fever.[31]

72. Timochares in winter had catarrhs, especially to the nose. When he had sexual activity all was dried up. Fatigue, fever came on; heavy head; much sweat from the head, but he used to sweat when in health. He was healthy the third day.[32]

73. After the Dogstar, fevers were accompanied with sweat, nor did they cool down entirely after the sweat. They got hot again, were significantly long, did not reach proper crises, did not produce much thirst. In a few cases they stopped in seven and nine days, and on the eleventh and fourteenth and seventeenth and twentieth they reached crises. For Polycrates, fever and the phenomena of the sweat were such as have been described. There was purging below after a drug, and the fever became mild. Again, small sweats at the temples and around the neck toward evening, then over the whole body. And he became fevered again. Around the twelfth and fourteenth days

[30] Cf. *Epid.* 7.67b.   [31] Cf. *Epid.* 7.82.
[32] Cf. *Epid.* 7.69.

[142] πάλιν . . . ὅλον] πάλιν καὶ περὶ τράχηλον καὶ κροτάφους ἱδρωτία. εἶτα ἐς ὅλον. V
[143] τέσσαρας V   [144] om. M

νεν ὁ πυρετός· καὶ ὑποχωρήσιες βραχεῖαι· ῥυφήμασι
μετὰ τὴν κάθαρσιν ἐχρήσατο. περὶ τὰς δεκαπέντε
γαστρὸς ἄλγημα κατὰ σπλῆνα καὶ κατὰ κενεῶνα
ἀριστερόν· θερμῶν προσθέσιες ἧσσον ἢ ψυχρῶν
προσωφέλεον· κλύσματι δὲ μαλακῷ χρησαμένῳ ἔληξεν ἡ ὀδύνη.

74. Τῷ ἐκ τοῦ μεγάλου πλοίου διόπῳ ἡ ἄγκυρα
λιχανὸν δάκτυλον καὶ τὸ κάτω ὀστέον ξυνέφλασε
δεξιῆς χειρός· φλεγμασίη[145] ἐπεγένετο, καὶ σφάκελος
καὶ πυρετός· ὑπεκαθάρθη μετρίως· θέρμαι ἤπιοι καὶ
ὀδύναι· δακτύλου τι ἀπέπεσεν. μετὰ τὰς ἑπτὰ ἐξῄει
ἰχὼρ ἐπιεικής. μετὰ ταῦτα, γλώσσης, οὐ πάντα ἔφη
248 δύνασθαι | ἑρμηνεύειν· προρρήσιος, ὅτι ὀπισθότο-
νος[146] ἥξει.[147] ξυνεφέροντο αἱ γνάθοι ξυνερειδόμεναι,
εἶτα ἐς τράχηλον· τριταῖος δὲ ὅλος[148] ἐσπᾶτο ἐς τοὐπίσω ξὺν ἱδρῶτι. ἑκταῖος ἀπὸ τῆς προρρήσιος ἔθανεν.

75. Ὁ δὲ Ἀρπάλου ἐκ τῆς ἀπελευθέρης Τηλεφάνης
στρέμμα κάτω μεγάλου δακτύλου ἔλαβεν· ἐφλέγμηνε,
καὶ ἐπώδυνος ἦν· καὶ ἐπεὶ ἀνῆκεν, ᾤχετο ἐς ἀγρόν.
ἀναχωρέων, ὀσφῦν ἤλγησεν· ἐλούσατο· αἱ γένυες ξυν-
ήπτοντο ἐς νύκτα, καὶ ὀπισθότονος παρῆν· τὸ σίελον
ἀφρῶδες μόγις διὰ τῶν ὀδόντων ἔξω διῄει·[149] τριταῖος
ἔθανεν.

145 φλέγμασιν ἢ V: -μασι ἢ M: corr. recc.
146 ὀπισθότονοι MV
147 ἵξει V
148 ὅλως V
149 ἐξωδιὴν MV (cf. Epid. 7.37)

the fever increased. Few feces. He took gruel after the purge. About the fifteenth day pain in the belly by the spleen and left abdomen. Applications of heat benefited less than cold. His pain was relieved when he used a soft clyster.[33]

74. The commander of the large ship; the anchor crushed his forefinger and the bone below it on the right hand. Inflammation developed, gangrene, and fever. He was purged moderately. Mild fevers and pain. Part of the finger fell away. After the seventh day satisfactory serum came out. After that, problems with the tongue: he said he could not articulate everything. Prediction made: that *opisthotonos* would come. His jaws became fixed together, then it went to the neck, on the third day he was entirely convulsed backward, with sweating. On the sixth day after the prediction he died.[34]

75. Telephanes, son of Harpalus and his freedwoman, got a sprain behind the thumb. It grew inflamed and was painful. When it desisted he went into the fields. On his way home he had pain in the lower back. He bathed. His jaws became fixed together towards night and *opisthotonos* developed. Saliva, frothy, passed out through the teeth with difficulty. He died on the third day.[35]

[33] Cf. *Epid.* 7.1.
[34] Cf. *Epid.* 7.36.
[35] Cf. *Epid.* 7.37.

76. Θρίνων ὁ τοῦ Δάμωνος,[150] περὶ κνήμης σφυρὸν ἕλκος κατὰ νεῦρον ἤδη καθαρόν·[151] τούτῳ δηχθέντι ὑπὸ φαρμάκου ξυνέβη ὀπισθοτόνῳ θανεῖν.

77. ῟Ηρά γε ἐν πᾶσι τοῖσιν ἐμπυήμασι, καὶ τοῖσι περὶ ὀφθαλμόν, ἐς νύκτα οἱ πόνοι;

78. Αἱ βῆχες χειμῶνος, μάλιστα δ' ἐν νοτίοισι. παχέα καὶ πολλὰ λευκὰ χρεμπτομένοισι πυρετοὶ ἐπεγίνοντο, ἐπιεικῶς καὶ πεμπταῖοι ἐπαύοντο· αἱ δὲ βῆχες περὶ τὰς τεσσαράκοντα, οἷον ῾Ηγησιπόλει.[152]

79. Εὐτυχίδης ἐκ χολερικῶν ἐπὶ τῶν σκελέων τετανώδεα· ἔληξεν ἅμα τῇ κάτω ὑποχωρήσει. κατακορέα χολὴν πολλὴν ἤμεσεν ἐπὶ τρεῖς ἡμέρας καὶ νύκτας, καὶ λίην ἐρυθρήν· καὶ ἀκρατὴς ἦν καὶ ἀσώδης· οὐδὲν δὲ κατέχειν ἐδύνατο οὔτε[153] ἐκ τῶν σιτίων· καὶ οὔρου πολλὴ σχέσις, καὶ τῆς κάτω διόδου· διὰ τοῦ ἐμέτου τρὺξ[154] μαλακὴ ἦλθε, καὶ κατέρριψε κάτω.

80. Ἀνδροθάλει[155] ἀφωνίη, λήρησις· λυθέντων δὲ
250 τούτων, περιῆν | ἔτη συχνά, καὶ ὑποστροφαὶ ἐγίνοντο· ἡ δὲ γλῶσσα διετέλει πάντα τὸν χρόνον ξηρή· καὶ εἰ μὴ διακλύζοιτο, διαλέγεσθαι οὐχ οἷος ἦν, καὶ πικρὴ λίην ἦν τὰ πολλά· ἔστι δ' ὅτε καὶ πρὸς καρδίην ὀδύνη. φλεβοτομίη ἔλυσεν· ταύτῃ ὑδροποσίη ἢ μελίκρητον ξυνήνεγκεν. ἐλλέβορον ἔπιε μέλανα, οὐδὲ τὸ χολῶδες διῄει, ἀλλ' ὀλίγον. τέλος δέ, χειμῶνος κατακλιθείς,

---

150 Δάνωνος V: δαίμωνος M (cf. *Epid.* 7.38)
151 om. V   152 ῾Ηγησίπολι mss. (cf. 7.58)
153 οὔτε ὁ M

76. Thrinon, son of Damon, a wound by the ankle of the lower leg, in the tendon; it was clean. When he was eroded by a drug, the event was that he died of *opisthotonos*.[36]

77. Is it true that in all suppurations, including those around the eye, the distress comes towards night?[37]

78. There were coughs in winter, especially in southerly weather. In people who brought up much thick white matter, fevers came on and stopped normally on the fifth day. But the coughs stopped about the fortieth day, as with Hegesipolis.[38]

79. Eutychides went from a choleric affection to convulsions in the legs. He got better at the same time as purgation through the bowel. He vomited much pure bile for three days and nights, exceedingly red. He was weak and nauseous. He could keep nothing down, and no solid food. Much blockage both of urine and of the bowels. Throughout the vomit soft "wine lees" came out and he cast them out below.[39]

80. Androthales had voicelessness, delirium. These passed; he survived many years and there was a relapse. His tongue was dry throughout; unless he rinsed his mouth he could not speak, and it was very bitter most of the time. Sometimes he had pain at the heart; which phlebotomy stopped. Drinking water or melicrat helped the heart. He drank black hellebore, but bilious matter did not pass, save a little. Finally, having taken to his bed in the winter,

---

[36] Cf. *Epid.* 7.38.   [37] Cf. *Epid.* 7.57.
[38] Cf. *Epid.* 7.58.   [39] Cf. *Epid.* 7.67.

---

[154] στρὺξ MV (cf. 7.58)
[155] Ἀνδροθαλὴ M

ΕΠΙΔΗΜΙΑΙ

ἔξω ἐγένετο, καὶ τὰ τῆς γλώσσης παθήματα ὅμοια,
θέρμη λεπτή, ἄπονος, γλῶσσα ἄχροος, φωνὴ περιπλευμονική, ἀπόσταξις· εἷμα ἀπεδύετο, καὶ ἐξάγειν
αὐτὸν ἐκέλευεν, οὐδὲν δὲ ἐδύνατο σάφα εἰπεῖν· ἐς
νύκτα ἐτελεύτα.

81. Τὸ Νικάνορος πάθος, ὁπότε ἐς ποτὸν ὁρμῷτο,[156]
φόβος τῆς αὐλητρίδος· ὁκότε φωνῆς αὐλοῦ ἀρχομένης
ἀκούσειεν αὐλεῖν ἐν ξυμποσίῳ, ὑπὸ δειμάτων ὄχλοι·
μόγις ὑπομένειν ἔφη, ὅτε εἴη[157] νύξ· ἡμέρης δὲ ἀκούων
οὐδὲν διετρέπετο· τοιαῦτα παρείπετο συχνὸν χρόνον.

82. Δημοκλῆς ὁ μετ᾽ ἐκείνου ἀμβλυώσσειν καὶ
λυσισωματεῖν[158] ἐδόκει, καὶ οὐκ ἂν παρῆλθε παρὰ
κρημνὸν οὐδὲ ἐπὶ γεφύρης οὐδὲ τοὐλάχιστον βάθος
τάφρου διαπορεύεσθαι, ἀλλὰ δι᾽ αὐτῆς τῆς τάφρου
οἷος ἦν· τοῦτο χρόνον τινὰ ξυνέβη αὐτῷ.

83. Τὸ[159] Φοίνικος, ἐκ τοῦ ὀφθαλμοῦ τοῦ δεξιοῦ τὰ
πολλὰ ὥσπερ ἀστραπὴν ἐδόκει ἐκλάμπειν· οὐ πολὺ δὲ
ἐπισχόντι ὀδύνη ἐς τὸν κρόταφον τὸν δεξιὸν ἐνέστη
δεινή, εἶτα ἐς ὅλην τὴν κεφαλὴν καὶ ἐς τράχηλον,
καθὸ δέδεται ἡ κεφαλὴ ὄπισθεν σπονδύλῳ·[160] | καὶ
252 ξύντασις, καὶ σκληρότης ἀμφὶ τοὺς ὀδόντας· καὶ διοί-
γειν[161] ἐπειρᾶτο ξυντεινόμενος. ἔμετοι, ὁκότε γενοίατο,
ἀπέτρεπον τὰς εἰρημένας ὀδύνας, καὶ ἠπιωτέρας ἐποί-

156 ὡρμῶτο MV   157 ἴη V
158 λυσισωματην δὲ (sic) M   159 τῷ V
160 ἐνέστη . . . σπονδύλῳ] ἐνεστηδεῖν ἢ σφονδύλῳ (om.
εἶτα . . . ὄπισθεν) M
161 διογεῖν M: διοιγεῖν V

194

he became delirious. The affections of the tongue were similar, light fever, no pain, tongue colorless, voice peripneumonic, nosebleeds. He tossed the covers off. He asked that someone take him out, but he could say nothing clearly. He died towards night.[40]

81. Nicanor's affection, when he went to a drinking party, was fear of the flute girl. Whenever he heard the voice of the flute begin to play at a symposium, masses of terrors rose up. He said that he could hardly bear it when it was night, but if he heard it in the daytime he was not affected. Such symptoms persisted over a long period of time.[41]

82. Democles, who was with him, seemed blind and powerless of body, and could not go along a cliff, nor on to a bridge to cross a ditch of the least depth, but he could go through the ditch itself. This affected him for some time.[42]

83. Phoenix's problem: he seemed to see flashes like lightning in his eye, usually the right. And when he had suffered that a short time a terrible pain developed towards his right temple, then in the whole head, and then into the part of the neck where the head is attached to the vertebra behind, and there was stretching and hardness around the teeth. He kept trying to open them, straining. Vomits, whenever they occurred, averted the pains I have described, and made them more gentle. Phlebotomy

[40] Cf. *Epid.* 7.85.
[41] Cf. *Epid.* 7.86.
[42] Cf. *Epid.* 7.87.

εον· φλεβοτομίη[162] ὠφέλει, καὶ ἐλλεβοροποσίη ἀνῆγε παντοδαπά, οὐχ ἥκιστα δὲ πρασοειδέα.

84. Παρμενίσκῳ καὶ πρότερον ἐνέπιπτον ἀθυμίαι καὶ ἀπαλλαγῆς βίου ἐπιθυμίη, ὁτὲ δὲ πάλιν εὐθυμίη.

85. Ἡ δὲ Κόνωνος θεράπαινα, ἐκ κεφαλῆς ὀδύνης ἀρξαμένη,[163] ἔκτοσθεν ἐγένετο· βοή, κλαυθμοὶ πολλοί, ὀλιγάκις ἡσυχίη. περὶ δὲ τὰς τεσσαράκοντα ἐτελεύτησεν· ὅτε δὲ ἔθνησκε,[164] δέκα ἡμέρας ἄφωνος καὶ σπασμώδης ἐγένετο.

86. Νεηνίσκος δέ τις πολὺν ἄκρητον πεπωκὼς ὕπτιος ἐκάθευδεν ἔν τινι σκηνῇ· τούτῳ ὄφις ἐς τὸ στόμα παρεισεδύετο ἀργής. καὶ δὴ ὅτε ᾔσθετο, οὐ δυνάμενος φράσασθαι, ἔβρυξε τοὺς ὀδόντας, καὶ παρέτραγε τοῦ ὄφιος,[165] καὶ ἀλγηδόνι μεγάλῃ εἴχετο, καὶ τὰς χεῖρας προσέφερεν ὡς ἀγχόμενος, καὶ ἐρρίπτει ἑωυτόν, καὶ σπασθεὶς ἔθανεν.

87. Καὶ ὁ τοῦ Τιμοχάριος θεράπων ἐκ μελαγχολικῶν δοκεόντων εἶναι καὶ τοιούτων καὶ τοσούτων, ἔθανεν ὁμοίως περὶ τὰς αὐτὰς ἡμέρας.

88. Τῷ Νικολάου περὶ ἡλίου τροπὰς χειμερινάς, ἐκ ποτῶν ἔφριξεν, ἐς νύκτα πυρετοί. τῇ ὑστεραίῃ, ἔμετος

---

162 καὶ φλεβ. M
163 ἀρξαμένης MV (cf. Epid. 7.90)
164 τὰς ὅτε ἔθνησκε δὲ M
165 τὴν ὄφιν V (in marg. λέγεται καὶ τῆς ὄφιος)

helped. A draught of hellebore produced variegated matter, not least leek-colored.[43]

84. Parmeniscus previously was visited by depressions and desire to end his life, but sometimes again with optimism.[44]

85. Conon's female servant, who began with a pain in the head, became delirious. Crying, much weeping, seldom quiet. She died about the fortieth day. When she died she had been for ten days voiceless and having convulsions.[45]

86. A youth who had drunk much undiluted wine was sleeping on his back in a tent. A shining snake went into his mouth.[46] When he felt it, unable to consider what to do, he ground his teeth together and bit off part of the snake. He was seized by a great pain and brought up his hands as though choking, tossed himself about, and died in convulsions.

87. Timochareus' male servant, after what appeared melancholic affections of that kind and degree, died similarly about the same day.[47]

88. Nicolaus' son, about the winter solstice, had shivering after drinking. Towards night, fever. The next day, bil-

---

[43] Cf. *Epid.* 7.88.

[44] Cf. *Epid.* 7.89.

[45] Cf. *Epid.* 7.90.

[46] ἀργῆς is interpreted as "white" by Erotian (frg. 22 N.). LSJ take the word to be the name of a snake, on the basis of Galen 19.89 K. This case is not paralleled in *Epid.* 7.

[47] Cf. *Epid.* 7.91. "Similarly" refers back to the case described in ch. 85, which in *Epid.* 7 immediately precedes this one.

χολώδης, ἄκρητος, ὀλίγος. τῇ τρίτῃ ἀγορῆς πλήρεος
ἐούσης ἱδρὼς ὅλου τοῦ σώματος· ἔληξεν. |

254   89. Τῇ Διοπείθεος ἀδελφεῇ, ἐν ἡμιτριταίῳ δεινὴ
καρδίη περὶ τὴν λῆψιν, καὶ ξυμπαρείπετο ὅλῃ τῇ
ἡμέρῃ· καὶ ἡ καρδιαλγίη, καὶ τῇσιν ἄλλῃσι παραπλησίως ὑπὸ Πληϊάδος δύσιν· ἀνδράσι σπανιώτερα ἐγένετο τὰ τοιαῦτα.

90. Τῇ Ἐπιχάρμου πρὸ τοῦ τεκεῖν δυσεντερίη,
πόνος,[166] ὑποχωρήματα ὕφαιμα, μυξώδεα· τεκοῦσα
παραχρῆμα ὑγίης.

91. Τῇ Πολεμάρχου ἐν ἀρθριτικοῖσιν, ἰσχίου ἀλγήματι δεινῷ ἐξ αἰτίης γυναικείων μὴ[167] γινομένων, ἡ
φωνὴ ἔσχετο νύκτα ἄχρι μέσον ἡμέρης· ἤκουε δὲ καὶ
ἐφρόνει, καὶ ἐσήμαινε τῇ χειρὶ[168] περὶ τὸ ἰσχίον εἶναι
τὸ ἄλγημα.

92. Ἐπιχάρμῳ, περὶ Πληϊάδων δύσιν, ὤμου[169] ὀδύνη, καὶ βάρος ἐς βραχίονα, νάρκη. ἔμετοι συχνοί,
ὑδροποσίη.[170]

93. Τῷ Εὐφάνορος παιδί, τὰ ἐξανθήματα οἷα ὑπὸ
κωνώπων, ὀλίγον δὲ χρόνον· τῇ ὑστεραίῃ ἐπεπυρέ-
τηνεν.

94. Αὐχμοὶ[171] μετὰ ζέφυρον ἐγένοντο μέχρι ἰσημερίης φθινοπωρινῆς· ὑπὸ κύνα, πνίγεα μεγάλα θερμά·

---

166 ὁ πόνος MV (cf. Epid. 7.99)
167 om. MV (cf. Epid. 7.100)
168 τῇ χειρὶ om. V
169 ὁμοῦ MV (cf. Epid. 7.103)
170 ὑδρωποσίη M   171 λύχμοι M

ious vomit, unmixed, small in quantity. On the third day while the agora was full,[48] sweat over his body. He was cured.

89. Diopeithes' sister in a semitertian fever had terrible heartburn at its onset and it continued all day.[49] The heartburn was similar for other women at the setting of the Pleiades. Such symptoms were more rare with men.

90. Epicharmus' wife before delivery had dysentery, fatigue, bloody feces with phlegm. When she had given birth she suddenly became healthy.[50]

91. Polemarchus' wife, in an arthritic condition, with a terrible pain in the hip joint, caused by the failure of her menses, lost her voice during the night and until midday. But she could hear, her mind was sound; she indicated with her hand that the pain was around the hip joint.[51]

92. Epicharmus, at the setting of the Pleiades, had pain in the shoulder, heaviness in the arm, loss of sensation. Frequent vomiting and water drinking.[52]

93. Euphanor's son had breaking out like the bites of mosquitoes for a short time. He was feverish besides on the following day.[53]

94. There was drought after the Zephyr until the fall equinox. At the Dogstar stifling, very hot weather. There

---

48 I.e., before noon. For this case, cf. *Epid.* 7.92.

49 *Cardie* and *cardialgia* can refer either to the heart or, as here, to the entrance of the stomach. For this case, cf. *Epid.* 7.95.

50 Cf. *Epid.* 7.99.

51 Cf. *Epid.* 7.100.

52 Cf. *Epid.* 7.103.

53 Cf. *Epid.* 7.104.

πυρετοὶ ἱδρώδεις· φύματα παρ' οὓς συχνοῖσιν ἐγένοντο.

95. Τύχων ἐν τῇ πολιορκίῃ τῇ[172] περὶ Δάτον ἐπλήγη ὑπὸ καταπέλτῃ[173] ἐς τὸ στῆθος, καὶ μετ' ὀλίγον γέλως ἦν περὶ αὐτὸν θορυβώδης· ἐδόκει δέ μοι ὁ ἰητρὸς ἐξαιρέων τὸ ξύλον ἐγκαταλιπεῖν τι τοῦ δόρατος κατὰ τὸ διάφραγμα, δοκέοντος δὲ αὐτοῦ. πρὸς τὴν ἑσπέρην ἔκλυσέ τε καὶ | ἐφαρμάκευσε κάτω. νύκτα διήγαγε τὴν πρώτην δυσφόρως· ἅμ' ἡμέρῃ δὲ ἐδόκει καὶ τῷ ἰητρῷ καὶ τοῖσιν ἄλλοισι βέλτιον ἔχειν· πρόρρησις, ὅτι σπασμοῦ γενομένου οὐ βραδέως ἀπολεῖται. τῇ ἐπιούσῃ νυκτὶ δύσφορος, ἄγρυπνος, ἐπὶ γαστέρα τὰ πολλὰ κλινόμενος. τῇ τρίτῃ ἅμ' ἡμέρῃ ἐσπᾶτο, καὶ περὶ μέσον[174] ἡμέρης ἐτελεύτησεν.

96. Τῷ Βίλλῳ πληγέντι ἐς τὸν νῶτον, τὸ πνεῦμα πολὺ κατὰ τὸ τρῶμα μετὰ ψόφου[175] ἐχώρειν· καὶ[176] ἡμορράγει· τῷ ἐναίμῳ καταδεθεὶς ὑγιής· καὶ τῷ Δυσλύτῃ ξυνέβη τωὐτό.

97. Τῷ τῆς Φίλης[177] παιδί, ψιλώματος ἐν μετώπῳ γενομένου, ἐναταίῳ πυρετός· ἐπελιάνθη τὸ ὀστέον· ἐτελεύτησεν. καὶ τῷ Φανίου καὶ τῷ Εὐεργέτου· πελιαινομένων δὲ τῶν ὀστέων, καὶ πυρεταινόντων, ἀφίσταται τὸ δέρμα ἀπὸ τοῦ ὀστέου, καὶ πύον ὑποφαίνεται.

256

[172] om. M
[173] κατα πελτην (sic) M
[174] om. M   [175] ψου M
[176] om. M
[177] Φίλλης V

were fevers with sweats, swellings developed by the ears of many.[54]

95. At the siege of Datum, Tychon was struck in the chest by a catapult. Shortly later he was overcome by a raucous laughter. It appeared to me that the physician who removed the wood left part of the shaft in the diaphragm, and the patient thought so. The physician gave him an enema towards evening and a drug by the bowel. He spent the first night in discomfort. At daybreak he seemed to the physician and others to be better. Prediction: spasms would come on and he would die quickly. On the subsequent night discomfort, sleeplessness. He lay on his stomach for the most part. Convulsions began with daybreak the third day and he died at midday.[55]

96. When Billus was wounded in the back, much breath came out through his wound noisily. And he hemorrhaged. Bandaged with drugs for stopping blood, he recovered. The same thing happened to Dyslytas.[56]

97. The son of Phile whose skull was laid bare on his forehead had fever on the ninth day. The bone became livid. He died. Also the sons of Phanias and Euergetes, whose bones became livid, had fever; their skin came away from the bone and pus showed from beneath.[57]

[54] Cf. *Epid.* 7.105.

[55] Cf. *Epid.* 7.121.

[56] Though this passage and 7.34 are versions of the same report, the names have become garbled. Meineke suggests that the original name in the first line was Ἄβδελος. He would suggest Δυσλώτας for the name of the second.

[57] Cf. *Epid.* 7.35.

98. Ἀρίστιππος ἐς τὴν κοιλίην ἐτοξεύθη ἄνω βίῃ χαλεπῶς· ἄλγος κοιλίης δεινόν· καὶ ἐπίμπρατο ταχέως· κάτω δὲ οὐ διεχώρει· ἀσώδης ἦν· χολώδεα κατακορέα· ὅτε ἀπήμεσεν,[178] ἐδόκει ῥῆϊον[179] εἶναι, μετ᾽ ὀλίγον δὲ πάλιν τὰ ἀλγήματα δεινά· καὶ ἡ κοιλίη ὡς ἐν εἰλεοῖσιν· θέρμαι, δίψαι· ἐν τῇσιν ἑπτὰ ἡμέρῃσιν ἐτελεύτησεν.

99. Ὁ δὲ Νεάπολις πληγεὶς ὁμοίως ταυτὰ[180] ἔπασχεν· κλυσθέντι δριμεῖ, κοιλίη κατερράγη· χρῶμα κατεχύθη λεπτόν, ὠχρόν, μελανόν,[181] ὄμματα αὐχμηρὰ καρώδεα[182] ἐνδεδινημένα ἀτενίζοντα.

100. Ἐν Καρδίῃ[183] τῷ Μητροδώρου παιδὶ ἐξ ὀδόντος ὀδύνη, | σφακελισμὸς τῆς γνάθου, καὶ οὔλων ὑπερσάρκωσις·[184] μετρίως ἐξεπύησεν· ἐξέπεσον οἱ γόμφιοι[185] καὶ ἡ σιηγών.

101. Γυναικί, ἐν Ἀβδήροισι καρκίνωμα ἐγένετο περὶ στῆθος, διὰ τῆς θηλῆς ἔρρει ἰχὼρ ὕφαιμος· ἐπιληφθείσης δὲ τῆς ῥύσιος ἔθανεν.

102. Ἐκ κατάρρου κατὰ τὸ ἥμισυ τῆς κεφαλῆς ἐπόνεον, καὶ κατὰ ῥῖνας ὑγροῦ χωρέοντος ἐπυρέταινον,[186] ἐπιεικῶς ἐν τῇσι πέντε ἡμέρῃσι περιεψύχοντο.

[178] ἐπήμεσεν MV (cf. *Epid.* 7.29)   [179] ῥήϊον M
[180] Lind.: ταῦτ᾽ mss.
[181] μέλαν ἐὸν MV: corr. Smith from *Epid.* 7.30
[182] καθαρώδεα M
[183] κραδίη MV (cf. *Epid.* 7.113)
[184] ὑπερσάρκησις MV (cf. *Epid.* 7.113)
[185] γόμφιοι M   [186] ἐπυρέτηνον V

98. Aristippus was severely wounded by being shot in the upper belly by an arrow. Terrible pain in the intestine. It was quickly inflamed, but no excrement passed below. He was nauseous; very bilious matter; when he had vomited he seemed better, but shortly later had the terrible pains again. His intestine as in intestinal obstructions. Fever, thirst. He died in seven days.[58]

99. Neapolis, similarly wounded, had the same experience. But after an enema with astringent medicine his bowel was freed up. A light color was diffused over him, yellow and black. Eyes dry, stuporous, rolling inwards, staring.[59]

100. In Cardia, Metrodorus' son had pain from the teeth, mortification of the jaw. Flesh grew over the gums. He was moderately purulent. The molars and the jawbone collapsed.[60]

101. A woman in Abdera had cancer on the chest and through her nipple a bloody serum flowed out. When the flow was interrupted, she died.[61]

102. After a catarrh they suffered in half the head. Moisture flowed from their noses, they were feverish. They grew properly cool again generally in five days.[62]

[58] Cf. *Epid.* 7.29.
[59] Cf. *Epid.* 7.30.
[60] Cf. *Epid.* 7.113.
[61] Cf. *Epid.* 7.116.
[62] Cf. *Epid.* 7.56.

103. Τῇ Σίμου[187] ἐν τόκῳ σεισθείσῃ, ἄλγημα περὶ στῆθος καὶ πλευρῶν ἀποχρέμψιες πυώδεις· φθισικὰ[188] κατέστη· ἐξ ἡμέρας οἱ πυρετοί· πάλιν διάρροια· παῦσις πυρετοῦ·[189] κοιλίη ἔστη, καὶ περὶ ἡμέρας ἑπτὰ ἔθανεν.

104. Ἡ κυναγχικὴ χεῖρα δεξιὴν καὶ σκέλος ἤλγησεν· πυρέτιον[190] ἐπεῖχε βληχρόν· ὁ πνιγμὸς τριταίην ἐχάλασεν. τετάρτῃ[191] σπασμώδης, ἄφωνος· ῥέγχος, ὀδόντων ξυνέρεισις, γνάθων ἔρευθος· ἔθανε πεμπταίη ἢ ἑκταίη· σημεῖον περὶ χεῖρα ὑποπέλιον.

105. Καὶ ἑτέρη ἐπὶ τοῦ ὑπερῴου ῥεγχώδης· γλῶσσα ξηρή, περιπλευμονική· ἔμφρων ἔθανεν.

106. Καὶ ὁ ἐν Ὀλύνθῳ ὑδρωπικός, ἐξαίφνης[192] ἄφωνος, ἔκφρων νύκτα καὶ ἡμέρην, ἔθανεν.

[187] τησίμου τοῦ Μ
[188] φθεὶς Μ
[189] πυρετοὶ ΜV (cf. Epid. 7.49)
[190] πυρετὸν ΜV (cf. Epid. 7.18)
[191] ἑβδόμη ΜV: corr. recc. CD
[192] ἐξαίφης Μ

103. Simus' wife, shaken in childbirth,[63] had a pain about the chest, and expectoration of pus from the lungs. Phthisis set in. Fever for six days.[64] Later diarrhea. Fever stopped. Bowels became stable, she died about seven days later.[65]

104. The woman with quinsy had pains in the right arm and leg. A low fever persisted. The choking relaxed on the third day. On the fourth she was convulsive, voiceless; rattling when she breathed; clenching of teeth; redness of cheeks. Death the fifth or sixth day. A livid mark on her hand.[66]

105. Another woman had rattling in the upper chest, dry tongue; peripneumonic. Died without delirium.[67]

106. And the hydropic man at Olynthus, suddenly voiceless, delirious a night and a day. He died.[68]

[63] This appears to indicate that there was some difficulty in the birth process, and that shaking or succussion of some sort was used to facilitate it.

[64] "Six months" in *Epid.* 7.49, probably correctly.

[65] Cf. *Epid.* 7.49, and cf. Simus' wife, who died from aborting, 5.53.

[66] Cf. *Epid.* 7.18.

[67] Cf. *Epid.* 7.15.

[68] Cf. *Epid.* 7.21.

ΕΠΙΔΗΜΙΩΝ ΤΟ ΕΚΤΟΝ

ΤΜΗΜΑ ΠΡΩΤΟΝ

V 266
Littré

1. Ὁκόσῃσιν ἐξ ἀποφθορῆς περὶ[1] ὑστέρην καὶ οἰδημάτων ἐς καρηβαρίην τρέπεται, κατὰ τὸ βρέγμα ὀδύναι μάλιστα, καὶ ὅσαι ἄλλαι ἀπὸ ὑστερέων· ταύτῃσιν ἐν ὀκτὼ ἢ δέκα μησὶν ἐς ἰσχίον τελευτᾷ.

2. Οἱ φοξοί, οἱ μὲν κραταύχενες,[2] ἰσχυροὶ καὶ τἄλλα καὶ ὀστέοισιν· οἱ δὲ κεφαλαλγεῖς, καὶ ὠτόρρυτοι, τούτοισιν ὑπερῷαι κοῖλαι, καὶ ὀδόντες παρηλλαγμένοι.

3. Ὁκόσοισιν ὀστέον ἀπὸ ὑπερῴης ἀπῆλθε, τούτοισι μέσῃ ἵζει ἡ ῥίς· οἷσιν ἔνθεν[3] οἱ ὀδόντες ἄκρη σιμοῦται. |

268

4. Αἱ τῶν νηπίων ἐκλάμψιες[4] ἅμα ἥβῃ· ἔστιν οἷσι μεταβολὰς ἴσχουσι καὶ ἄλλας.

5. Ἀτὰρ[5] καὶ ἐς νεφρὸν ὀδύνη βαρείη ὅταν πληρῶνται τοῦ σίτου ἢ πότου,[6] ἐμέουσί τε φλέγμα, ὅταν δὲ

---

mss. MV, recentiores HIR
[1] καὶ περὶ M
[2] κρατεραύχενες M Gal.
[3] ὅθεν (perhaps correctly) M: δὲ ὅθεν Gal.

206

EPIDEMICS 6

SECTION 1

1. In women in whom, after miscarriage and after swellings about the uterus, it turns to headaches, there is pain mostly around the front of the head, along with whatever other pains come from the uterus: for these women it terminates at the hip in the eighth or tenth month.

2. People with pointed heads: if they have strong necks they are strong elsewhere and in the bones. But if they have headaches and running ears, in them palates are hollow and teeth uneven.

3. People whose bone has gone away from the palate have the nose settling in the middle; if where the teeth are, they have the tip of the nose upturned.[1]

4. The blossoming of infants at puberty, for some, contains other changes as well.

5. However, there is also severe pain in a kidney when they are filled with food or drink; they vomit phlegm and,

---

[1] *Epid.* 4.19 deals with related subject matter.

---

4 ἐπιλήμψιες M (v.l.) Gal.

5 om. M

6 τοῦ σίτου ἢ πότου] σίτου M

πλεονάζωσιν αἱ ὀδύναι,[7] ἰώδεα· καὶ ῥάους μὲν γίνονται, λύονται δέ, ὅταν σίτων[8] κενωθῶσιν· ψαμμία τε
πυρρὰ ὑφίσταται, αἱματῶδές τε οὐρέουσιν·[9] νάρκη
μηροῦ τοῦ κατ᾽ ἴξιν. ἐλινύειν οὐ ξυμφέρει, ἀλλὰ γυμ-
νάσια· μὴ ἐμπίπλασθαι· τοὺς νέους ἐλλεβορίζειν,
ἰγνύην τάμνειν, οὐρητικοῖσι καθαίρειν, λεπτῦναι καὶ
ἁπαλῦναι.

6. Γυναικεῖα τῇσιν ὑδαταινούσῃσιν ἐπιπολὺ παρα-
μένει· ὅταν δὲ μὴ ταχὺ ἴῃ ἐποιδεῖ.

7. Ἐν Κρανῶνι, αἱ παλαιαὶ ὀδύναι ψυχραί, αἱ δὲ
νεαραὶ θερμαί, αἱματίαι[10] αἱ πλεῖσται· καὶ τὰ ἀπὸ
ἰσχίου ψυχρά. |

8. Τὰ ἐς ῥίγεα ἰσχυρὰ ἰόντα,[11] οὐ πάνυ τι προ-
πρηΰνεται, ἀλλ᾽ ἐγγὺς τῆς ἀκμῆς. πρὸ ῥίγεος αἱ[12]
σχέσιες τῶν οὔρων, ἢν ἐκ χρηστῶν ἴωσι, καὶ κοιλίη
ἢν[13] ὑποδιέλθῃ καὶ ὕπνοι ἐνέωσιν· ἴσως δὲ[14] καὶ ὁ
τρόπος τοῦ πυρετοῦ· ἴσως δὲ καὶ τὰ ἐκ κόπων. ἀποστάσιες οὐ μάλα οἷσι ῥίγεα.

9. Αἱ[15] τῶν σκελέων ἐκθηλύνσιες, οἷον ἢ πρὸ νούσου ὁδοιπορήσαντι, ἢ ἐκ νούσου αὐτίκα,[16] διότι ἴσως
τὸ ἐκκόπτον[17] ἐς ἄρθρα ἀπέστη, διὸ καὶ τῶν σκελέων
ἐκθηλύνσιες.

10. Φύματα ἔξω ἐξοιδέοντα, καὶ τὰ ἀπόξη καὶ
κορυφώδεα, καὶ τὰ ὁμαλῶς ξυμπεπαινόμενα, καὶ μὴ

---

[7] αἱ ὀδύναι om. M   [8] σῖτον M
[9] ῥέουσι M   [10] αἵματι δὲ Gal.
[11] Gal.: ἔοντα MV (v.l.) Gal.   [12] om. M

when the pains increase, greenish material. They become easier and are relieved when they are emptied of food. Reddish sand develops and they urinate bloody matter. Loss of sensation in the thigh on the same side. Rest is not helpful, rather exercises. Avoid repletion; give hellebore to young people, bleed at the groin, purge with diuretics, thin them down and soften them.

6. In moist women the menses last long. But when they do not come quickly there is swelling.

7. In Crannon, old pains were cold, but recent ones warm, and mostly engorged with blood. The parts near the hip were cold.

8. Affections that develop to powerful shivers do not much lessen, except near their acme. Before the shivers, retention of urine, if they come from favorable developments, and if the bowel is open and if there is sleep. Perhaps also the manner of fever (makes a difference). Perhaps also affections from fatigue. There are no apostases in those who have shivering.

9. Weakening of the legs, for example in one who has made a journey before the disease, or just after the commencement of the disease: perhaps because the material from fatigue settled in the joints, wherefore the weakening of the legs.

10. Swellings which go outwards, which taper to a peak, and are ripened evenly, which are not hard, not pendent,

---

13 κοιλίην ἦν M
14 ἴσος τε V
15 ἐκ V
16 om. V
17 ἐκ κόπων M

περίσκληρα, καὶ κατάρροπα, καὶ μὴ δίκραια, ἀμείνω·
τὰ δὲ ἐναντία κακά, καὶ ὅσα πλείστῳ ἐναντία, κάκι-
στα. |

272   11. Τὸ θηριῶδες φθινοπώρου, καὶ αἱ καρδιαλγικαί,
καὶ τὸ φρικῶδες, καὶ τὸ μελαγχολικόν. πρὸς τὰς
ἀρχὰς τοὺς παροξυσμοὺς σκέπτεσθαι, καὶ ἐν ἁπάσῃ
τῇ νούσῳ· οἷον τὸ ἐς δείλην παροξύνεσθαι, καὶ ὁ
ἐνιαυτὸς ἐς δείλην· καὶ αἱ ἀσκαρίδες.

12. Νηπίοισι βηχίον ξὺν γαστρὸς ταραχῇ καὶ
πυρετῷ ξυνεχεῖ διμηνιαίῳ, τὸ ξύμπαν σημαίνει μετὰ
κρίσιν[18] εἰκοσταίῳ καὶ οἰδήματα ἐς ἄρθρα· καὶ ἢν μὲν
κάτω τοῦ ὀμφαλοῦ καταστῇ, τὰ ἄνω ἐν τοῖσι κάτω
ἄρθροισιν, ἀγαθόν· ἢν δὲ ἄνω, οὐχ ὁμοίως λύει τὴν
νοῦσον, ἢν μὴ ἐκπυήσῃ· τὰ δὲ ἐν ὤμοισιν ἐκπυεῦντα
τοῖσι τηλικούτοισι γαλιάγκωνας ποιεῖ· λύσειε δ᾽ ἂν
καὶ ἑλκυδρίων κάτω ἔκχυσις, ἢν μὴ στρογγύλα καὶ
βαθέα ᾖ, τὰ δὲ τοιαῦτα ὀλέθρια καὶ ἄλλως παιδίοισιν·
λύσειε δ᾽ ἂν καὶ αἷμα ῥαγέν·[19] μᾶλλον δὲ τοῖσι τελειο-
τέροισιν ἐπιφαίνεται.

13. Δάκρυον ἐν τοῖσιν ὀξέσι τῶν φλαύρως ἐχόντων,
ἑκόντων μὲν χρηστόν, ἀκόντων δὲ παραρρέον κακόν·
καὶ οἷσι περιτείνεται βλέφαρα, κακόν· κακὸν δὲ καὶ τὸ
274   ἐπιξηραινόμενον οἷον ἄχνη, καὶ τὸ | ἀμαυρὸν κακὸν
καὶ αὐχμηρόν. καὶ οἱ ῥυτιδούμενοι ἔνδοθεν, καὶ οἱ
πεπηγότες, καὶ οἱ μόγις στρεφόμενοι, καὶ οἱ ἐνδεδινη-
μένοι, καὶ τἄλλα ὅσα παρεῖται.

and not split, are best. The opposite are bad. The most opposite are the worst.

11. In autumn there are worms and cardialgic ailments (heartburn), shivering, melancholy. One should watch for paroxysms at the onset; also in the whole disease: as is the exacerbation at evening, so is the year at its evening. Intestinal worms also.[2]

12. In infants, a cough with stomach upset and continuous fever in the second month indicates generally that there will also be swellings in the joints twenty days after the crisis. If the material from above settles below the navel in the lower joints, it is good. But if in the upper ones, it does not similarly resolve the disease unless there be festering. Festering in the shoulder for infants of that age makes them weasel-armed. Eruption of sores below can resolve that, if they are not round and deep, but that kind are even otherwise injurious to children. A flow of blood, too, can resolve it, but is more common in older children.

13. Weeping in patients who are badly off in acute diseases: good if they are voluntary, but flowing involuntarily, bad. Bad if the eyelids are stretched out. Dryness like chaff on them is bad, and dimming of sight and drying of the eye. If there is wrinkling in the eye's interior, if it is fixed or rotates with difficulty, or if it whirls. I pass over other phenomena.

[2] Cf. *Epid.* 2.1.3 for similar material.

---

[18] ξυνεχεῖ σημ. μετὰ κρίσιν διμην. τὸ ξύμπαν mss.: I have transposed on the basis of Gal. comm. ms. U

[19] ῥυέν Gal.

14. Πυρετοί, οἱ μὲν δακνώδεις τῇ χειρί, οἱ δὲ πρη-
εῖς·[20] οἱ δ' οὐ δακνώδεις μέν, ἐπαναδιδόντες δέ· οἱ δ'
ὀξεῖς μέν, ἡσσώμενοι δὲ τῇ χειρί.[21] οἱ δὲ περικαεῖς
εὐθέως, οἱ δὲ διὰ παντὸς βληχροί· ξηροί· οἱ δὲ ἁλμυ-
ρώδεις· οἱ δὲ πεμφιγώδεις ἰδεῖν δεινοί· οἱ δὲ πρὸς τὴν
χεῖρα νοτιώδεις· οἱ δὲ ἐξέρυθροι· οἱ δὲ πελιοί· οἱ δὲ
ἔξωχροι· καὶ τἄλλα τοιουτότροπα.

15. Αἱ ξυντάσιες τοῦ σώματος, καὶ οἱ σκληρυσμοὶ
τῶν ἄρθρων, κακόν· καὶ αὐτὸς διαλελυμένος, κακόν·
276 καὶ αἱ κατακλάσιες τῶν | ἄρθρων, κακαί. ὄμματος
θράσος, παρακρουστικόν· καὶ ἔρριψις καὶ κατάκλα-
σις, κακόν.

ΤΜΗΜΑ ΔΕΥΤΕΡΟΝ

1. Εὐρῦναι, στενυγρῶσαι, τὰ μέν, τὰ δὲ μή. χυμούς,
τοὺς μέν ἐξῶσαι,[22] τοὺς δὲ ξηρᾶναι, τοὺς δὲ ἐνθεῖναι,
καὶ τῇ μέν, τῇ δὲ μή. λεπτῦναι, παχῦναι τεῦχος,
δέρμα, σάρκας, τἄλλα, καὶ τὰ μέν, τὰ δὲ μή. λειῆναι,
τρηχῦναι, σκληρῦναι, μαλθάξαι, τὰ δὲ μή. ἐπεγεῖραι,
ναρκῶσαι· καὶ τἄλλα ὅσα τοιαῦτα. παροχετεύειν,
ὑπείξαντα ἀντισπᾶν αὐτίκα, ἀντιτείναντα ὑπεῖξαι.
ἄλλον χυμόν, μὴ τὸν ἰόντα, ἄγειν, τὸν δὲ ἰόντα ξυν-

---

[20] πρηῆες MV (perhaps correctly)
[21] τῆς χειρός M
[22] στεγνῶσαι V

14. Fevers: some are pungent to the touch, some gentle. Some are not pungent but increasing. Some are sharp but decreasing to the touch, some are straightway burning hot, and some are faint throughout. Some dry, some salty, some with blisters dreadful to see. Some damp to the touch. Some are red, some livid, some yellow. And so on.

15. Stretching of the body and hardening of the joints is bad. The joint itself being loosened is bad. Breakings of the joints are bad. Boldness of eye shows delirium. Both restlessness and drooping are bad.

SECTION 2

1. To dilate or to constrict, in some cases yes, in some, no. Some humors should be expelled, some dried, some injected, sometimes, but sometimes not. To reduce or increase the body, the skin, flesh, and so on; here, too, some yes, some no. To smoothe, roughen, harden, soften. Some no. To wake up, to put to sleep. And other things of the sort. Pump out, stretch again immediately what is relaxed, relax what is tight. Induce another humor, not the one running, help evacuate the one running, produce a similar

3 This and similar lists of actions hint, in *Epid.* 6 as in *Epid.* 2 (e.g., 2.3.8), at the objectives of the physician: to be in control of bodily processes (as he imagines them) in order to assure that the process of disease is successfully carried through. The implications of drastic intervention appear to contradict the expectative posture of *Epid.* 6.5.1 below. But the "wisdom of Nature" described in 6.5.1 is what the active, interventionist physician strives to imitate.

εκχυμοῦν,²³ ἐργάσασθαι²⁴ τὸ ὅμοιον, οἷον ὀδύνην²⁵ παύει τὰ ἀνόμοια. ᾗ²⁶ ῥέπει ἄνωθεν | ἀρθέντα, κάτωθεν

278 λύειν, καὶ τὰ ἐναντία τὸ αὐτό,²⁷ οἷον κεφαλῆς κάθαρσις, φλεβοτομίη, ὅτε οὐκ εἰκῆ ἀφαιρεῖται.

2. Αἱ ἀποστάσιες, οἷον βουβῶνες, σημεῖον μὲν τῶν τὰ βλαστήματα ἐχόντων, ἀτὰρ καὶ ἄλλων, μάλιστα δὲ περὶ τὰ²⁸ σπλάγχνα, κακοήθεις δὲ οὗτοι.

3. Πνεύματα, σμικρά, πυκνά· μεγάλα, ἀραιά· σμικρά, ἀραιά· πυκνά, μεγάλα·²⁹ ἔξω μέγα, ἔσω σμικρόν· ἔσω μέγα, ἔξω σμικρόν· τὸ μὲν ἐκτεῖνον, τὸ δὲ κατεπεῖγον· διπλῆ ἔσω, ἐπανάκλησις, οἷον ἐπεισπνέουσι, θερμόν, ψυχρόν.

4. Ἰητήριον ξυνεχέων χασμέων, μακρόπνους, ἐν τοῖσιν ἀπότοισι καὶ μόγις, βραχύπνους.

5. Κατ' ἴξιν καὶ πλευρέων ὀδύνη, καὶ ξυντάσιες ὑποχονδρίων, καὶ σπληνὸς ἐπάρσιες, καὶ ἐκ ῥινῶν ῥήξιες, καὶ ὦτα κατ' ἴξιν τούτων τὰ πλεῖστα· ταὐτὰ καὶ ἐς ὀφθάλμους. πότερον ἦρα πάντα, ἢ τὰ μὲν κάτωθεν ἄνω, οἷα τὰ παρὰ γνάθους ἢ παρ' ὀφθαλμὸν καὶ οὖς, τὰ δὲ ἄνωθεν κάτω, οὐ κατ' ἴξιν; καίτοι καὶ τὰ

280 ξυναγχικὰ ἐρυθή|ματα καὶ πλευρέων ἀλγήματα κατ' ἴξιν·³⁰ ἢ καὶ τὰ κάτω ἥπατος ἄνωθεν διαδιδόντα, οἷον τὰ ἐς ὄρχιας καὶ κιρσούς; σκεπτέα³¹ ταῦτα, ὅπη καὶ ὅθεν καὶ διότι.

---

²³ μὴ συνεκχυμοῦν V   ²⁴ ὀργάσασθαι Gal. (v.ll. ὀργίσασθαι, ἐργάσασθαι)   ²⁵ ὀδύνην] ὀδύνη ὀδύνην Gal.
²⁶ εἰ Gal.

condition, just as dissimilars stop pain. Where it inclines upward, being elevated, resolve it below, and in the opposite case the same thing, for example, purging the head, phlebotomy, when the removal is not random.[3]

2. Apostases, such as buboes, are indications about the parts which have the excrescences, and of other parts as well, particularly the intestinal area. Those people are of bad habitude.

3. Breathing: shallow and rapid; deep and intermittent; shallow and intermittent; rapid and deep. Out large, in small; in large, out small. One prolonged, another hurried. Double inspiration, like people breathing in successively, hot, cold.[4]

4. A cure for constant yawning is deep breathing; in those who cannot drink, shallow breathing.

5. On the same side occur pains in the side, stretching of the hypochondria, swelling of the spleen, eruptions from the nostrils; also ear problems generally on the corresponding side with these. The same things coming to the eyes as well. Is this true for all cases, or only when what is below rises, as in affections by the jaws or the eye and ear, while what goes down from above does not do so according to side? Still, flush of quinsy and pains in the ribs are by sides. Is it also material below the liver distributed upward, like affections of testicles and varicose veins? This must be investigated, where, whence, and why.

[4] *Epid.* 2.3.7 is related to this passage.

27 τὸ αὐτό Gal. comm.: ταὐτά M: ταῦτα V Gal. lemma
28 om. V   29 ἀραιά . . . μεγάλα om. V
30 καίτοι . . . ἵξιν Gal.: om. mss.   31 σκεπτέον V

6. Φλέβες κροτάφων οὐχ ἱδρυμέναι, οὐδὲ χλώ-
ρασμα λαμπρόν. ἢν πνεῦμα ἐγκαταλείπηται, ἢ βὴξ
ξηρή, μὴ θηριώδης, ἐς ἄρθρα στήριξιν προσδέχεσθαι
δεῖ,[32] κατ᾽ ἴξιν τῶν ἐντασίων τῶν κατὰ κοιλίην ὡς
ἐπιτοπολύ· ἔχουσι δὲ οὗτοι οἱ πλεῖστοι καὶ ἐξέρυθρα,
καὶ τῇ φύσει τοῦ λευκοχρωτέρου τρόπου, καὶ οὐχ
αἱμορραγέουσι ῥῖνες, ἢ σμικρὰ αἱμορραγέουσιν· καὶ
ἢν μὲν ῥυέντων ἐγκαταλείπηται, ἕτοιμον· δίψα ἐγ-
καταλειφθεῖσα καὶ στόματος ἐπιξηρασίη[33] καὶ ἀη-
δίη[34] καὶ ἀποσιτίη τοῦτον τὸν τρόπον· πυρετοὶ δὲ οὐκ
ὀξεῖς οἱ τοιοίδε, ὑποστροφώδεις δέ.|

282   7. Τὰ ἐγκαταλιμπανόμενα μετὰ κρίσιν, ὑποστρο-
φώδεα. τὸ γοῦν πρῶτον σπληνῶν ἐπάρσιες, ἢν μὴ ἐς
ἄρθρα τελευτήσῃ, ἢ αἱμορραγίη γένηται· ἢ δεξιοῦ
ὑποχονδρίου ἔντασις, ἢν μὴ ἐξοδεύῃ οὖρα· αὕτη γὰρ ἡ
ἐγκατάληψις ἀμφοτέρων, καὶ αἱ ὑποστροφαὶ τούτων
εἰκότως. ἀπόστασιν οὖν ποιεῖσθαι αὐτὸν μὴ γινομένας, τὰς δὲ ἐκκλίνειν γινομένας,[35] τὰς δὲ ἀποδέχεσθαι[36] ἢν ἴωσιν οἷα δεῖ καὶ ᾗ δεῖ· ὁπόσαι δὲ μή,
σφόδρα ξυνδρᾶν· τὰς δ᾽ ἀποτρέπειν ἢν πάντη ἀξύμ-
φοροι ἔωσι, μάλιστα δὲ ταύτας μελλούσας, εἰ δὲ[37] μή,
ἀρχομένας ἄρτι.

8. Αἱ τεταρταῖαι αἱμορραγίαι, δύσκριτοι.

[32] χρή V   [33] ἐπὶ ξηρασίην M
[34] ἠδίη M
[35] τὰς ... γινομένας om. V
[36] ἀπὸ ἀποδέχεσθαι M
[37] om. V

6. Blood vessels in the temples unstable, color bad; if there be breathing trouble or a dry cough (not too violent[5]), you must expect a deposit on the joints, generally on the side where the intestines are distended. These people generally are ruddy, even if they are of the whiter type; they do not have nosebleeds, or very little. If in nosebleeds there is a residuum, it is ready to cause trouble. Persistent thirst and residual dryness of the mouth, and indifference and aversion to food are of that sort. Such fevers are not acute, though they tend to relapse.

7. Material left after a crisis tends toward relapses. First there is swelling of the spleen, unless it terminate in the joints or hemorrhage occur. Or a stretching of the right hypochondrium unless there be exit of urine. For this is the blockage of both, and the relapse occurs predictably. One should oneself cause an apostasis if they do not happen, deflect them when they are occurring, accept them if they come of the right kind in the right place. If they do not, offer vigorous assistance, but turn them back if they are totally inappropriate, especially when they are about to occur, but otherwise just after they begin.[6]

8. Fourth day hemorrhages indicate unfavorable crises.[7]

[5] Θηριώδης, here and in 11 below, I interpret as "fierce." It could mean, as often, "indicating worms." Galen tells us that the ancient commentators disagreed about it.

[6] *Epid.* 2.3.8 offers a related discussion of controlling apostases.

[7] = *Epid.* 2.3.9.

9. Οἱ διαλείποντες[38] μίαν τῇ ἑτέρῃ ἐπιρριγέουσιν ἅμα κρίσει ἐκ τῶν πέντε εἰς τὰς ἑπτά.

10. Ὅσοι τριταιοφυεῖς, τούτοισιν ἡ νὺξ δύσφορος ἡ πρὸ τοῦ παροξυσμοῦ· ἡ δὲ ἐπιοῦσα, εὐφορωτέρη ὡς ἐπὶ τὸ πολύ.

11. Βῆχες ξηραὶ βραχὺ ἐρεθίζουσαι ἀπὸ πυρετοῦ πυρικαέος, οὐ κατὰ λόγον διψώδεις, οὐδὲ γλῶσσαι καταπεφρυγμέναι, οὐ τῷ θηριώδει, ἀλλὰ τῷ πνεύματι, δῆλον δέ· ὅταν γὰρ διαλέγωνται ἢ χασκῶσι, τότε βήσσουσιν· ὅταν δὲ μή, οὔ· τοῦτο ἐν τοῖσι κοπιώδεσι μάλιστα πυρετοῖσι[39] γίνεται. |

284   12. Μηδὲν εἰκῆ, μηδὲν ὑπερορᾶν. ἐκ προσαγωγῆς τἀναντία, ἃ[40] προσάγειν, καὶ διαναπαύειν.

13. Τὰ ὄπισθεν κεφαλῆς ὀδυνωμένῳ, ἐν μετώπῳ ὀρθὴ ἡ φλὲψ τμηθεῖσα ὠφέλησεν.

14. Αἱ διαδέξιες τῶν ὑποχονδρίων, ἐξ οἵων, οἷα ἄλλοισι[41] καὶ τῶν σπλάγχνων τῶν φλεγμονῶν οἷα δύνανται, εἴτ᾽ ἐξ ἥπατος σπληνί, καὶ τἀναντία, καὶ ὅσα τοιαῦτα. ἀντισπᾶν ἢν μὴ ᾖ δεῖ[42] ῥέπῃ·[43] ἢν δὲ ὅπη δεῖ, τούτοισι δὲ στομοῦν οἵως ἕκαστα ῥέπει.

15. Τὰ πλατέα ἐξανθήματα, οὐ πάνυ τι κνησμώδεα, οἷα Σίμων εἶχε χειμῶνος· [ὃς][44] ὅτε πρὸς πῦρ ἀλείψαιτο ἢ θερμῷ λούσαιτο ἀνίστατο· ἔμετοι οὐκ ὠφέλευν. οἶμαι εἴ τις ἐξεπυρία ἀνιέναι ἄν.

---

[38] διαλιπόντες M   [39] πυρετοῖσι μάλιστα V
[40] om. Gal.   [41] ἐξ οἵων οἷα ἀλλοιοῦσι (v.l.) Gal. (i.e., "what causes alteration")

9. Fevers that remit one day have shivering on the other, along with crisis between the fifth and seventh day.[8]

10. For those with tertians, the night before the paroxysm is difficult, the subsequent one generally more comfortable.

11. There are dry coughs without much irritation after ardent fever; less than reasonable thirst, tongue not dry; they are not from malignity but from the breathing problems, clearly, for when they talk or yawn they cough, otherwise not. This happens mostly in fevers from fatigue.

12. Nothing at random. Overlook nothing. Opposites by gradual addition: add them, pause between.

13. For a pain in the back of the head it helps to cut the straight vein on the forehead.

14. Receptions into the hypochondria, from what kind of things, what they are, what effect they have on the other inflamed viscera, whether they are from the liver on to the spleen, or the opposite, and similar things. Draw them back, unless they are inclining where they should. If where they should, open the channel for them in the way they each are tending.

15. Broad exanthemas, not very itchy, such as Simon had in winter. Whenever he oiled himself by the fire or bathed in warm water, they broke out. Vomiting was no help. I think one could have brought them out if one had treated him with heat.

[8] There is a similar text at *Epid.* 2.3.10.

---

[42] ἢ δεῖ] ἴδει (sic) M
[43] ῥέπει M
[44] del. Li.

16. Ὅσα πεπαίνεσθαι δεῖ, κατακεκλεῖσθαι δεῖ. τἀναντία [ἃ]⁴⁵ δὲ ξηραίνειν ἢ ἀνεῷχθαι. [οἷον] ὀμμά-
286 των ῥοωδέων,⁴⁶ ἢν ἄλλως φαίνηται | ξυμφέρειν, ἀντι-σπᾶν ἐς φάρυγγα καὶ ὅπη ἔρευξις λυσιτελεῖ καὶ ἄλλα τοιαῦτα. τὰς⁴⁷ ἐφόδους ἀνεστομῶσθαι, οἷον ῥῖνας, καὶ τὰς ἄλλας, ὧν δεῖ καὶ οὗ⁴⁸ δεῖ καὶ ὅτε καὶ ὅσον δεῖ, οἷον ἱδρῶτας καὶ τἄλλα δὴ⁴⁹ πάντα.

17. Ἐπὶ τοῖσι μεγάλοισι κακοῖσι πρόσωπον ἢν ᾖ χρηστόν, σημεῖον χρηστόν· ἐπὶ δὲ τοῖσι σμικροῖσι τἀναντία σημαῖνον ἢ εὖ σημεῖον⁵⁰ κακόν.

18. Παρὰ τὸ μέγα, οὗ ἡ γυνὴ ὄπισθεν τοῦ Ἡρῴου, ἰκτερώδεος ἐπιγενομένου παρέμενεν αὐτῇ . . .

19. Ὁ παρὰ Τιμένεω ἀδελφιδῇ, οὗτος μελάγχρως, ἐν Περίνθῳ· τὸ γονοειδὲς τὸ τοιοῦτον ὅτι κρίσιμον, καὶ
288 τῶν ἤτρων τὰ τοιαῦτα· | ὅτι αἱ οὐρήσιες ῥύονται, ὅτι οὔτε φύσης πολλῆς, οὔτε κόπρου πολλῆς, γλίσχρης δέ, διελθούσης ἐμαλάσσετο· οὐ γὰρ δὴ μέγα ἦν τὸ ὑποχόνδριον· κράμβην ἑβδομαῖος ἔφαγεν, ἔτι δύσ-πνους ἐών, ἐπὶ τὸ ἦτρον ἐμαλάσσετο, εὐθύπνους ἐγέ-νετο.

20. Περὶ τοῦ αἵματος τοῦ ἰχωροειδέος, ὅτι ἐν τοῖσι πτοιώδεσι⁵¹ τὸ τοιοῦτον ἢ τοῖσιν ἠγρυπνηκόσι, καὶ εἴτε φλαῦρον, εἴτε χρηστόν. οἷσιν ὁ σπλήν ἐστι

45 om. Gal., secl. Corn.   46 ξηραίνει ἢ ἀνεῷχθαι οἷον ὀμμάτων ῥυωδέων MV: corr. Li.   47 τὰς δὲ V
48 οὐ MV: ὅπη Gal.   49 δεῖ M
50 ἢ εὖ σημεῖον] τῇ εὐσημείῃ Gal.
51 Erot. (70.10 N) Gal.: πτυώδεσι MV

16. (Sores) If they need to achieve coction, they should be covered; but for the opposite, dry them or open them up. When the eyes run, if it seems that other therapy is good, draw backwards into the gullet, where eructation and similar things help. Open the exits, such as nostrils, and the others, those that are needed where they are needed, when, and how much, such as sweating and all the rest.

17. In severe illness if the face looks very good it is a very good indication. But in minor ones, when it gives other than a favorable sign it is bad.

18. Beside the big building where the woman lived behind the Heroon, when jaundice developed, there persisted for her. . . .

19. The man at the home of Timenes' niece in Perinthus, with the dark skin. Urine like semen, the sort that is critical, and comparable symptoms of the lower abdomen. Because the urine flowed, because there was softening after some little flatulence, after passing of not much feces, but glutinous ones. His hypochondrium was not enlarged. He ate cabbage on the seventh day while still having breathing trouble. He grew softer in the lower abdomen; his breathing got easy.[9]

20. Concerning serum-like blood, that it occurs in frightened people and insomniacs, and whether it is good or bad. Those whose spleen has descended have warm

[9] There is much use of logical particles in the report of this case, referring to assumptions confirmed or denied about the course of the disease, apparently.

κατάρροπος, πόδες καὶ γούνατα καὶ χεῖρες θερμά, ῥίς,
ὦτα ἀεὶ ψυχρά. ἦρα διὰ τοῦτο λεπτὸν τὸ αἷμα; ἦρα καὶ
φύσει[52] τοιοῦτον οὗτοι ἔχουσιν;

21. Ἡ ἐν τοῖσιν ἐμπυήμασιν ὅρος·[53] οἷσι μέλλου-
σιν ἐκπυεῖν, αἱ κοιλίαι ἐκταράσσονται. |

290   22. Σπλὴν σκληρὸς οὐ τὰ ἄνω, κάτω στρογγύλος,
πλατύς, παχύς, λεπτός, μακρός.

23. Ἧσσον τοῖσιν ἀπὸ κεφαλῆς κορυζώδεσιν. . . .

24. Ἡ περὶ τὸν νοσέοντα οἰκονομίη, καὶ ἐς τὴν
νοῦσον ἐρώτησις·[54] ἃ διηγεῖται, οἷα, ὡς ἀποδεκτέον, οἱ
λόγοι· τὰ πρὸς τὸν νοσέοντα, πρὸς τοὺς παρεόντας,
πρὸς τοὺς ἔξω.

25. Ὅτι ἐν[55] θερμοτέρῳ[56] τὸ ἐν τοῖσι δεξιοῖσι, καὶ
μελανθὲς[57] διὰ τοῦτο, καὶ ἔξω αἱ φλέβες μᾶλλον.
ξυνεκρίθη, ξυνέστη ὀξύτερον, κινηθέν, ἐμωλύνθη, καὶ
βραδύτερον αὔξεται καὶ ἐπὶ πλείω χρόνον. ὅτι ἐστε-
ρεώθη καὶ χολωδέστερόν τε καὶ ἐναιμότερον, ᾗ τοῦτο
θερμότερόν ἐστι τὸ χωρίον τῶν ζώων. |

ΤΜΗΜΑ ΤΡΙΤΟΝ

292   1. Ἡ δέρματος ἀραιότης, ἡ κοιλίης πυκνότης· ἡ
δέρματος ξύνδεσις, ἡ σαρκῶν αὔξησις· ἡ κοιλίης
νάρκωσις, ἡ τῶν ἄλλων[58] ξύγχυσις· ἡ τῶν ἀγγείων

---

[52] φύσις MV   [53] ὀμφαλος ὅρος Gal.
[54] καὶ ἐρώτησις M   [55] om. V
[56] θερμοτέρῳ στερεωτέρῳ Gal. (v.l. στερεώτεροι)
[57] Smith: μέλανες mss.   [58] ἀλῶν M: ὅλων Gal.

feet, knees, hands. Noses and ears always cold. Is the blood thin because of that? Do they have the condition naturally?

21. Could this be a defining characteristic in empyema? Those who are going to develop suppuration have upset bowels.

22. The spleen: hard, not on the top, round below, flat, thick, long, thin.

23. Less for people with *coryzas* (running noses) from head problems.[10]

24. Arrangements for the sick person and inquiry about the disease: what is explained, what kind of things, how it must be accepted; the reasoning; what relates to the patient, what relates to those who are present, and to people elsewhere.

25. Because what is on the right is in a hotter place it is darker because of that, and its blood vessels are more external. It congeals more quickly, is composed more quickly, moves, becomes softer and grows more slowly and for a longer time. Because it is solidified it is more bilious and more blooded, to the extent that that is the warmer area in animals.

SECTION 3

1. Looseness of the skin: denseness of the intestines; binding up of skin: increase of flesh; lack of feeling in the intestine: collapse of the rest; lack of cleansing of the hol-

---

[10] Taken by ancient commentators to refer to the preceding, but cf. below 6.3.3.

ἀκαθαρσίη, ἡ ἐγκεφάλου ἀνάλωσις, διὸ καὶ φαλακρό-
της, ἡ τῶν ὀργάνων κατάτριψις. καθαίρεσις, δρόμοισιν, πάλῃσι, ἡσυχίῃσιν, πολλοῖσι περιπάτοισι τάχε-
σιν, οἷσιν ἐφθὴ μάζα τὸ πλεῖστον, ἄρτος ὀλίγος.
καθαιρέσιος[59] σημεῖον τὴν αὐτὴν ὥρην τῆς ἡμέρης
φυλάσσειν· ἐξαπίνης γὰρ ἐρείπεται·[60] ὑφιέναι τῶν
πόνων, ᾗ ῥύεται, ὁμοίως γὰρ ὅλον ξυμπίπτει· ὅταν δὲ
δὴ ξυμπέσωσι, προσάγειν ὕεια ὀπτά· ὅταν δὲ πληρῶνται, σημεῖον, αὖτις τὸ σῶμα ἀνθηρὸν γίνεται. ἐν
γυμνασίοισι σημεῖον ὁ ἱδρὼς ὁ ῥέων στάγδην ἔξεισιν
ὥσπερ ἐξ ὀχετῶν, ἢ ξύμπτωσις ἐξ ἐπάρσιος.

2. Ἡ γυνὴ ἣν πρῶτον ἐθεράπευσα ἐν Κρανῶνι,
294 σπλὴν φύσει | μέγας· πυρετὸς καυσώδης· ἐξέρυθρος·
πνεῦμα, δεκάτῃ· ἱδρὼς τὰ πολλὰ ἄνω, ἀτάρ τι καὶ
κάτω τεσσαρεσκαιδεκάτῃ.

3. Ἧσσον τοῖσιν ἀπὸ κεφαλῆς κορυζώδεσι καὶ
βραγχώδεσιν ἐπιπυρετήνασιν, ὡς οἶμαι, ὑποστροφαί.

4. Πᾶν τὸ ἐκπυέον ἀνυπόστροφον· οὗτος γὰρ πε-
πασμὸς καὶ κρίσις ἅμα καὶ ἀπόστασις.

5. Ἔστιν οἷσιν ὅταν ἀφροδισιάζωσι φυσᾶται ἡ
γαστήρ, ὡς Δαμναγόρᾳ, οἷσι δ᾽ ἐν τούτῳ[61] ψόφος.
Ἀρκεσιλάῳ δὲ καὶ ᾧδει. τὸ φυσῶδες ξυναίτιον τοῖσι
πτερυγώδεσι, καὶ γὰρ εἰσι φυσώδεις.

6. Τὸ ψυχρὸν πάνυ φλεβῶν ῥηκτικὸν καὶ βηχῶδες
296 οἷον χιών, | κρύσταλλος, καὶ ξυστρεπτικὸν οἷον τὰ
φήρεα, καὶ αἱ γογγρῶναι· ξυναίτιον καὶ αἱ σκληρό-
τητες.

low vessels: wasting of the brain (whence baldness also), destruction of the organs. Purging: by running, by wrestling, by quiescence, by frequent fast walks (and with them barley gruel mostly, little bread). Watch for signs of purging at the same time each day, for the collapse is sudden. Reduce exercise to the extent that there is contraction. For the whole collapses similarly. And when there is collapse, give roast pork. When they are full, the sign is that the body looks blooming again. A sign in exercises: the sweat which flows comes out in drops, as from water pipes; or a collapse after swelling.

2. The woman I had first treated in Crannon: a naturally large spleen. Burning fever; ruddy; breathing difficulty on the tenth. Sweat generally above, and also some below on the fourteenth day.

3. There are fewer relapses, I think, in fevers with coryzas from the head and with throat involvement.

4. All suppuration makes relapse unlikely, for it is at the same time ripening, crisis, and apostasis.

5. Some people get intestinal gas from sexual activity, like Damnagoras, and some from it, noise. Arcesilaus swelled up besides. Intestinal gas is contributory to protruding shoulder blades, for such people are flatulent.

6. Very cold water will rupture vessels and cause coughs, like snow and ice, and it causes bulging, as of the parotid glands, and swellings on the neck. Hardnesses are also contributory.

---

59 Gal.: καθαίρεσις MV
60 εἰρύεται Gal.
61 οἶσι δ᾽ ἐν τούτῳ] ὅθεν τουτέοισιν V

7. Τὸ μετ᾿ οὔρησιν σύναγμα, παιδίοισι μᾶλλον· ἦρ᾿ ὅτι θερμότερα;

8. Τὰ σχήματα τὰ ῥηΐζοντα, οἷον ὁ τὰ κλήματα τῇ χειρὶ πλέκων ἢ στρέφων, ὑπεροδυνέων, κατακείμενος, λαβόμενος πασσάλου ἄκρον ὑπερπεπηγότος εἴχετο, καὶ ἐρρήϊσεν.

9. Ὃν ἐξ ὀροιτυπίης[62] παρὰ τὴν γέφυραν εἶδον ἐγὼ ῥιπτεῦντα σκέλεα, κνήμην ἑτέρην ἥκιστα ἐλεπτύνετο, μηροὶ[63] δὲ κάρτα· οὖρα καὶ γονὴ οὐκ ἔσχετο.

10. Ὅσαι πτερυγώδεις φύσιες πλευρέων δι᾿ ἀδυναμίην τῆς ἀφορμῆς, ἐπὶ τοῖσι κατάρροισι τοῖσι κακοήθεσιν, εἰ ἔκκρισις εἴη καὶ εἰ[64] μὴ εἴη, κακόν.

11. Ῥίγεα ἄρχεται γυναιξὶ μὲν μᾶλλον ἀπ᾿ ὀσφύος καὶ διὰ νώτου, τότε ἐς κεφαλήν· ἀτὰρ καὶ ἀνδράσιν ὄπισθεν μᾶλλον ἢ ἔμπροσθεν· φρίσσομεν τὰ ἔξωθεν μᾶλλον ἢ τὰ ἔνδοθεν[65] τοῦ σώματος, οἷον πήχεων, μηρῶν· ἀτὰρ καὶ τὸ δέρμα ἀραιότερον, δηλοῖ δὲ ἡ θρίξ· ἀφ᾿ ὧν δὲ ἄλλων ῥιγέουσιν ἴσως ἑλκέων, ἄρχεται ἀπὸ τῶν ἀγγείων.

12. Κεφάλαιον ἐκ τῆς γενέσιος καὶ ἀφορμῆς καὶ πλείστων λόγων καὶ κατὰ σμικρὰ γινωσκομένων ξυνάγοντα καὶ καταμανθάνοντα εἰ ὁμοιά ἐστιν ἀλλήλοισιν, ⟨αὖτις τὰς ἀνομοιότητας τούτοισιν⟩,[66] εἰ[67] ὅμοιαι ἀλλήλῃσιν εἰσίν, ὡς ἐκ τῶν ἀνομοιοτήτων ὁμοιότης

---

[62] ὀροτυπίης MV: ὀρειτυπίης Gal.   [63] μηροὺς Gal.
[64] εἰ . . . καὶ εἰ Gal.: ἢν . . . κ᾿ ἢν MV
[65] ἔσωθεν M   [66] Gal.: om. mss.

7. Sedimentation after urination is more frequent in children. Is it because they are warmer?

8. The postures which offer more relief: for example a man twining or twisting vine poles with his hand was overcome with pain and lay down. He took hold of the tip of a peg that had been fixed above, hung on, and was eased.

9. The man I saw beside the bridge: he had ruined his legs with working in the mountains.[11] He was not much wasted in one calf, but his upper legs were seriously so. He was incontinent of urine and semen.

10. Those physiques that have winglike shoulder blades because of weakness of impulse from the lungs: if there are morbid flows it is bad, whether they achieve secretion or not.

11. Shivering begins for women more from the loins and along the back, then to the head. In men it is more behind than in front. We shiver more on the outside than the inside of the body, e.g. on the forearms and thighs. But the skin is less dense, as the hair shows. In other lesions whence there is shivering it perhaps begins from the blood vessels.[12]

12. The summary conclusion comes from the origin and the going forth, and from very many accounts and things learned little by little, when one gathers them together and studies them thoroughly, whether the things are like one another; again whether the dissimilarities in them are like each other, so that from dissimilarities there

---

11 Perhaps "falling down a mountain."

12 This material recurs at *Epid.* 2.3.16 and *Aph.* 5.69, in neither of which places is there indication of the potential usefulness of the observations.

γένηται μία· οὕτως ἂν ἡ ὁδός· οὕτω καὶ τῶν ὀρθῶς
ἐχόντων δοκιμασίη, καὶ τῶν μή, ἔλεγχος.

13. Αἱμορραγίη ἐκ ῥινῶν ἢ τοῖσιν ὑποχλωρομέ-
λασιν, ἢ τοῖσιν ἐρυθροχλώροισιν,[68] ἢ τοῖσιν ὑπο-
χλώροισιν· βραχέα ὑφέντα, παχῦναι ξηρῷ.[69] τοῖσι δὲ
ἑτέροισι, παχυσμοῖς ἧσσον.[70] ξηρῷ δὲ δεῖ λευκὰ οἷον
κηκίς.[71] |

300   14. Ἐπὴν ἀφροδισιάζειν ἄρξωνται ἢ τραγίζειν αἱ-
μορραγέουσιν. ἐν τῇσι προσόδοισιν ἔστιν οἳ ἀπο-
ψοφέουσιν, οἷον Ἀρκεσίλαος·[72] οἱ δὲ μέλλοντες ῥιγοῦ-
σι[73] ῥικνώδεις·[74] οἱ δ' ἐπὴν προσέλθωσι, φυσῶνται
κοιλίην, οἷον Δαμναγόρας.

15. Αἱ μεταβολαὶ φυλακτέαι· ὀλιγοσιτίη, ἄκοπον,
ἄδιψον πίνοντι.[75]

16. Πᾶς[76] λεπτυσμὸς χαλᾷ τὸ δέρμα, ἔπειτα περι-
τείνεται· ἀνάθρεψις τἀναντία· χρωτὸς ῥίκνωσις ξυμ-
πίπτοντος, ἔκτασις ἀνατρεφομένου· τὸ ῥικνῶδες,[77] τὸ
λεῖον, ἑκατέρου σημεῖον, τὸ ὑπόχολον, τὸ ὑπέρυθρον·
οὕτω τὸ κατασπᾶσθαι μαζούς, ἰσχνοὺς δὲ ἀνεσπά-

---

[67] εἰ . . . εἰ] ἢ . . . ἢ V: ἢ . . . καὶ ἢ M
[68] Gal.: ἐρυθροχόλοισιν mss.   [69] ξηρῶς M Gal.
[70] ἧσσον παχυσμοῖς V   [71] κικίς V
[72] Gal.: Ἀρκεσίλλος MV
[73] Capito (ap. Gal.): ῥιγώσειν MV   [74] φρικώδεες Gal.
[75] om. Gal.: πεινῶντι (v.l.) Gal.   [76] πᾶς δὲ V

---

[13] Notable in this attempt to formulate an inductive methodology is the sophisticated approach to what would later be called
empirical method, but without technical, philosophical language.

arises one similarity. This would be the road (i.e., method). In this way develop verification of correct accounts and refutation of erroneous ones.[13]

13. Hemorrhage from the nostrils in the pale dark-complexioned or in the pale ruddy-complexioned or in the pale-complexioned: having removed a little, pack it tight with dry substance. In the rest use less packing. For dry substance you need white drugs, formed like an oak gall.[14]

14. When they commence sexual activity or their voices change, they have nosebleeds. In the sexual act some have rumbling noises, like Arcesilaus. And some, in anticipation, shiver and have corrugation of the skin. And some, in the sexual act, have wind in the intestines, like Damnagoras.[15]

15. Guard against changes: reduce food, avoid fatigue, avoid drinking to fullness.

16. All thinning down relaxes the skin, later tightens it. Feeding up does the opposite. Corrugation is of the collapsing skin, stretching of the well nourished. The shriveling, the smoothness, a sign of each: biliousness, ruddiness. So with the sagging of the breasts, but thin ones elevated

---

The methodology is self-conscious, as the word "road," "method," shows. Δοκιμασίη, "verification," is a word for auditing of accounts of officeholders at the end of their terms, ἔλεγχος, "refutation," has some flavor from testimony in legal proceedings.

[14] This is a prescription for something like the "white root," a medicine employed as a styptic in the gynecological works; here it is to be applied in a round pellet, perhaps with wool (cf. *Nat. Fem.* 32, 7.352.16 Li.). While the gall, an excrescence on the oak, was used medically and for ink, it is referred to here for its size and shape.   [15] Cf. *Epid.* 6.3.5.

σθαι καὶ περιτετάσθαι· καίτοι οὐκ ἄν τις οἴοιτο διὰ τοῦτο, ἀλλὰ σαρκωθέντος τοῦτο γενέσθαι.

17. Ἡ ἄγαν πλήρωσις περιφανής, φλέβες διαφανεῖς. |

18. Ἡρόδικος τοὺς πυρεταίνοντας ἔκτεινε περιόδοισι, πάλῃσι πολλῇσι, πυρίῃ· κακόν· τὸ πυρετῶδες πολέμιον πάλῃσι, περιόδοισι, δρόμοισιν, ἀνατρίψει· πόνῳ πόνον αὐτοῖσιν· ὄγκοι⁷⁸ φλεβῶν, ἔρευθος,⁷⁹ πελίωσις, χλωρότης⁸⁰ πλευρέων ὀδύναι λαπαραί.⁸¹

19. Ὅτε ἐχρῆν ἄδιψον, ξυνέχειν στόμα, σιγᾶν, ἄνεμον ξὺν τῷ ποτῷ ψυχρὸν εἰσάγειν.

20. Τὰς ἀφορμάς, ὁκόθεν ἤρξαντο κάμνειν σκεπτέον, εἴτε κεφαλῆς ὀδύνη, εἴτε⁸² ὠτός, εἴτε πλευροῦ. σημεῖον, οἱ ὀδόντες, καὶ ἐφ᾽ οἷσι βουβῶνες.

21. Καὶ τὰ γινόμενα ἕλκεα, καὶ φύματα, κρίνοντα πυρετούς· οἷσι ταῦτα μὴ παραγίνεται, ἀκρισίη·⁸³ οἶ-

---

⁷⁷ φρικῶδες Gal.
⁷⁸ Gal. (v.l.): ὅτε mss.: om. Gal. com.
⁷⁹ Gal.: ἔρευσιν M: εὖρες V (εὑρέσεως C ex V)
⁸⁰ χωλώτης V
⁸¹ Gal.: λαπάρης (v.l.) Gal.: λαπάραι MV
⁸² ψυχρὸν . . . εἴτε Galen: om. MV
⁸³ ἀκρασίη mss. (cf. Epid. 2)

---

16 Galen says that this reading, which our manuscripts carry, was introduced by Capito, who substituted it for κλεῖς περιφανέες, "When collarbones are obvious." I do not credit Galen's statement that Capito did so arbitrarily, without basis in ancient evidence.

and tightened. Indeed, one might not expect that it was from that cause, but would expect it to come from increase of flesh.

17. Overfullness is apparent,[16] the blood vessels are visible.

18. Herodicus killed fever patients with running, much wrestling, hot baths. A bad procedure. Fever is inimical to wrestling, walks, running, massage; that is trouble on trouble for them. Swelling of the blood vessels, redness, lividness, pallor, soft pains in the ribs.[17]

19. When it is necessary to prevent thirst, keep the mouth closed, do not speak, inhale cold wind with drink.

20. The point of departure should be studied, whence they first began to be ill, whether pain in the head, or the ear, or the side; an indication is the teeth, and the parts in which the glands swell.

21. Developing sores and swellings bring fevers to their crises. Those in which they do not occur are without crisis.

---

[17] Galen, in his commentary, says that he will not discuss whether Hippocrates refers here to Herodicus of Selymbria or of Leontini. Apparently there was discussion in antiquity. Plato, in *Republic* 406a, satirizes Herodicus as the inventor of contemporary "pampering" medicine: he was an athletic trainer with a chronic disease. He mingled medicine and gymnastic and spent his life tending his disease and keeping himself alive. In *Protagoras* 316e Plato speaks of the gymnastic trainer Herodicus, "now of Selymbria, formerly of Megara," a "first-rate sophist." It is possible that these various notices fit together, but difficult to see how.

σιν ἐγκαταλείπεται[84] βεβαιόταται καὶ τάχισται ὑπο-
στροφαί. |

304   22. Τὰ στρογγυλλόμενα πτύαλα παρακρουστικά.

23. Οἱ αἱμορροΐδας ἔχοντες,[85] οὔτε πλευρίτιδι, οὔτε
περιπλευμονίῃ, οὔτε φαγεδαίνῃ, οὔτε δοθιῆσιν, ‹οὔτε
τερμίνθοισιν›,[86] ἴσως δὲ οὐδὲ λέπρῃσιν· ἴσως δὲ οὐδὲ
ἄλλοισιν· ἰητρευθέντες γε μὴν ἀκαίρως συχνοὶ τοῖσι
τοιούτοισιν οὐ βραδέως ἑάλωσαν, καὶ ὀλέθρια οὕτω·
καὶ ὅσαι ἄλλαι ἀποστάσιες,[87] οἷον σύριγγες, ἢ ἕτεραι·
ἀπόσκηψις,[88] ἐφ᾽ οἷσι γινομένη[89] ῥύεται, τούτων προ-
γενομένη[90] κωλύει. ἄλλου δὲ τόποι οὗτοι οἱ δεξάμενοι
ἢ πόνῳ ἢ βάρει ἢ ἄλλῳ[91] τινὶ ῥύονται· ἄλλοισιν αἱ
κοινωνίαι.

24. Διὰ τὴν ῥοπὴν[92] οὐκ ἔτι αἷμα ἔρχεται, ἀλλὰ
κατὰ τοῦ χυμοῦ τὴν ξυγγένειαν τοιαῦτ᾽ ἀποπτύουσιν.
ἔστιν οἷσιν αἷμα ἀφαιρεῖσθαι[93] ἐν καιρῷ ἐπὶ τού-
τοισιν· ἐπ᾽ ἄλλοισι δέ, ὥσπερ ἐπὶ τούτοισι, τοῦτο οὐκ
εἰκός· κώλυσις· ἐπὶ τοῖσιν αἱματώδεα πτύουσιν ὥρῃσι,
πλευρῖτις, χολή.[94] |

84 ἐγκαταλείπηται Μ
85 ἔχοντας V
86 Gal. *Hum.* 20: om. mss.
87 *Hum.* Gal.: ὑπο- mss.
88 σκῆψις Gal.
89 (v.l.) Gal.: γινόμενα MV Gal.
90 (v.l.) Gal.: προγενόμενα MV Gal.
91 ἄλλο Μ   92 τρόπην (v.l.) Gal.
93 Gal.: ἀφίεσθαι MV   94 ὥρῃ πλευρῖτις χολή Gal.:
ὥρῃ πλευρίτιδι χολῇ (v.l.) Gal.

Those in which they persist have the surest and quickest relapses.[18]

22. Globular expectoration portends delirium.[19]

23. People with hemorrhoids are not affected with pleuritis or peripneumonia or spreading sores or boils or terminth-like swellings.[20] Also, perhaps, not with lepra (white, flaking skin), and perhaps not with other things. Indeed, many people whose hemorrhoids are healed untimely are soon seized by these kinds of diseases, which, coming so, are often deadly.[21] That is true also of other apostases, like fistulas and others. Drainage cures conditions on which it supervenes, prevents them when it precedes. These places which receive with pain, heaviness or anything else, give protection from another affection. The communion with other things.

24. Because of the tilt of the balance blood does not continue to come, but according to the relationship of the humor they expectorate those sorts of things. In some, blood should be let at the right time in those diseases. But in others, as in these people, it is not proper. Negative indication in addition to the seasons, for those spitting bloody matter: pleuritis, bile.

[18] This and the preceding chapter are paralleled in *Epid.* 2.1.11.   [19] Cf. *Epid.* 6.6.9.

[20] The terminth-like swelling is apparently similar in shape to the fruit of the terebinth (turpentine) tree. This simile is not in our mss. of *Epid.* 6, but is in *Humors* ch. 20 (Li. V 500.9), a passage drawn from *Epid.* 6, and was in the version of *Epid.* 6 that Galen read.

[21] These judgments are paralleled in *Epid.* 4.58 and Aph. 6.11–12.

ΕΠΙΔΗΜΙΑΙ

ΤΜΗΜΑ ΤΕΤΑΡΤΟΝ

1. Τὰ παρ᾽ οὖς οἷς ⟨ἂν⟩[95] ἀμφὶ κρίσιν γινόμενα μὴ ἐκπυήσῃ, τούτου λαπασσομένου, ὑποστροφὴ γίνεται κατὰ λόγον τῶν ὑποστροφέων. τῆς ὑποστροφῆς γενομένης αὖτις αἴρεται καὶ παραμένει ὥσπερ αἱ τῶν πυρετῶν ὑποστροφαί, ἐν ὁμοίῃ περιόδῳ· ἐπὶ τούτοισιν ἐλπὶς ἐς ἄρθρα ἀφίστασθαι.

2. Οὖρον παχύ, λευκόν, οἷον τὸ[96] τοῦ Ἀμφιγένεος,[97] ἐπὶ τοῖσι κοπιώδεσι τεταρταίοισιν ἔστιν ὅτε ἔρχεται καὶ ῥύεται τῆς ἀποστάσιος, ἢν δὲ καὶ πρὸς τούτῳ αἱμορραγήσῃ ἀπὸ ῥινῶν ἱκανῶς, καὶ πάνυ.

3. Ὧι[98] τὸ ἔντερον[99] ἐπὶ δεξιὰ ἀρθριτικὸς ἐγένετο, ἢν δὲ ἡσυχώτερος, καὶ ἐπεὶ[100] ἰητρεύθη ἐπιπονώτερος.

4. Ἡ Ἀγάσιος, κόρη μὲν ἐοῦσα, πυκνοπνεύματος ἦν· γυνὴ δὲ γενομένη, ἐκ τόκου πάλαι ἐπίπονος ἐοῦσα ἐπιπολαίως, ἦρεν ἄχθος μέγα· αὐτίκα μὲν ψοφῆσαί τι ἐδόκει κατὰ τὸ στῆθος· τῇ δ᾽ | ὑστεραίῃ ἆσθμά τε εἶχε καὶ ἤλγει ἰσχίον τὸ δεξιόν· ὁπότε τοῦτο πονέοι, ἐπόνει,[101] τότε καὶ τὸ ἆσθμα εἶχε, παυσαμένου δέ,[102] ἐπαύσατο· ἔπτυσεν ἀφρώδεα, ἀρχομένη δὲ ἀνθηρά, κατασταθὲν δὲ ἐμέσματι χολώδει[103] ἐῴκει λεπτῷ· οἱ πόνοι μάλιστα μὲν ὁπότε πονοίη τῇ χειρὶ ταύτῃ·[104] εἴργεσθαι σκορόδου, χοιρίου, ὄϊος, βοός, ἐν δὲ τοῖσι ποιευμένοισι, βοῆς, ὀξυθυμίης.

---

95 add. Erm. 96 τῷ M Gal. 97 MV: Ἀντιγένεος Gal.
Hum. 20 98 Gal. Hum.: ὁ MV 99 ἕτερον (v.l.) Gal.

SECTION 4

1. People in whom swellings by the ear, appearing towards the crisis, do not suppurate: when that area becomes soft they will have a relapse according to the pattern of relapses. When the relapse occurs the swelling rises again and persists like the relapses in fevers, in a similar period. In these swellings there is hope that they will depart towards the joints.

2. In affections with prostration from fatigue, white, thick urine, like that of Antigenes, comes forth sometimes on the fourth day, and wards off the apostasis. If there is also an adequate hemorrhage from the nose, it is sure to.

3. The man with intestinal problems on the right: he became arthritic and was relieved. When he was cured, his distress increased.

4. Agasis' wife had breathing difficulties as a child. After she was married, having had for a long time following parturition a superficial pain, she developed intense suffering. Straightway she seemed to have noise in the chest. On the following day, she had asthma and pain in the right loin. Whenever she had the one pain she had the other, and at the same time she had the asthma also. When the pain stopped so did the asthma. Foamy expectorant, at the beginning bright colored, but when it sat it was like thin, bilious vomit. The suffering was worst when she worked with that hand. Avoidance of garlic, pork, mutton, beef, and, in her activities, shouting, passion.

---

100 καὶ ἐπεὶ V: ἐπὶ δὲ τούτῳ M: ἐπεὶ δὲ τοῦτο Gal. *Hum.*
101 om. Gal.   102 οὐδὲ V   103 om. V
104 ταύτην (v.l.) Gal.

5. Ὦι ἐν τῇ κεφαλῇ ἐνέμετο,[105] ᾧ πρῶτον ἡ στυπτη-ρίη ἡ κεκαυμένη[106] ἐνήρμοσεν, εἶχεν ἄλλη[107] ἀπό-στασιν, ἴσως ὅτι ὀστέον ἔμελλεν ἀποστήσεσθαι· ἀπέ-στη ἑξηκοσταῖον· ὑπὲρ τοῦ ὠτὸς ἄνω πρὸς κορυφῇ[108] τὸ τρῶμα ἦν.

6. Τὰ κόλα ἔχει οἷα κυνός, μέζω[109] δέ· ἤρτηται ἐκ τῶν μεσοκόλων· ταῦτα δὲ ἐκ νεύρων ἀπὸ τῆς ῥάχιος ὑπὸ τὴν γαστέρα.

7. Αἱ τοῖσι κάμνουσι χάριτες, οἷον τὸ καθαρείως δρᾶν ἢ ποτὰ ἢ βρωτὰ ἢ ἃ[110] ἂν ὁρᾷ, μαλακῶς ὅσα ψαύει·[111] ἃ μὴ μέγα[112] βλάπτει, ἢ εὐανάληπτα, οἷον ψυχρόν, ὅπου τοῦτο δεῖ· εἴσοδοι, λόγοι· σχῆμα, ἐσθὴς τῷ νοσέοντι, κουρή, ὄνυχες, ὀδμή.

8. Ὕδωρ ἀφεψηθέν, τὸ μὲν ὡς δέχηται τὸν ἠέρα· τὸ δὲ μή. ἔμπλεον[113] εἶναι καὶ ἐπίθημα ἔχειν. |

310   9. Ὅτι ἐξ αἱμορραγιῶν ἐξυδεροῦνται.

[105] Gal.: ἐννίμετο Μ: ἐννείμετο V
[106] καυμένη Μ
[107] ἄλλην Gal. (lemma)
[108] κορυφὴν Gal.
[109] Gal.: μείζων ΜV
[110] om. V
[111] ψαύει ἄλλαι Μ Gal. (ἄλλα (v.l.) Gal.)
[112] μεγάλα Gal.
[113] ἔμπλεων V

22 This description of courtesy to the patient drew much atten-tion from the ancient commentators. As Galen's commentary in-dicates, their attention produced considerable variation in the

5. The one who was corroding on the head, to whom at the beginning the burnt stiptic ointment was applied, had an apostasis elsewhere, perhaps because the bone was going to come away. It came away on the sixtieth day. His sore was above the ear at the crown.

6. He has intestines like a dog's, but larger, hung from the mesocolon. And that is hung from tendons from the backbone behind the stomach.

7. Kindnesses to those who are ill. For example to do in a clean way his food or drink or whatever he sees, softly what he touches. Things that do no great harm and are easily got, such as cool drink where it is needed. Entrance, conversation. Position and clothing for the sick person, hair, nails, scents.[22]

8. Boiled water: on the one hand so it receives the air, on the other, not. Make it full and have a lid.[23]

9. That after hemorrhage they become watery.

wording of the transmitted text. While Galen takes the last part, after "Entrance," to relate entirely to the grooming, smell, and comportment of the medical man, and offers interesting anecdotes of boorish physicians and of the Roman emperor's feeling about haircuts, I have preferred the interpretation offered here. Cf. Manetti and Roselli, ad loc., and on the tradition of the physician's indulgence of the patient, K. Deichgräber, *Medicus Gratiosus*, Abh., Akad. der Wiss. und der Lit., Mainz 1970, Nr. 3. For *eisodos* meaning "manner of entrance" or "visit," cf. *Decorum* chs. 11–13, Loeb *Hippocrates* vol. 2, p. 295.

[23] Galen informs us that this passage baffled commentators. Some attached it to the preceding, as describing how to prepare warm drinking water to please patients, while others took it as a discussion of the best way to improve water for medical purposes.

10. Ἦν οἷα δεῖ καθαίρεσθαι[114] καθαίρωνται, καὶ[115] εὐφόρως φέρουσιν.

11. Ἐν Αἴνῳ ἐν λίμῳ,[116] ὀσπριοφαγέοντες ξυνεχέως, θήλεα, ἄρσενα, σκελέων ἀκρατεῖς ἐγένοντο καὶ διετέλεον, ἀτὰρ καὶ ὀροβοφαγέοντες γουναλγεῖς.

12. Ἐμφανέως ἐγρηγορὼς θερμότερος τὰ ἔξω, τὰ ἔσω δὲ ψυχρότερος, καθεύδων τἀναντία.

13. Ἐνθέρμῳ φύσει, ψύξις, ποτὸν ὕδωρ, ἐλινύειν.

14. Ὕπνος ἐν ψύχει ἐπιβεβλημένῳ.

15. Ὕπνος ἑδραῖος, ὀρθονυσταγμός.[117]

16. Αἱ ἀσθενεῖς δίαιται ψυχραί· αἱ δὲ ἰσχυραὶ θερμαί.

17. Ὑδάτων ἀτεχνέων, τὸ μὲν ἀπὸ τοῦ αἰθέρος ἀποκριθὲν βρονταῖον ὡραῖον, τὸ δὲ λαιλαπῶδες κακόν. |

312   18. Ὕδωρ βορόν, καὶ ἀγρυπνίη βορόν. ἐνθέρμῳ φύσει καὶ θερμῇ ὥρῃ, κοίτη ἐν ψύχει παχύνει, ἐν θερμῷ λεπτύνει. ἄσκησις ὑγιής,[118] ἀκορίη τροφῆς, ἀοκνίη πόνων. ἐν τῷ ἐγρηγορέναι δίψης ἐπιπολαίου ὕπνος ἄκος, τῷ[119] δὲ ἐξ ὕπνου ἔγερσις.[120]

19. Οἷσι πλεῖστον τὸ θερμόν, μεγαλοφωνότατοι· καὶ γὰρ ψυχρὸς ἀὴρ πλεῖστος· δύο δὲ μεγάλων μεγάλα τὰ ἔκγονα γίνεται. οἱ θερμοκοίλιοι, ψυχρόσαρκοι καὶ λεπτοί· οὗτοι ἐπίφλεβοι, καὶ ὀξυθυμότεροι.

114 om. M   115 om. V (v.l.) Gal.
116 ἐν Αἴνῳ ἐν λίμῳ Gal.: ἐν Αἴνῳ MV: ἐν λίμῳ Zeuxis et alii (Gal.)

10. If what ought to be purged is purged, they bear that too more easily.

11. In Aenus, in a famine, those who always ate beans, men and women, became weak in the legs and continued so. And those who ate vetch had knee troubles.[24]

12. Obviously when one is awake his exterior is warmer, his interior cooler, when asleep the opposite.

13. For a warm nature: cooling, water for drinking, inactivity.

14. Sleep covered in the cold.

15. Sleep while sitting up: drowsing erect.

16. Weak regimens are cold. Strong ones, warm.

17. Of natural waters: what comes from the aether (dry upper air), from thunder, is good. What comes out of a hurricane, bad.

18. Water makes hunger, sleeplessness makes hunger. For a hot nature and in a hot season, sleep in the cold fattens, in the heat, thins. Healthy discipline, not gluttonizing, not avoiding work. Sleep after being awake cures superficial thirst. In arousal from sleep, waking cures.

19. Those with most heat have very big voices: that is because cold air is most abundant. The products of two great things are great. Those with hot intestines have cold skin and are thin. They have prominent veins and are quick to passion.

---

[24] Cf. *Epid.* 2.4.3.

---

[117] ὄρθῳ νυσταγμός M Gal.
[118] ὑγιείης Gal. Erot. (70.16 N)
[119] τῇ Gal.
[120] ἐγρήγορσις ἐνίοις Gal.

20. Αὐχμοῦ ἐπὶ γῆς, οἰωνῶν γένος εὐθηνεῖ.

21. Τράγος, ὁπότερος ἂν φανῇ ἔξω ὄρχις, δεξιός, ἄρσεν, εὐώνυμος, θῆλυ.

22. Ὀφθαλμοί, ὡς ἂν ἰσχύωσιν, οὕτω καὶ γυῖον· καὶ χροιὴ ἐπὶ τὸ κάκιον ἢ ἄμεινον ἐπιδιδοῖ· δίκαιον δέ, ὡς ἂν ἔχῃ ἡ τροφή, οὕτω καὶ τὸ ἔξω ἔπεσθαι. σημεῖα θανατώδεα, ἀνὰ ῥινὸν[121] | θερμότατος[122] ἀτμός· πρότερον δὲ ῥὶς ψυχρὸν πνεῦμα ἀφίησιν· τὰ ζωτικὰ ἐναντία.

23. Πρὸς ὑγιείην πόνοι σιτίων ἡγείσθωσαν.[123]

314

ΤΜΗΜΑ ΠΕΜΠΤΟΝ

1. Νούσων φύσιες ἰητροί. ἀνευρίσκει ἡ φύσις ἑωυ- τῇ[124] τὰς ἐφόδους, οὐκ ἐκ διανοίης, οἷον τὸ σκαρδα- μύσσειν, ἢ[125] γλῶσσα δὲ ὑπουργεῖ, καὶ ὅσα ἄλλα τοιαῦτα· εὐπαίδευτος[126] ἡ φύσις ἑκοῦσα οὐ[127] μαθοῦσα τὰ δέοντα ποιεῖ.[128] δάκρυα, ῥινῶν ὑγρότης, πταρμοί, ὠτὸς ῥύπος, στόματος σιάλου[129] ἀναγωγή, πνεύματος εἴσοδος, ἔξοδος, χάσμη, βήξ, λύγξ, οὐ τοῦ αὐτοῦ

---

121 ῥινῶν V
122 θερμὸς (v.l.) Gal.
123 πόνος . . . ἡγείσθω Gal.
124 αὐτὴ ἑωυτῇ Gal.
125 τὰ δὲ καὶ ἡ Gal.
126 ἀπαίδευτος (v.l.) Gal.
127 ἑκοῦσα οὐ] ἐκτουσαου (sic) M: ἐοῦσα καὶ οὐ Gal.
128 Gal.: ποιέειν MV
129 σίαλον Gal.

20. In a dry land the tribe of birds thrives.[25]

21. Lubriciousness: whichever testicle appears outside; if right, male, if left, female.[26]

22. To the degree that the eyes are strong, so too is the body. Color tends towards the better or worse. And it is proper that the state of nutrition is followed by the exterior. Fatal signs: through the skin an extremely hot exhalation. And before, the nose sends out cold breath. Vital signs, the opposite.

23. Exercise before food for health.

SECTION 5

1.[27] The body's nature is the physician in disease. Nature finds the way for herself, not from thought. For example, blinking, and the tongue offers its assistance, and all similar things. Well trained, readily and without instruction, nature does what is needed. Tears, moisture of the nostrils, sneezing, ear wax, production of saliva in the mouth, the intake of breath, exhalation, yawning, cough-

[25] The ancients were intrigued by the fact that birds have no urinary bladder. [26] I have interpreted the word τράγος, "goat," as lubriciousness, the urge to sexual activity, for which cf. τραγίζειν, *Epid.* 6.3.14. Galen and others take the word to refer to pubescence. In any case, this aphorism appears to deal with predicting the sex of the child that will be produced.

[27] This is the famous statement that nature heals, to which commentators add that the physician is only nature's auxiliary. The text of the manuscripts shows signs of commentators' interference. This passage plays on the double meaning of *physis*: "the physique of the body," and "nature" in a more general sense, as opposed to culture and education.

παντάπασι τρόπου. οὔρου ἄφοδος καὶ φύσης[130] καὶ[131]
ταύτης τῆς ἑκατέρης,[132] τροφῆς καὶ πνοιῆς, καὶ τοῖσι
θήλεσιν ἃ τούτοισι, καὶ κατὰ τὸ ἄλλο σῶμα, ἱδρῶτες,
κνησμοί, σκορδινισμοί,[133] ὅσα[134] τοιαῦτα.

2. Ἀνθρώπου ψυχή[135] φύεται μέχρι θανάτου· ἢν δὲ
ἐκπυρωθῇ ἅμα τῇ νούσῳ καὶ ἡ ψυχή, τὸ σῶμα φέρ-
βεται. |

316   3. Νοῦσοι ξύντροφοι ἐν γήραϊ λείπουσι[136] καὶ διὰ
πεπασμόν, καὶ διὰ λύσιν, καὶ διὰ ἀραίωσιν.

4. Ἴησις ἀντίνοον, μὴ ὁμονοεῖν τῷ πάθει· τὸ ψυχρὸν καὶ ἐπικουρεῖ καὶ κτείνει.[137] ὁκόσα δ᾽ ἐκ θερ-
μοῦ[138] ταὐτά.[139]

5. Ὀξυθυμίη ἀνασπᾷ καὶ καρδίην καὶ πλεύμονα ἐς
ἑωυτά, καὶ ἐς κεφαλὴν τὰ θερμὰ καὶ τὸ ὑγρόν· ἡ δ᾽
εὐθυμίη ἀφίει καρδίην καὶ[140] ταῦτα. πόνος τοῖσιν
ἄρθροισι καὶ σαρκὶ σῖτος, ὕπνος σπλάγχνοισιν. ψυ-
χῆς περίπατος[141] φροντὶς ἀνθρώποισιν.

6. Ἐν τοῖσι τρώμασι τὸ αἷμα ξυντρέχει, βοηθητέον
ὡς τὸ κενὸν πληρώσῃς.[142] |

318   7. Ἢν οὖς ἀλγέῃ, εἰρίον περὶ τὸν δάκτυλον ἑλί-
ξας,[143] ἐγχεῖν ἄλειφα θερμόν, ἔπειτα ἐπιθεὶς ἔσω ἐν
τῷ θέναρι τὸ εἰρίον τὸ οὖς ὑπερθεῖναι ὡς δοκέῃ τί οἱ
ἐξιέναι, ἔπειτα ἐπὶ πῦρ ἐπιβάλλειν· ἀπάτη.

8. Γλῶσσα οὖρον[144] σημαίνει· γλῶσσαι χλωραὶ
χολώδεις, τὸ δὲ χολῶδες ἀπὸ πίονος· ἐρυθραὶ δὲ ἀφ᾽
αἵματος· μέλαιναι δὲ ἀπὸ μελαίνης χολῆς· αὖαι δὲ

---

130 φύσις MV   131 om. V

ing, hiccough, in a variety of ways. The excretion of urine and wind (wind of both kinds, from food and breath), and in women, the things characteristic of them, and in the rest of the body sweat, itchings, stretching, and so on.

2. The soul of man grows until death. If the soul be burnt up with a disease it consumes the body.

3. Diseases that grow old with us leave us in old age by concoction, by resolution, and by rarifaction.

4. Healing is dispute with a disease, not agreement. Cold both helps and kills. The same for the effects of heat.

5. Anger contracts the heart and the lungs and draws the hot and the moist substances into the head. Contentment releases the heart and those substances. Labor is food for the joints and the flesh, sleep for the intestines. Intellection is a stroll for the soul in men.

6. In wounds blood collects. You must help to fill the empty place.

7. If the ear aches, wrap wool around your fingers, pour on warm oil, then put the wool in the palm of the hand and put it over the ear so that something will seem to him to come out. Then throw it in the fire. A deception.

8. The tongue indicates the urine. Greenish tongues are bilious. Biliousness is from fat. Ruddy ones are from blood. Black ones are from black bile. Dry ones are from

132 Man.-Ros.: ἑτέρης mss. Gal.   133 κορδινισμοί V
134 καὶ ὅσα Gal.   135 ψυχὴ αἰεὶ Gal.
136 λίπουσιν M   137 ἐκτείνει MV   138 θυμοῦ V
139 ταῦτα MV: ὅκοσα . . . ταὐτά om. Gal.
140 om. M   141 περὶ παντός (v.l.) Gal.
142 πλησθῆναι Gal.   143 ἐλίξασα MV
144 ὀρὸν Jouanna, perhaps rightly

ἀπὸ ἐκκαύσιος λιγνυώδεος καὶ μητρῴου μορίου· λευκαὶ δὲ ἀπὸ φλέγματος.

9. Οὖρον ὁμόχροον σώματι καὶ πόματι, καὶ ὡς
ἔσωθεν ἐὸν[145] ποτοῦ ὑγροῦ ξύντηξις.

10. Γλῶσσα ὁμόχροος τῇσι προσστάσεσι,[146] διὸ
ταύτῃ γινώσκομεν τοὺς χυμούς. ἢν ἁλμυραὶ σάρκες
γευομένῳ, περισσώσιος.

11. Ἢν τῶν μαζῶν αἱ θηλαὶ καὶ τὸ ἐρυθρὸν χλωρὸν
ᾖ, νοσῶδες τὸ ἄγγος.

12. Ἀνθρώποισιν ὁ ἐν τοῖσιν ὠσὶ ῥύπος, ὁ μὲν
γλυκὺς θανάσιμος, ὁ δὲ πικρὸς οὔ.

13. Γῆν μεταμείβειν ξύντροφον ἐπὶ τοῖσι μακροῖσι
νοσήμασιν.

14. Τὰ ἀσθενέστατα[147] σιτία[148] ὀλιγοχρόνιον βιο-
τὴν ἔχει. |

320   15. Κεδμάτων, τὰς ἐπὶ τοῖσιν ὠσὶ ὄπισθεν φλέβας
σχάζειν. λαγνείη τῶν ἀπὸ φλέγματος νούσων ὠφέλι-
μον. θερμοκοιλίοισιν ἰσχυρὰ ποτὰ ἢ βρωτὰ ταρακτι-
κά. μελαίνης χολῆς, ἐς ὅμοιον, αἱμορροΐδι. τὰς ἐπαυ-
ξέας νούσους μίξις ψύχει.[149] ψύξις[150] τὰ κατὰ κοιλίην
σκληρύνει. ἐλλέβορον πιόντα θᾶσσον καθαίρειν ἢν
θέλῃς,[151] λούειν ἢ φαγεῖν. τὸ αἷμα ἐν ὕπνῳ ἔσω
μᾶλλον φεύγει. ῥῖγος ἀπὸ τῆς ἄνω κοιλίης, πῦρ δὲ

145 ἐὸν M   146 Man.-Ros.: προσστάσεσι mss. Gal.
147 ἀσθενέστερα Gal.   148 σώματα (v.l.) Gal.
149 Diosc. et Capito (Gal.): ψύξει MV: om. Gal.
150 μίξις (v.l.) Gal.
151 MV repeat θᾶσσον after θέλῃς

smoky burning and from the area of the womb. White ones are from phlegm.

9. Urine matches the body and drink in color, and, being within the watery drink, is a melting (of the body).[28]

10. The tongue takes the color of what touches it, whence we know the humors by it. If flesh is salty to the taste it is from excess.

11. If the nipples of the breasts and the areola are pale, the hollow space is ill.[29]

12. Wax in people's ears: if sweet, a mortal sign, if bitter, not.

13. It is supportive to go to another land in long illnesses.

14. Very weak foods contain short-lived sustenance.

15. For *kedmata*, cut the veins behind the ears.[30] Venery helps diseases from phlegm. For those with heated intestines strong food or drink is upsetting. A condition of black bile is normalized by hemorrhoids. Sexual intercourse chills incipient illnesses. Cold hardens the area of the intestine. If you want to speed up the purging action of hellebore in drink, prescribe bathing or eating. Blood in sleep retreats more to the interior. Shivering is from the upper gut, fever more from the lower. Rapid intake of air, if

[28] The meaning of this sentence is dubious. Probably the text is corrupt.

[29] Perhaps this refers to the womb, as Galen says, perhaps to the breast.

[30] *Kedmata* is a word that puzzled the ancient commentators. It probably signifies an affection of the joints in the inguinal area. Cf. *Epid.* 7.122.

ἀπὸ τῆς κάτω μᾶλλον. ἐπισπασμός, ἢν[152] πλεύμων
ξηρὸς[153] ἢ [ὑγρὸν][154] καῦμα. ὑπέρινον[155] ἰσχναίνει καὶ
322 ὕπνος πολύς. ψυχρότατον βρῶμα, | φακοί, κέγχροι,
κολοκύντη. ἕλκεα ἐκθύουσιν, ἢν ἀκάθαρτος ἐὼν πονή-
σῃ. γυναικὶ ἐλατήριον[156] ἢ[157] σίκυον ἄγριον βεβρωκυίῃσι[158] παιδίοισι κάθαρσις. ἐνθέρμῳ βρωθὲν ἔσωθεν ψύξις, ἔξωθεν πόνος, ἡλίῳ, πυρί, ἐσθῆτι, ἐν ὥρῃ
θερινῇ· τῷ δὲ ἐναντίῳ ὡς ἐναντίως.[159] βρώματα τὰ μὲν
ταχέως κρατεῖται, τὰ δὲ ἐναντίως.

ΤΜΗΜΑ ΕΚΤΟΝ

1. Σάρκες ὁλκοὶ καὶ ἐκ κοιλίης καὶ ἔξωθεν· δῆλον
αἴσθησις,[160] ἐν πόνῳ[161] καὶ ἔκπνοον.[162] ἐνθερμότερον
φλέβιον αἵματος πλήθει ἀνίσχει τὸ καυσῶδες, καὶ
εὐθὺς ἀποκρίνει. καὶ οἷσι τὸ μὲν πῖον, χολὴν ξανθήν,
τὸ δ᾽ αἷμα, μέλαιναν.

2. Γνώμης, μνήμης, ὀδμῆς, τῶν ἄλλων, καὶ πεί-
324 ρης,[163] ὀργάνων | ἄσκησις. πόνοι, σιτία, ποτά,

---

152 om. Gal.   153 ξηρὸν (v.l.) Gal.
154 ὑγρὸν MV Capito (Gal.): om. Gal.
155 Gal.: ὑπερινώμενος Erot.: ὑπὲρ ῥινῶν mss.
156 ἐλατηρίων V   157 om. MV
158 γυναικὶ . . . βεβρωκυίῃσι Smith: γυνὴ αἴξ (ἐξ V) . . .
βεβρωκυῖα (-αι Gal.) MV Gal.
159 MV Capito (Gal.): βραδέως Gal.
160 Gal.: αἰσθήσιος MV: αἰσθήσει (v.l.) Gal.
161 ἐμ πόνο M   162 ἐν πόνῳ καὶ ἔκπνοον] ὡς ἔκπνοον
καὶ εἴσπνοον ὅλον τὸ σῶμα Gal.

the lung is dry or the weather hot and dry.[31] Much sleep reduces one who has been emptied by purge. The coldest foods: lentils, millet, cucumber. Sores break out if one exercises unpurged. For women who eat *elaterion* or wild cucumber, there is purgation for their infants. Food eaten by a hot nature, cold inside, suffering outside, from sun, fire, clothing, in summer weather, and for the opposite the opposite.[32] Some foods are digested quickly, others slowly.

SECTION 6

1. The flesh draws both from the intestine and from the exterior. Obvious in exercise is the perception that there is exhalation also. A small blood vessel, heated with fullness of blood, raises up burning heat and straightway separates it off; in those in which there is fat, yellow bile; in which blood, black bile.[33]

2. Intelligence, memory, smell, the others, and experience: exercise of the faculties. Exertion, food, drink, sleep,

[31] This sentence was variously written, punctuated, and interpreted in antiquity, as Galen tells us. I have interpreted the reading of mss. M and V, but have omitted ὑγρόν, which Galen attributes to Capito.   [32] Neither the grammatical structure nor the intended meaning of this sentence is clear.

[33] This is apparently an interpretation of the aspect of the skin's surface, its varicolored vessels, and its absorption of, e.g., olive oil, and excretion of sweat, etc. Galen's commentary and other citations offer an expanded text of this passage adapted to ancient discussions of breathing: "The perception is clear that the whole body inhales and exhales."

[163] Man.-Ros.: πείνης (πίνης M) mss. Gal.

ὕπνος,[164] ἀφροδίσια, μέτρια. ὁ ἐμψυχρότερος ἐν ψυ-
χρῇ χώρῃ, ὥρῃ, ἐνθερμότερος ἔσται.

3. Ὀδυνέων, τὴν ἐγγύτατα κοιλίην καθαίρειν, †αἵ-
ματος δὲ κοιλίην διαιρεῖν,† καῦσις, τομή, θάλψις,
ψύξις, πταρμοί, φυτῶν[165] χυμοί, ἐφ' ὧν τὴν δύναμιν
ἔχουσι, καὶ κυκεών· κακούργων,[166] σκόροδον, γάλα,
οἶνος ἀπεζεσμένος, ὄξος, ἄλες.

4. Ἄνθρωπος, ἐκ κόπων ἐξ ὁδοῦ ἀδυναμίη καὶ
βάρος, ἀνέπτυεν· ἔβησσε γὰρ ἐκ κορυφῆς· πυρετὸς
πρὸς χεῖρα ὀξύς, ὑποδάκνων. δευτεραίῳ δὲ καρη-
βαρίη, γλῶσσα ἐπεκαύθη· ῥὶς ὀνυχογραφηθεῖσα οὐχ
ἡμορράγησεν· ἀριστερὸς σπλὴν μέγας καὶ σκληρός,
ὠδυνᾶτο.

5. Οἱ ὑπὸ[167] τεταρταίου ἁλισκόμενοι[168] ὑπὸ τῆς
μεγάλης νούσου οὐχ ἁλίσκονται· ἢν δ' ἁλίσκωνται
πρότερον καὶ ἐπιγένηται τεταρταῖος, παύονται. ἀνθ'[169]
οἵων αἱ νοῦσοι. ἡ χολή, οἷον εἶπον περὶ τῶν | ὀρνίθων,
ὅτι χολώδεις. ἡ θερμότης δριμύτητος σημεῖον. οἱ
ὄχλοι, αἱ δίοδοι·[170] ὅτι τοῖσι παρακρούουσι λήγουσιν
ὀδύναι πλευρέων· ἔστι δ' οἷσι καὶ πυρετοί· ἔστι δ'
οἷσιν οὔ, ἀλλὰ ξὺν ἱδρῶσιν· ἔστι δ' οἷσι σὺν ὄχλῳ·
ἔστι δ' οἷσι καρφαλέον καὶ περιτεταμένον τὸ δέρμα
καὶ ἁλμυρῶδες. αἱ ναρκώσιες, οἷαι ἐξ οἵων[171] ᾧ τὸ
ἰσχίον. δι' οὐάτων, ἐξ οὐάτων, τὰ πολλὰ θνήσκει

326

---

164 ὕπνοι V   165 Corn.: φυσῶν mss.
166 κακοῦργον V   167 πο Μ   168 αὐλισκόμενοι Μ
169 ἂν Μ   170 αἱ δίοδοι] ἐδίοδοι Μ

sexual activity, in moderation. The man who is cooler in a cold land and season will be warmer.

3. In pains: purge the nearest part of the intestine. [Separate the intestine from blood.][34] Cautery, excision, heating, cooling, sneezing, the juices of plants for things they affect, and cyceon.[35] Harmful things: garlic, milk, boiled wine, vinegar, salt.

4. A man, from fatigue from a journey: weakness and heaviness, expectoration. He had a cough from the head. A fever, acute to the touch, somewhat biting. Heaviness in the head on the second day, his tongue parched. His nose did not bleed when scratched with a fingernail. On the left, the spleen enlarged, hard, painful.

5. People seized with quartan fevers are not seized with epilepsy. If they have it already and a quartan fever supervenes they are cured. Diseases are responses to what kinds of things? As I said about birds, bile because they are bilious. Heat is a sign of acridness. Consider blockage, free passage. The fact that pains in the ribs stop in delirious people. And fever for some, but some not, but with sweats. And some with blockage. And in some drying and stretching of the skin, and saltiness. Loss of sensation, what kind from what in one with hip troubles. Through the ears, from the ears: most deaths on the third day. Those with

---

[34] This sentence, well attested in the mss. and commentaries, is not good Greek nor good ancient medicine, but I have not found emendation.

[35] *Cyceon* is a traditional therapeutic drink, usually barley mush, cheese, wine, possibly onions, and herbs.

---

171 οἴων] οὐάτων καὶ οἷαι Gal.: οἴων καὶ οἷαι M

τριταῖα.[172] οἷσι δέρματα περιτείνεται καρφαλέα καὶ
σκληρά, ἄνευ ἱδρῶτος, οἷσι δὲ χαλαρά, ξὺν ἱδρῶτι
τελευτῶσιν. ἐν τοῖσι παλιμβόλοισιν αἱ μεταβολαὶ
ὠφελέουσι, τούτοισι μεταβάλλειν, πρὶν κακοῦσθαι, ἐς
τὰ πρέποντα, οἷον Χαιρίωνι, τὰ ἐρεθιζόμενα ἐξ οἵων
τὰ κερχνώδεα.

6. Ὅριον, οἷσι μὲν ὑγιὲς καταλείπεται κάτω ὑφῃρη-
μένης τῆς προφάσιος, ἢ καθαίρων, ἢ ἀποδέων, ἢ
ἐκβάλλων, ἢ ἀποτάμνων, ἢ ἀποκαίων· οἷσι δὲ μή, οὔ.

7. Οἷς αἷμα ῥεῖ πολὺ καὶ πολλάκις ἐκ ῥινῶν, οἷσι
328   μὲν ἄχροιαι, | ἄκρητοι,[173] τούτοισιν ὀλίγα ἀρήγουσιν·
οἷσι δὲ ἐξέρυθροι [χρῶτες],[174] οὐχ ὁμοίως. καὶ οἷσι
κεφαλαὶ εὔφοροι, ἄκρητος ἀρήγει, οἷσι δὲ μή, οὔ.

8. Οἷσι ῥῖνες ὑγρότεραι φύσει καὶ ἡ γονὴ ὑγροτέρη
καὶ πλείων, ὑγιαίνουσιν. νοσηλότεροι δὲ οἷσι τἀναντία.[175]

9. Τὰ στρογγυλούμενα πτύαλα παρακρουστικά,
οἷον τῷ ἐν Πλινθίῳ· τούτῳ ἡμορράγησεν ἐξ ἀριστεροῦ
πεμπταίῳ,[176] καὶ ἐλύθη.

10. Οὖρον πολλὴν ὑπόστασιν ἔχον ῥύεται τὰς
παρακρούσιας, οἷον καὶ τὸ τοῦ Δεξίππου μετὰ μάθη-
σιν.

11. Οὐ πρόσω ἐνιαυτοῦ τεταρταῖος.

12. Ὦτα[177] τοῦ θέρεος, ῥήξιες πεμπταίοισιν, ἔστι δ'
ὅτε καὶ μακρότερα· τὰ παρὰ τὰ οὖλα ‹καὶ γλῶσσαν

172 mss. Dioscurides (Gal.): om. Gal.
173 ἄκριτοι V   174 om. Gal.: del. Man.-Ros.

250

stretched dry hard skin reach the end without sweat; those with slack skin die with sweat. In unstable conditions change helps; they should change to the appropriate things before deterioration, as with Chaerion, the irritations that brought on hoarseness.

6. A distinction: if a healthy area remains when you have removed the cause below, purge, bind off, expel, excise, or cauterise. If it does not, not.

7. In people who have frequent copious nosebleeds, if they are of uniform bad color, uniform, they are rarely helpful to them. If they are bright red it is otherwise. For those with healthy heads undiluted wine helps; those without, not.

8. Those whose noses are moist by nature and whose semen is moister and more copious: they are healthy. But those with the opposite condition tend to illness.

9. Globular expectoration portends delirium, as with the man in Plinthius. He had hemorrhage from the left nostril and, on the fifth day, was cured.[36]

10. Urine which leaves a large deposit helps delirium, as in Dexippus' case after he became bald.

11. A quartan fever does not extend beyond a year.

12. In ears in the summer, eruptions on the fifth days, sometimes at greater intervals. In affections of the gums and the tongue, suppuration on seventh days, but espe-

---

36 Cf. *Epid.* 6.3.22.

---

175 νοσηλότεροι κτλ.] οὗτοι νοσηλότερον τοῖσι πλείστοισι δὲ οἷσιν ὑπὸ νούσου τἀναντία M   176 Zeuxis placed
πεμπταίῳ here, the mss. and Galen after ἐλύθη
177 ᾧ τὰ M

ἀποπνεῖ ἑβδομαίοισι, μάλιστα δὲ τὰ κατὰ τὰ οὖλα>,[178] καὶ αἱ κατὰ ῥῖνας ἐκπυήσιες.

13. Οἷσιν ἐπὶ ὀδόντων ὀδύνης ἀπὸ ῥινὸς[179] λεπτὰ ἔρχεται, τούτοισιν ἀπὸ πεπέρεως εὖ ἐνερεισθέντος παχύτερα τῇ ὑστεραίῃ ἔρχεται, ἢν καὶ τὰ ἄλλα μὴ κρατήσῃ· Ἡγησίππῳ γὰρ τὸ ὑπνικὸν | ἐντεθὲν οὐκ ἐκράτησε, μᾶλλον δέ τι καὶ προσεσκαλεύθη βιαιό-

330 τερον.

14. Τὴν ἀπὸ κεφαλῆς ὀστέων φύσιν, ἔπειτα νεύρων, καὶ φλεβῶν, καὶ σαρκῶν, καὶ τῶν ἄλλων χυμῶν, καὶ τῶν ἄνω καὶ τῶν κάτω κοιλιῶν, καὶ γνώμης, καὶ τρόπων, καὶ τῶν κατ' ἐνιαυτὸν γινομένων ὥρῃ τινί, τὸ[180] ἐπὶ πρωϊαίτερον τοῦ ἔτεος οἷον ἐξανθήματα καὶ τὰ τοιαῦτα, ὅμοιον τοῖσι καθ' ἡμέρην πρωϊαίτερον λαμβανομένοισιν, ἢ τὸ ὀψιαίτερον ὡσαύτως. τὸ ἐπίχολον καὶ ἔναιμον σῶμα[181] μελαγχολικόν, μὴ ἔχον ἐξαρύσιας.[182]

15. Λυκίνῳ τὰ ὕστατα σπλὴν[183] μέγας, ὀδυνώδης ἐν τῇ τετάρτῃ ἢ[184] πέμπτῃ.

ΤΜΗΜΑ ΕΒΔΟΜΟΝ

1. Βῆχες ἤρξαντο περὶ ἡλίου τροπὰς τὰς χειμερινὰς ἢ πέμπτῃ καὶ δεκάτῃ ἢ εἰκοστῇ ἡμέρῃ ἐκ μεταβολῆς πυκνῆς νοτίων ἢ βορείων καὶ χιονωδέων· ἐκ τούτων τὰ μὲν βραχύτερα, τὰ δὲ μακρότερα ἐγίνετο· καὶ περιπλευμονικὰ συχνὰ μετὰ ταῦτα. πρὸ ἰσημερίης αὖτις ὑπέστρεφε τοὺς πλείστους, ὡς ἐπὶ τὸ πολύ,

cially on the gums, and purulent gatherings in the nostrils.

13. For people who, with pain of the teeth, have thin discharges from the nose, pepper packed in will produce thickening on the following day, unless other symptoms prevail. For in Hegesippus a soporific was applied but failed; rather there was more powerful erosion.

14. The nature of the bones from the head, then of the tendons, blood vessels, flesh and the other humors, and of the upper and lower intestines, the intellect, habits, and the things that occur each year at a certain time in the early part of the year such as breaking out and the like (similar to the things got in the early part of each day) or those in the later part likewise. The bilious and sanguineous body is melancholic when it lacks evacuation.

15. At the end Lycinus had a large, painful spleen on the fourth or fifth day.

SECTION 7

1. Coughs began around the winter solstice, on the fifteenth or twentieth day after frequent change between southerly weather and northerly with snow. Some affections were shorter, some longer. Pneumonia frequently followed the longer ones. Before the equinox in most affections there was relapse, generally around forty days

---

178 Gal. Pall.: om. mss.
179 Smith (cf. *Epid.* 4.40): ὀφρύος mss. Gal.
180 τότ᾽ M   181 χρῶμα (v.l.) Gal.
182 Li. from Gal. *Glos.*: ἐξερώσιας MV
183 πλὴν M   184 καὶ V

τεσσαρακοσταίους ἀπὸ τῆς ἀρχῆς· καὶ τοῖσι μὲν πάνυ
332 βραχέα καὶ εὔκριτα[185] | ἐγένετο· τοῖσι δὲ φάρυγγες
ἐφλέγμηναν, τοῖσι δὲ κυνάγχαι· τοῖσι δὲ παραπλη-
γικά· τοῖσι δὲ νυκτάλωπες, μᾶλλον δὲ παιδίοισιν·
περιπλευμονικὰ δὲ πάνυ βραχέα ἐγένετο. νυκτάλωπες
μὲν οὖν οὐδὲν βήξασι[186] τὸ ὕστερον ἢ πάνυ βραχὺ
ἀντὶ τῆς βηχὸς ἐγίνοντο, φάρυγγες δὲ βραχέαι, μᾶλλον δὲ νυκταλώπων. κυνάγχαι δὲ καὶ παραπληγικά, ἢ
σκληρὰ καὶ ξηρά, ἢ σμικρὰ καὶ ὀλιγάκις ἀνάγουσαι
πέπονα, ἔστι δ' οἷσι καὶ κάρτα. οἱ μὲν οὖν ἢ φωνῇσι
πλέον ταλαιπωρήσαντες, ἢ ῥιγώσαντες, ἐς κυνάγχας
μᾶλλον ἐτρέποντο.[187] οἱ δὲ τῇ χειρὶ πονήσαντες ἐς
χεῖρας μοῦνον παραπληγικοί, οἱ δ' ἱππεύσαντες ἢ
πλείω ὁδὸν πορευόμενοι ἢ ἄλλο τι τοῖσι σκέλεσι
ταλαιπωρήσαντες τούτοισιν[188] ἐς ὀσφῦν ἢ σκέλεα
ἀκρασίαι παραπληγικαί, καὶ ἐς μηροὺς καὶ κνήμας
κόπος καὶ πόνος· σκληρόταται δὲ καὶ βιαιόταται αἱ[189]
ἐς τὰ παραπληγικὰ ἄγουσαι. πάντα δὲ ταῦτα ἐπὶ
334 τῇσιν ὑποστροφῇσιν | ἐγένετο, ἐν ἀρχῇσι δὲ οὐ μάλα.
πολλοῖσι δὲ τούτων ἀνῆκαν μὲν αἱ βῆχες ἐν τῷ μέσῳ,
ἐξέλιπον δὲ τελέως οὔ· ἀλλὰ ξυνῆσαν τῇ ὑποστροφῇ.
οἷσι φωναὶ ἀπερρήγνυντο ἐς τὸ βηχῶδες, τούτων οἱ
πλεῖστοι οὐδὲ ἐπυρέτηνον, οἱ δέ τινες βραχέα· ἀτὰρ
οὐδὲ περιπλευμονικὰ ἐγίνετο τούτων οὐδενὶ οὐδὲ
παραπληγικὰ οὐδὲ ἄλλο οὐδὲν ἐσημάνθη, ἀλλ' ἐν τῇ
φωνῇ ἐκρίνετο. τὰ δὲ νυκταλωπικὰ ἱδρύετο ὡς καὶ τὰ
ἐξ ἄλλων προφασίων γινόμενα· ἐγίνετο δὲ νυκταλωπι-

from the commencement. Some had brief affections with
successful crises. Others had inflammations of the throat
and others quinsy, others paralyses, others, primarily children, night blindness. The pneumonias were very brief.
Night blindness did not develop subsequently in those
with coughs or it quickly replaced the coughs, and the
inflammations of the throat were brief, more so than the
night blindness. The quinsy and paralyses: coughs produced hard, dry matter or rarely small amounts of concocted matter, sometimes very rarely. Those who were
more affected in the voice, or had more chills, more frequently ended in quinsy. Those who labored with their
hands had paralyses only in them, those who rode horses or
did more walking or other exertion with their legs had par-
alytic weakness in the hip or legs and pain and fatigue
in hams and shanks. Weakness that led to paralysis was
particularly harsh and violent. All these complications developed in the relapses, not much at the commencement.
For many patients the coughs relented in the meantime
but did not stop entirely. But they were there at the
relapse. Those whose speech would break into coughing
were mostly not feverish, but some had brief fevers. None
of these developed pneumonia or paralyses, nor was there
anything else exhibited, but it reached a crisis in the voice.
The night blindness became established just as the affections from other causes. Night blindness problems devel-

---

185 εὔκρητα M   186 νυκτάλωπες . . . βήξασι V: φάρυγγες δὲ (1/3 line blank) βήξασι M
187 V Gal.: ἐτελεύτων M (v.l.) Gal.
188 τούτοισιν δὲ M
189 Gal.: om. mss.

κὰ τοῖσι παιδίοισι μάλιστα· ὀμμάτων δὲ τὰ μέλανα
ὑποποίκιλα, ὅσα τὰς μὲν κόρρας[190] σμικρὰς ἔχει, τὸ
δὲ ξύμπαν μέλαν ὡς ἐπὶ τὸ πολύ· μεγαλόφθαλμοι δὲ
μᾶλλον, καὶ οὐ σμικρόφθαλμοι, καὶ ἰθύτριχες οἱ πλεῖστοι καὶ μελανότριχες.

Γυναῖκες δὲ οὐχ ὁμοίως ἐπόνησαν ὑπὸ τῆς βηχός,
ἀλλ᾽ ὀλίγαι τε ἐπυρέτηναν, καὶ τούτων πάνυ ὀλίγαι ἐς
τὸ περιπλευμονικὸν ἦλθον, καὶ αὗται πρεσβύτεραι,
καὶ πᾶσαι περιεγένοντο. ἠτιώμην [καὶ][191] τοῦτο καὶ τὸ
μὴ ἐξιέναι ὁμοίως ἀνδράσι καὶ ὅτι οὐδ᾽ ἄλλως ὁμοίως
ἁλίσκονται[192] ἀνδράσιν. κυνάγχαι δὲ ἐγίνοντο μὲν καὶ
ἐλευθέρῃσι δισσῇσι, καὶ αὗται τοῦ εὐηθεστάτου τρόπου, περισσοτέρως δὲ δούλῃσιν, ὅσῃσί τε ἐγίνοντο
βιαιόταται καὶ ταχύτατα[193] ἀπώλλυντο. ἀνδράσι δὲ
πολλοῖσιν ἐγίνοντο καὶ οἱ μὲν διέφυγον, οἱ δὲ ἀπώλλυντο. τὸ δὲ ξύμπαν οἱ μὲν μὴ δυνάμενοι καταπίνειν[194]
μοῦνον πάνυ εὐήθη καὶ εὔφορα, οἱ δὲ καὶ διαλεγόμενοι
πρὸς τούτοισιν ἀσαφέως καὶ ὀχλωδέστερα καὶ χρονι-
336 ώτερα· οἷσι δὲ | καὶ φλέβες αἱ[195] περὶ κρόταφον καὶ
αὐχένα ἐπήροντο ὑποπόνηρα· οἷσι δὲ καὶ πνεῦμα
ξυνεμετεωρίζετο κάκιστον, οὗτοι γὰρ καὶ ἐπεχλιαί-
νοντο.

Ὡς γὰρ γέγραπται, οὕτως αἱ ξυγκληρίαι τῶν πα-
θημάτων ἦσαν· τὰ μὲν πρῶτα γεγραμμένα καὶ ἄνευ
τῶν ὕστερον γεγραμμένων ἐγίνετο· τὰ δ᾽ ὕστερον οὐκ
ἄνευ τῶν πρότερον· τάχιστα δ᾽ ἔθνησκον ὅτ᾽ ἐπιρρι-
γώσειαν πυρετώδει ῥίγει. τούτους οὔτε ἀναστάσει πιε-
ζομένους οὐδὲν ἄξιον λόγου ὠφέλει, οὔτε γαστρὸς

oped mostly in children. The dark parts of the eyes which have the small pupil were varicolored, and so was the whole dark part generally. Patients were large-eyed more often, not small-eyed, and with straight hair and dark hair.

Women did not suffer similarly from the cough, but few of them had fever, and of those very few went into pneumonia, and those the older. All survived. I attributed this to their not going out as the men did and because they were not otherwise susceptible like the men. Two free women got quinsy, and that was of the mildest sort. Slave women got it in a more extreme way, and those with very violent cases died very quickly. But many men got it; some survived, some died. For the most part if they only could not drink it was mild and bearable. Those who also spoke unintelligibly had more troublesome and longer cases. Those whose blood vessels swelled at the temple and neck had painful cases. Those whose breathing was elevated had it worst since they also became hot externally.

As I have written, these were the relationships of the affections. The first described occurred also without the later, but the later ones not without the former. Patients died most quickly when they were chilled with a feverish chill. Nothing I tried worth notice helped these, not when they were pressed to evacuate the bowels, not roiling the

---

190 κόρραν V
191 om. Gal.: del. Li.
192 καὶ ὅτι . . . ἁλίσκονται om. V
193 ταχύταται V
194 κταπίνειν V
195 οἱ V

ταραχή, οὔτε φλεβοτομίη, ὅσα ἐπειράθην· ἔταμον δὲ
καὶ ὑπὸ γλῶσσαν· οὓς δὲ ἄνω ἐφαρμάκευσα. ταῦτα
μὲν οὖν καὶ διὰ παντὸς ἐν τῷ θέρει, ὡς δὲ καὶ τὰ
ἐπιρρηγνύμενα πάμπολλα. πρῶτον μὲν ἐν τοῖσιν αὐ-
χμοῖσιν ὀφθαλμίαι ἐπεδήμησαν ὀδυνώδεις.

2. Αἵματος φλεβῶν στάσιες, λειποθυμίη, σχῆμα,
338 ἄλλη ἀπόληψις, μοτώματος ξυστροφή, πρόσθεσις,
ἐπίδεσις, ἐπίπλασις.[196] βουβωνοῦται τὰ πλείω διότι
ἡπατῖτις· ἦν δὲ καὶ ἀπὸ ἀρτηρίης κακωθείσης κακὸν
σημεῖον οἵως[197] Ποσειδωνίη. οἱ αἱμορραγέοντες τελευ-
τῶντες[198] οὐκ ἐφίδρωσαν μέτωπον, ἀλλ᾽ οἷα ξυμπε-
πτωκότες· καὶ οἱ πνευματίαι καὶ οἱ ὑπὸ ἱδρώτων ὀλλύ-
μενοι, πονηρόν. τῶν γαστέρων αἱ εὐφορίαι ταραχήν,
οἷον Ποσειδωνίη, καὶ τὰ θηρία οἷα ἐνεποίει· ἐν τῷ
λεπτυσμῷ ἡ περίτασις πρὸ τῆς τελευτῆς, καὶ ὁ ὀμφαλὸς πρόμακρος εἱλκύσθη αὐτῇ, καὶ οὔλων ἐφελκώ-
σιες τῶν ἐπιόντων ἐπὶ ὀδόντα.

3. Ὅτι πολλὰ περὶ ἑκάστου ἐστὶν ὀρθῶς ἐντείλα-
σθαι, τὰ μὲν ταῦτα δυνάμενα, τὰ δὲ οὔ· οἷόν ἐστι τὰ
τοιάδε, διαχυθῆναι, καὶ πιληθῆναι, καὶ ἐξαχθῆναι, καὶ
340 σκληρυνθῆναι, καὶ πεπανθῆναι, καὶ | ὅπῃ κλίνειν
διώσασθαι. τοὺς ἀτολμέοντας δέον μεταβολῆς ἀνεγεί-
ρειν κατανεναρκωμένους.[199]

4. Ἃ ὑστερέουσιν· ὑδατώδεας θᾶσσον τάμνειν, φθί-
νοντας καίειν, αὐτίκα πρίειν κεφαλάς, τὰ τοιαῦτα· τῶν

[196] ἐπίθεσις ἐπίπλασις V: ἐπίδεσις (om. ἐπιπλ.) M Gal.
[197] οἷον V

stomach, not phlebotomy. And I cut the vein under the tongue, and tried emetics on some. Those affections continued for the whole summer as did the outbreaks generally. Initially with the dry weather painful ophthalmias were epidemic.

2. Stopping blood from the veins; fainting; posture; withdrawing it elsewhere; wrapping in lint; application; bandaging; plastering. Most cases have swollen glands because of the hepatic vein. There was also a bad indication from degeneration of the artery (windpipe?) as in Posidonia. In death those who hemorrhaged did not sweat on the forehead, but, as it were, collapsed. Those with lung affections and those laid low by sweat, unfavorable. Easiness of the intestine foretold upset, as with Posidonia, and effects such as worms produced. Along with loss of flesh, stretching before death, and the navel was drawn out large by it. There were ulcerations of the gums where they came over the teeth.

3. The fact that one can prescribe many things properly for each person: some are effective, some not. For example, this sort of thing: to be scattered and to be compressed; to be hardened and to be softened and ripened; and to force them to where they incline. The diffident, change being necessary, must be roused when they grow torpid.

4. Areas of neglect: incise the dropsical quickly, cauterize the consumptive, immediately trephine heads, and so

ὑδατουμένων μὴ ψαύειν ἤτρου, μηδὲ τῶν ἔσω· ὅμοιον
γὰρ τοῖσι πολλοῖσι γούνασιν.

5. Τὰ παρὰ καρδίην Ξενάρχῳ· καὶ θερμὸν ἅλες
ἐσπνεῖν²⁰⁰ ἐς τὸ ἕλκος· ἀντὶ τῆς κενώσιος θάλπειν.

6. Ἀρχῆθεν σημεῖον τῆς ὀργῆς καὶ τῶν τοιούτων·
φωνὴ οἵη γίνεται ὀργιζομένοισιν, ἢν τοιαύτη ᾖ μὴ
ὀργιζομένῳ φύσει, ἢ καὶ ὄμματα οἷσιν²⁰¹ ἂν ᾖ φύσει
ταραχώδεα οἷα ὅταν ὀργίζωνται²⁰² οἱ μὴ τοιοῦτοι, καὶ
τἆλλα κατὰ λόγον· καὶ νούσων,²⁰³ οἷον τὸ φθινῶδες
ποιεῖ τὸ εἶδος, ἢν τοιοῦτος φύσει ὑπάρχῃ,²⁰⁴ ἐς τοιοῦτον νόσημα παρέσται, καὶ τἆλλα οὕτως.

7. Αἱ βῆχες κοπώδεις καὶ ἅπτονται τῶν σιναρῶν,
ἀτὰρ καὶ μάλιστα ἄρθρων· ἀτὰρ καὶ ἐν τοῖσι κοπιώδεσι πυρετοῖσι βῆχες ξηραὶ γίνονται· αἱ ξηραὶ βῆχες
<ἐς>²⁰⁵ ἄρθρα²⁰⁶ στηρίζουσι ξὺν πυρετῷ ἢν ἐγκατα-
λίπωνται. |

342   8. Τὰ πνεύματα τοῖσι φθινώδεσι τὰ ἄσημα, κακά,
καὶ τοῖσιν ἀτόκοισι, καὶ ὅσα ἄλλα τοιαῦτα, ἀπὸ τῆς
αὐτῆς καταστάσιος.

9. Τοῖσι φθίνουσι τὸ φθινόπωρον κακόν· κακὸν δὲ
καὶ ἔαρ ὅταν τὰ τῆς συκῆς φύλλα κορώνης ποσὶν
ἴκελα ᾖ.

10. Ἐν Περίνθῳ ἔαρος οἱ πλεῖστοι, ξυναίτιον δὲ
βὴξ χειμερινὴ ἐπιδημήσασα, καὶ τοῖσιν ἄλλοισιν

---

²⁰⁰ εὐπνεῖν M   ²⁰¹ οἷ M   ²⁰² ὀργίζονται M
²⁰³ νοῦς V   ²⁰⁴ ὑπάρχει M
²⁰⁵ add. Lind.   ²⁰⁶ ἄθρα V

on. Do not palpate the liver of the dropsical, nor any of the insides. It is similar with most knees.

5. In Xenarchus' affection by the heart: inhaling adequate warmth into the sore; warm to counter the evacuation.

6. From the outset, a sign of derangement and the like: the quality of the voice in people in passion: whether it is that way by nature when he is not angry. Or the eyes, for those in whom they are by nature disturbed like those of normal people who are angry, and the other things on the same rationale. And of diseases, the way consumption makes the body: if he is that way naturally he will come into that kind of disease. And the rest similarly.

7. Exhausting coughs also attach themselves to damaged areas, especially the joints. In exhausting fevers, however, coughs become dry. Dry coughs become fixed in the joints with fever if they settle there.

8. Breathing that is without symptoms in consumptive people is bad. So, too, in the barren, and all such things that come from the same condition.

9. For the consumptive the fall of the year is bad. And the spring is bad when the fig leaves are like a crow's feet.[37]

10. In Perinthus most of them in spring; an epidemic winter cough was a contributing cause, and for the rest

---

[37] The author is punning on "phthisis" and "phthinoporon" (autumn). For the generic statement that diseases are worse in the fall, see *Epid.* 2.1.4 and *Aph.* 3.9–10. In the *Works and Days* (679–82) Hesiod sets the time for spring sailing when the topmost fig leaves are like the crow's foot.

ΕΠΙΔΗΜΙΑΙ

ὅσα χρόνια, καὶ γὰρ τοῖσιν ἐνδοιαστοῖσιν ἐβεβαίω-
σαν· ἔστι δ' οἷσι τῶν χρονίων οὐκ ἐγένετο, οἷον τοῖσι
τὰς νεφρικὰς ὀδύνας ἔχουσιν· ἀτὰρ καὶ τοῖσιν ἄλλοισιν, οἷον ὁ ἄνθρωπος πρὸς ὃν ὁ Κυνίσκος ἤγαγέ με.

11. Τῶν ὀδυνέων καὶ ἐν πλευρῆσι καὶ στήθει καὶ
τοῖσιν ἄλλοισι τὰς ὥρας εἰ μέγα, διαφέρουσι κατα-
μαθητέον, ὅτι ὅταν βέλτιον ἴσχωσιν²⁰⁷ αὖτις κάκιον
ἴσχουσιν οὐχ ἁμαρτάνοντες.²⁰⁸

ΤΜΗΜΑ ΟΓΔΟΟΝ

1. Ἐν τῆσι μακρῆσι δυσεντερίῃσιν²⁰⁹ αἱ ἀποσιτίαι,
κακόν, ἄλλως τε καὶ ἢν ἐπιπυρεταίνωσιν.²¹⁰

2. Τὰ περιμάδαρα ἕλκεα κακοήθεα. |

344   3. Ὀσφῦν ἀλγέοντι, ἀναδρομὴ ἐς τὸ πλευρόν, καὶ
ἐκφύματα ἃ σῆψ καλεῖται.

4. Τὰ νεφριτικὰ οὐκ εἶδον²¹¹ ὑγιασθέντα ὑπὲρ
πεντήκοντα ἔτεα.

5. Τὰ ἐν τοῖσιν ὕπνοισι παροξυνόμενα, καὶ ὅσοις
ἄκρεα περιψύχεται, καὶ ἡ γνώμη ταράσσεται, καὶ
τἄλλα ὅσα²¹² περὶ ὕπνον τοιαῦτα, καὶ οἷσι τἀναντία.

6. Ὅσῃσι μὲν οὐδὲν ἔσω τοῦ τεταγμένου χρόνου,
ἑκάστῃσι τὰ τικτόμενα ἀπόγονα²¹³ γίνεται. τὰ²¹⁴ ἐπιφαινόμενα ἐν οἷσι μησὶ γίνεται· οἱ πόνοι ἐν περιόδοι-
σιν· ὅτι ἐν ἑπτὰ κινεῖται ἐν τριπλασίῃ τελειοῦται, καὶ

²⁰⁷ ἴσχουσι M
²⁰⁸ οὐδὲν ἐξαμαρτάνοντες V

as many diseases as were chronic, for they were powerful in ambiguous conditions. But it did not happen in some chronic diseases, for example in those with kidney pains, but did for the rest, for example the man to whom Cyniscus brought me.

11. One must note the seasons of pains in the sides, chest, and elsewhere, whether they differ greatly, because when patients are better they are again worse though they have done nothing wrong.

SECTION 8

1. In lengthy intestinal affections aversion to food is bad, especially if there is fever.

2. Ulcers saturated with fluid are of evil nature.

3. In a patient with an affection of the loins, there was an incursion to the ribs and eruptions that are called *seps*.

4. I did not see kidney infections get better beyond fifty years.

5. Affections that exacerbate in sleep, and people whose extremities are cold and mind disordered, and all similar things in relation to sleep, and people subject to the opposites.

6. Women to whom nothing happens within the prescribed time have viable babies. Additional symptoms, in what months they occur: pains come in cycles. What moves in seven is perfected in thrice that, what moves in nine is

---

209 "Blutfluss" Gal.   210 ἐπιπυρετήνωσιν M
211 ἴδον M   212 om. M
213 *Epid. 2, Gal. Gloss.*: ἄγονα MV: γόνιμα Pall.
214 τὰ δ' V

ὅτι ἐν ἐννέα κινεῖται, ἐν τριπλασίῃ τελειοῦται.²¹⁵ ὅτι
μετὰ τὰ γυναικεῖα τὰ δεξιὰ τὰ δ᾽ ἀριστερὰ χάσκων,²¹⁶
ὑγρότης διὰ τῶν ἀπιόντων, διαίτης ξηρότης. ὅτι δὲ τὸ
θᾶσσον διακριθέν, κινηθέν, αὖτις αὔξεται βραδύτερον, ἐπὶ πλείονα δὲ χρόνον. οἱ πόνοι, τρίτῳ, πέμπτῳ,
ἑβδόμῳ, ἐνάτῳ μηνί, δευτέρῳ, τετάρτῳ, ἕκτῳ.

7. Τὰ ἐκ τοῦ σμικροῦ πινακιδίου σκεπτέα. δίαιτα
γίνεται πλησμονῇ, κενώσει, βρωμάτων, πομάτων·
μεταβολαὶ τούτων, οἷα ἐξ οἵων, ὡς ἔχει. ὀδμαὶ τέρπου-
346 σαι, λυποῦσαι, πιμπλῶσαι,²¹⁷ πειθό|μεναι· μεταβολαὶ
ἐξ οἵων οἵως ἔχουσιν. τὰ ἐσπίπτοντα ἢ ἐξιόντα πνεύματα, ἢ καὶ σώματα. ἀκοαὶ κρείσσονες, αἱ δὲ λυποῦ-
σαι. καὶ γλώσσης, ἐξ οἵων οἷα προκαλεῖται. πνεῦμα,
τὸ ταύτῃ θερμότερον, ψυχρότερον, παχύτερον, λεπτό-
τερον, ξηρότερον, ὑγρότερον, πεπληρωμένον, μεῖόν τε
καὶ τὸ πλεῖον· ἀφ᾽ ὧν αἱ μεταβολαί, οἷαι ἐξ οἵων, ὡς
ἔχουσιν. τὰ ἴσχοντα, ἢ ὁρμῶντα, ἢ ἐνισχόμενα σώματα. λόγοι, σιγή,²¹⁸ εἰπεῖν ἃ βούλεται· λόγοι,²¹⁹ οἷσι
λέγει, ἢ μέγα, ἢ πολλοί, ἀτρεκεῖς, ἢ πλαστοί.

8. Τὰ ἀπιόντα· ἱδρώς, ὅθεν²²⁰ ἤρξατο, ἢ ὅπη ἐπαύσατο, ἢ ὅσοισι διεσπᾶτο, χρώματά τε, οἷσιν ἢ θερμότατα, ἢ ἁλμυρότατα, ἢ γλυκύτατα, ἢ λεπτότατα, ἢ

---

²¹⁵ καὶ . . . τελειοῦται om. V, perhaps correctly
²¹⁶ χασκῶν Diosc. ap. Gal. *Gloss.*
²¹⁷ πιμπλάσαι M
²¹⁸ λόγοι σιγή M: λόγοισι δεῖ V
²¹⁹ Li.: λόγοις mss.
²²⁰ ἱδρὼς ὅθεν] ἱδρωσθὲν M

perfected in thrice that. That after the menses there is gaping on the right and the left. Dampness because of what passes out: dryness of regimen. That what is quickly separated, disturbed, afterwards grows more slowly and for a longer time. Pains at the third, fifth, seventh, ninth month, second, fourth, sixth.[38]

7.[39] Things from the small tablet, to be observed. Regimen consists in repletion and evacuation of foods and drinks. Changes of these: what from what, how it is. Odors: pleasant, noxious, filling, persuading. Alterations, from what kinds of things, how they are. The airs that come in or go out, solid bodies also. Better sounds, and those that harm. Of the tongue, what things are called forth by what. Breath, what is hotter to the tongue, colder, thicker, thinner, dryer, wetter, filled up, less and greater. From what come changes, what kind out of what kinds of things, how they are. Bodies that restrain or stimulate, or are restrained. Speech, silence, saying what he wishes. The words with which he speaks: loudly or many, unerring or molded.

8. Excretions: sweat, whence it started or where it stopped, or by how many things it was drawn out; and the colors, those according to which they are extremely hot, salty, sweet, thin, thick, uniformly or not. What kinds of

---

38 With a few differences in the text this appears also in *Epid.* 2.3.17.

39 This list of basic subjects for research "from the small tablet," beginning here in ch. 7, had already by Galen's time generated much discussion as to what it might tell us about the mode of composition of *Epid.* 6, and of *Epid.* 2 and 4 as well.

παχύτατα, ὁμαλῶς ἢ ἀνωμάλως, τὸ σῶμα, τοῦ χρό-
νου²²¹ αἱ μεταβολαί, οἷαι ἐξ οἵων [ἔχουσιν].²²² ἤρ-
ξατο²²³ ἔρυξιν,²²⁴ οὐκ ἐκράτησεν. δάκρυα, ἑκόντι, ἀέ-
κοντι, πολλά, ὀλίγα, θερμά, ψυχρά, πάχος, γεῦσις.
πτύαλον αὐτόθεν ἀναχρεμπτόμενον, ἢ ἀναβήσσοντα,
ἔμετος.

9. Ἡλίου θάλπος, ψῦχος, τέγξις, ξηρότης, μεταβολὴ διὰ²²⁵ οἷα, ἐξ οἵων, ἐς οἷα ὡς²²⁶ ἔχει. πόνοι,
ἀργίαι, ὕπνοι, ἀγρυπνίαι. τὰ ἐν ὕπνῳ· ἐνύπνια, κοῖται,
καὶ ἐν οἷσι, καὶ ὑφ' οἵων. |

348   10. Καὶ τῆς γνώμης ξύννοια, αὐτὴ καθ' ἑωυτήν,
χωρὶς τῶν ὀργάνων καὶ τῶν πρηγμάτων, ἄχθεται²²⁷
καὶ ἥδεται, καὶ φοβεῖται καὶ θαρσεῖ, καὶ ἐλπίζει καὶ²²⁸
ἀδοξεῖ, οἷον ἡ Ἱπποθόου οἰκουρὸς τῆς γνώμης αὐτῆς
καθ' ἑωυτὴν²²⁹ ἐπίστημος ἐοῦσα τῶν ἐν τῇ νούσῳ
ἐπιγενομένων.

11. Ἡλικίην μὲν ἡλίκος, καὶ ἡλικίην ἢ πρότερον ἢ
ὕστερον τοῦ δέοντος, οἷον εἰ παρελύθη ἐν χειμῶνι καὶ
γέροντι τὴν ἡλικίην ἢ νηπίῳ καὶ θερίης, ἢ πρότερον ἢ
ὕστερον τοῦ δέοντος ὀδόντων ἐκβολαί, τριχῶν φύσιες,
γόνος, τὸ μᾶλλον καὶ ἧσσον τριχῶν αὔξησις, παχυ-
σμός, κρατυσμός, μινύθησις.

221 Zeuxis (Gal.): τὸν χρόνον mss.
222 om. Gal. Zeuxis
223 ἤρξαντο V
224 ἔρυξιν V: ερυξι (sic) M: οὐκ ἠρύξατο Gal. Zeuxis
225 δ' V
226 (v.l.) Gal.: om. mss. Gal.

alterations from what kinds of conditions possess the body, within what time. It commences belching, it does not conquer it.[40] Tears, voluntary, involuntary, many, few, hot, cold; thickness, taste. Expectorant, hawking it right out, or coughing it up. Vomit.

9. Sun's heat, cold, dampness, dryness, the nature of alteration, on account of what it occurs, from what to what. Pains, lassitude, sleep, restlessness. Phenomena of sleep: dreams, going to bed, both in what circumstances and from what cause.

10. Even the mind's consciousness, itself by itself, distinct from the organs and events, feels misery and joy, is fearful and optimistic, feels hope and despair.[41] Like the servant of Hippothous, although by herself in her mind she was conscious of the things that followed on her disease.

11. Of an age with one's age, or earlier or later than is proper for the age: such as, if there was loss of faculties in winter for one who was, in age, an old man, or for an infant, and in the summer. Or earlier or later than appropriate, the eruption of teeth, growth of hair, semen; excess and defective growth of hair, thickness, toughness, diminution.

---

[40] Ancient commentators tried unsuccessfully to make sense of this sentence. The text is probably corrupt.

[41] ἀδοξέω does not mean "despair" except here. It generally means, as Galen and his predecessors note, "think badly of a person." I think that we should add the meaning "despair, feel diffidence" to the lexicon.

---

227 ἄρχεται V
228 τε καὶ M
229 mss.: κατ᾽ ἐνιαυτόν (v.l.) Gal.

12. Τὸ ξυγγενές, καὶ τὸ καθ᾽ ἑωυτό, ὅσῳ μᾶλλον καὶ ἧσσον.

13. Τοῦ ἔτεος ἡ ὥρη ἐν ἧτε καὶ πρωΐτερον ἢ ὀψίτερον ἡ ὥρη ἐν ᾗ ἐγεγόνει· ἔπομβρος ἢ αὐχμός, ψυχρὴ ⟨ἢ⟩²³⁰ θερμή, νήνεμος ἢ ἐπάνεμος,²³¹ καὶ οἵων ἀνέμων, τῆς ὥρης ἐν ἀρχῇ, ἢ μέσα, ἢ ἔσχατον,²³² ἢ διὰ παντός, παροιχομένη²³³ ἢ παρεοῦσα ὥρη.

14. Τῆς νούσου, ὁ χρόνος, τὰ ἐπιγινόμενα, αἱ περίοδοι, καὶ τῶν περιόδων αἱ μέζονες, καὶ εἰ διὰ πλείονος, αἱ ἐπιδόσιες, καὶ ἡ τῆς ἄλλης νούσου ἐπίδοσις, ἡ χάλασις, ἡ ἀκμὴ καὶ τὸ μᾶλλον καὶ τὸ ἧσσον ἀποτελέουσα, καὶ ὅτε, καὶ ὁποίως, καὶ ἐν οἵῃ ὥρῃ καὶ ἡλικίῃ.

15. Τῶν ἐπιδημεουσέων νούσων οἱ τρόποι, καὶ εἴ τις τῶν ἀρχομένων ἄρξαιτο²³⁴ ἀνήμετος, οἷον ἢ πιόντες τι, ἢ κατισχόντες, ἢ βραχὺ καθαιρόμενοι. |

350   16. Πυρετώδεις ἴσως οἱ ἔκλευκοι, ὧν καὶ χείλεα· οἷος ὁ τρόπος, οἱ χρόνοι.

17. Τὸ σῶμα ἔργον ἐς τὴν σκέψιν ἄγειν, ὄψις, ἀκοή, ῥίς, ἁφή, γλῶσσα, λογισμὸς καταμανθάνει.²³⁵

18. Τὰ ἐν τοῖσι βλεφάροισι τοῖς ἄνω καταλειπόμενα οἰδήματα, τῶν ἄλλων περισχναινομένων, ὑποστροφαί·²³⁶ ἐν δ᾽ ἄκρῳ ὑπέρυθρα σκληρά, καὶ πάνυ τούτοισι, γλίσχρα, καὶ ἀνιστάμενα, καὶ ἐνεχόμενα

---

²³⁰ Gal.: om. MV
²³¹ Gal., recc.: ἀνήνεμος MV Pall.
²³² ἔσχατα M   ²³³ παροιχομένης M

12. The congenital and that in and of itself, to what extent more or less.

13. The season of the year in which the time in which it developed was also earlier or later: rainy or with drought, cold or hot, calm or windy, with what sort of winds at the beginning of the season or middle or end or throughout, incidental or persistent.

14. The time of the disease, the things that follow on it, the periods, and of the periods the longer ones and whether they are increasing, and the increments, and, of the disease generally, the increment, slackening, the acme, and whether it is more or less perfect, when, how, and in what season and age.

15. The types of epidemic diseases. And whether at their coming on anyone commenced vomitless: for example, either when they drank something or retained it or were slightly purged.

16. Possibly feverish, the excessively pale and those with (cracking) lips.[42] As is the type of disease, so are its time periods.

17. The task is to bring the body under consideration. Vision, hearing, nose, touch, tongue, reasoning arrive at knowledge.

18. Residual swellings on the upper eyelids when the others have subsided: relapse. If they are hard and red on their tips, especially so. The viscous ones and those that protrude are included among them, as in the case of

[42] Cf. *Epid.* 2.1.10.

---

234 recc.: ἄρξετο M: ἄρξεται V
235 om. Gal.   236 ὑποστροφήν (v.l.) Gal.

ἐν τούτοισιν, οἷον Φαρσάλῳ, Πολυμήδει. τὰ καταρ-
ρηγνύμενα οἰδήματα, ἢ πελιώματα, ἢ ἐπ' ὀφθαλμί-
ῃσιν ἢ ἕλκεσιν· ἄλλα γὰρ τὰ ἐκ τῶν φυμάτων καὶ ἐμ-
πυημάτων καταφερόμενα ὅτι σημεῖον ἀμφίτμητον.²³⁷

19. Στρόφοι περὶ τὸν ὀμφαλὸν καὶ ὀδύναι ἔστιν
οἷσιν²³⁸ ἀπὸ πράσων· ἀτὰρ καὶ σκορόδων· τούτοισιν
ὕστερον· ἐρυσίπελάς τ' ἐστὶν οἷσιν ἀπὸ τραχήλου καὶ
περὶ ταῦτα ἐν προσώπῳ ἐκρίνετο· τὰ μέλανα, καὶ οἷσι
τράχηλοι, κακά, καὶ εἰ φλύκταιναι· κακὰ²³⁹ καὶ οἷσι
ταρακτικά. |

352 20. Ὁ μελαγχολικὸς ὁ Ἀδάμαντος ἀπὸ πεπλίων
πλειόνων ἤμεσέ ποτε μέλανα, ἄλλοτε ἀπὸ κρομμύων.

21. Οἱ πυρετοὶ ἐν οἷσιν ἐφελκοῦται χείλεα ἴσως
διαλείποντες²⁴⁰ μὲν καὶ τριταίοισι ψύξιες, οἱ δὲ περικαεῖς αὐτίκα πρὸς τὴν χεῖρα λυόμενοι αἰεί.

22. Οἷον ἔνεστιν ἐν τοῖσιν ἄρθροισι σκεπτέον· ἦρα
οὐκ ἐκπυοῦνται;

23. Ἔθος δέ, ἐξ ὧν ὑγιαίνομεν, διαίτῃσι, σκέπῃσι,
πόνοισιν, ὕπνοισιν, ἀφροδισίοισι, γνώμῃ.

24. Ὅπως τὰς νούσους, ἀπὸ τίνων τίνα σχήματα,
ἐφ' οὓς τόπους ἐτράποντο, ἤρχοντο, παρῆσαν, ἐπαύ-
σαντο. ἐν οἷσιν ἀκρατὴς φοβερός. τὰ ἐναντία ἐν τῇ
νούσῳ διαιτήματα. τὸ εὔφορον, τὸ δύσφορον.

---

²³⁷ (v.l.) Gal.: ἀμφίδμητον MV: ἀμφιμητήριον Diosc. (Gal.):
ἀμφὶ ἄμητον Capito (Gal.): ἀμφίβλητον Rufus (Gal.)
²³⁸ Gal.: om. MV
²³⁹ καὶ εἰ φλύκταιναι κακά om. V
²⁴⁰ διαλιπόντες M

Pharsalus and Polymedes. Swellings that rupture or are livid belong either to inflammation of the eyes or to ulcers, for those are different that are developed from abscesses and empyemas because their significance is ambiguous.[43]

19. Cramps and pains around the navel sometimes from onions, also from garlic, the latter later. Red eruptions are sometimes resolved from the neck and thereabouts on the face. Dark ones, also for those whose necks are affected, are bad, also if there are blisters; they are bad also for those for whom they cause systemic upset.

20. The melancholic Adamantus once vomited black matter after too much purslane, at another time after onions.

21. Fevers in which the lips are ulcerated are perhaps remittent, with chills on the third day, but fevers that are immediately very hot to the touch are always resolved.

22. One must consider how it is in the joints: are they not purulent?

23. And the habitude, things from which we are healthy: in diet, covering, exercise, sleep, sexual activity, mental activity.

24. How (to know) diseases, from what things come what patterns, towards what places they turn, originate, settle, stop. Those in which one is powerless, fearful. Opposite diet in disease. Easily borne, borne with difficulty.

---

[43] I have no confidence in this text and translation of ch. 18, which seems to address swellings that do or do not come to a head. Galen's commentary indicates considerable confusion about it in antiquity.

25. Αἱ δίαιται, ὅσον γνῶναι μὴ ἐξειδῆσαι, ξυμφοραὶ γὰρ πολλαί.

26. Ἀγαθοῖσι δὲ ἰητροῖσιν, αἱ ὁμοιότητες πλάνας καὶ ἀπορίας, ἀλλὰ τἀναντία. ἡ πρόφασις, οἵη· ὅτι χαλεπόν ἐστιν ἐκλογίσασθαι εἰδότα τὰς ὁδούς· οἷον εἰ φοξός, εἰ σιμός, ὑπόξυρος, | χολώδης, δυσήμετος, χολώδης μέλας, νέος εἰκῇ βεβιωκώς, ἅμα ταῦτα πρὸς ἄλληλα ξυνομολογήσασθαι χαλεπόν.[241]

354

27. Ὧι τὸ συρίγγιον ἐπανερρήγνυτο,[242] βηχία ἐκώλυε διαμένειν.

28. Ὧι ὁ λοβὸς τοῦ ἥπατος ἐπεπτύχθη, διέσεισα, ἐξαίφνης ὁ πόνος ἐπαύσατο.

29. Σάτυρος, ἐν Θάσῳ, παρωνύμιον ἐκαλεῖτο γρυπαλώπηξ, περὶ ἔτεα ἐὼν πέντε καὶ εἴκοσιν, ἐξωνείρωσσε πολλάκις·[243] προήει δ᾽ αὐτῷ καὶ δι᾽ ἡμέρης πλεονάκις· γενόμενος δὲ περὶ ἔτεα τριήκοντα, φθινώδης ἐγένετο καὶ ἀπέθανεν.

30. Ἐν Ἀβδήροισιν ὁ παλαιστροφύλαξ, ὁ Κλεισθένεος[244] < . . . >[245] γενόμενος, παλαίσας πλείω πρὸς ἰσχυρότερον καὶ πεσὼν ἐπὶ κεφαλήν, ἀπελθὼν ἔπιε ψυχρὸν ὕδωρ πολύ· μετὰ δὲ ταῦτα, ἐκείνης τῆς νυκτός, ἀγρυπνίη, δυσφορίη, ἄκρεα ψυχρά. τῇ δὲ ὑστεραίῃ εἰσῆλθον εἰς οἶκον·[246] κοιλίη βαλάνου προστεθείσης οὐχ ὑπῆλθεν, οὔρησε δὲ σμικρόν, πρότερον δὲ οὐδὲν

---

[241] mss. Capito (Gal.): Galen omits χ. and adds καὶ ἐπὶ μᾶλλον καὶ ἧττον   [242] πανερρήγνυντο M
[243] πλεονάκις M   [244] K. Gal.: καὶ σθένεος MV
[245] Man.-Ros. conjecture a lacuna here

25. Regimen, how far to know is not to comprehend, for there are many accidents.

26. For good physicians similarities cause wanderings and uncertainty, but so do opposites. It has to be considered what kind of explanation one can give, and that reasoning is difficult even if one knows the method. For example, if a man has a pointed head and flat nose, is sharp-nosed, bilious, vomits with difficulty, full of black bile, young and has lived at random: it is hard for all these to be in concord with one another.[44]

27. The one in whom a fistula broke through above: coughs kept him from staying still.[45]

28. The one whose liver lobe was folded, I shook him and suddenly the pain stopped.

29. Satyrus in Thasos was nicknamed Griffinfox. He was near twenty-five years, had frequent wet dreams, and it also occurred often in the daytime. As he approached thirty he became consumptive and died.

30. In Abdera the wrestling master, son of Cleisthenes . . . , having wrestled too much with a stronger man and fallen on his head, withdrew and drank a large quantity of cold water. Afterwards, during that night, restlessness, discomfort, cold extremities. On the next day I went to his house. When a suppository was applied to his bowel there was no movement; he urinated a little, having urinated

[44] Galen's interpretation reads "Evaluate all this together, taking into account also the quantitative degrees."

[45] The wording suggests that this is related to the fuller description of the case of Deinias' child, *Epid.* 7.117.

---

246 εἰσῆλθε εἰς βάλνιον Gal.

οὐρήκει· ἐς νύκτα ἐλούσατο· οὐδὲν ἧσσον[247] ἀγρυπνίη
καὶ δυσφορίη, παρέκρουσεν. ἐόντι δὲ τριταίῳ κατά-
ψυξις ἀκρέων· ἐκθερμανθεὶς ἵδρωσε πιὼν μελίκρητον·
ἀπέθανε τριταῖος.

31. Οἱ μελαγχολικοὶ καὶ ἐπιληπτικοὶ εἰώθασι γίνε-
356 σθαι ὡς ἐπὶ | τὸ πολύ, καὶ οἱ ἐπίληπτοι μελαγχολικοί·
τούτων δὲ ἑκάτερον μᾶλλον γίνεται, ἐφ᾽ ὁπότερα ἂν
ῥέψῃ τὸ ἀρρώστημα· ἢν μὲν ἐς τὸ σῶμα ἐπίληπτοι, ἢν
δὲ ἐπὶ τὴν διάνοιαν μελαγχολικοί.

32. Ἐν Ἀβδήροισι Φαέθουσα ἡ Πυθέου γυνὴ οἰ-
κουρός, ἐπίτοκος ἐοῦσα τοῦ ἔμπροσθεν χρόνου, τοῦ δὲ
ἀνδρὸς αὐτῆς φυγόντος, τὰ γυναικεῖα ἀπελήφθη χρόνον πολύν· μετὰ δὲ ἐς ἄρθρα πόνοι καὶ ἐρυθήματα·
τούτων[248] δὲ ξυμβάντων τό τε σῶμα ἠνδρώθη καὶ
ἐδασύνθη πάντα, καὶ πώγωνα ἔφυσε, καὶ φωνὴ τρηχέη ἐγενήθη,[249] καὶ πάντα πραγματευσαμένων ἡμέων
ὅσα ἦν πρὸς τὸ τὰ γυναικεῖα κατασπάσαι οὐκ ἦλθεν,
ἀλλ᾽ ἀπέθανεν οὐ πολὺν μετέπειτα χρόνον βιώσασα.
ξυνέβη δὲ καὶ Ναννοῖ[250] τῇ Γοργίππου γυναικὶ ἐν
Θάσῳ τωὐτόν· ἐδόκει δὲ πᾶσι τοῖσιν ἰητροῖσιν οἷσι
κἀγὼ ἐνέτυχον μία ἐλπὶς εἶναι τοῦ γυναικωθῆναι, εἰ
τὰ κατὰ φύσιν ἔλθοι· ἀλλὰ καὶ ταύτῃ οὐκ ἐδυνήθη
πάντα ποιούντων γενέσθαι, ἀλλ᾽ ἐτελεύτησεν οὐ βραδέως.

[247] om. V   [248] τοῦτο M
[249] ἐγένετο V
[250] Li.: Nanno, Nano Gal.: Ναννύι M: Ναννύη V

274

nothing before. He bathed toward night. No less restlessness, discomfort; he became delirious. On his third day, chill of the extremities. He grew feverish and sweated after having drunk melicrat. He died on the third day.

31. Melancholics tend to become epileptic generally and epileptics melancholic. Each of these develops more according to what the weakness inclines towards: if towards the body, epileptics, if towards the mind, melancholics.

32. In Abdera, Phaëthusa the wife of Pytheas, who kept at home, having borne children in the preceding time, when her husband was exiled stopped menstruating for a long time. Afterwards pains and reddening in the joints. When that happened her body was masculinized and grew hairy all over, she grew a beard, her voice became harsh, and though we did everything we could to bring forth menses they did not come, but she died after surviving a short time. The same thing happened to Nanno, Gorgippus' wife, in Thasos. All the physicians I met thought that there was one hope of feminizing her, if normal menstruation occurred. But in her case, too, it was not possible, though we did everything, but she died quickly.

ΕΠΙΔΗΜΙΩΝ ΤΟ ΕΒΔΟΜΟΝ

V 364
Littré

1. Μετὰ κύνα οἱ πυρετοὶ ἐγένοντο ἰδρώδεις, καὶ οὐ περιεψύχοντο παντάπασι μετὰ τὸν ἰδρῶτα· πάλιν δὲ ἐπεθερμαίνοντο, καὶ μακροὶ ἐπιεικῶς καὶ δύσκριτοι καὶ οὐ πάνυ διψώδεις· ὀλίγοισιν ἐν ἑπτὰ καὶ ἐννέα ἐπαύοντο, ἄλλοισιν ἕνδεκα, καὶ τεσσαρεσκαίδεκα, καὶ ἑπτακαίδεκα, καὶ εἰκοσιδύω.[1] Πολυκράτει πυρετὸς καὶ τὰ τοῦ ἱδρῶτος οἷα γέγραπται· ἀπὸ φαρμάκου κάθαρσις κάτω σφοδρὴ[2] ἐγένετο, καὶ τὰ τοῦ πυρετοῦ οὕτως ἤπια ὡς ἄδηλα πλὴν ἐν κροτάφοισιν· καὶ τὰ ἰδρώτια πρὸς δείλην, περὶ κεφαλήν, τράχηλον, στήθεα, εἶτ᾽ αὖτις καὶ ἐς ὅλην κοιλίην, καὶ πάλιν ἐπεθερμαίνετο. περὶ δὲ τὰς[3] δώδεκα καὶ τεσσαρεσκαίδεκα ἐπέτεινεν ὁ πυρετός· καὶ ὑποχωρήσιες βραχέαι· ῥυφήμασι δὲ μετὰ τὴν κάθαρσιν ἐχρήσατο. περὶ δὲ τὰς πεντεκαίδεκα ἀλγήματα γαστρὸς κατὰ σπλῆνα καὶ κενεῶνα ἀριστερόν· θερμῶν προσθέσιες ἦσσον ἢ ψυχρῶν[4] ξυνέφερον· κλύσματι μαλακῷ χρησαμένῳ ἔληξεν ἡ ὀδύνη. τὸ αὐτὸ δὲ καὶ Κλεοκύδει ξυνήνεγκε πρὸς

mss MV, recentiores HIR

[1] καὶ ἑπτακαίδεκα καὶ εἰκοσιδύω] καὶ εἰς δευτέρην καὶ εἰκοστὴν καὶ ἐς ἑπτὰ καὶ ἐς δέκα mss. (cf. Epid. 5.73)

EPIDEMICS 7

1. After the Dogstar, fevers were accompanied with sweat, nor did they cool down entirely after the sweats. They got hot again, were significantly long, did not reach proper crises, did not produce much thirst. In a few cases they stopped in seven and nine days, in other cases in eleven and fourteen and seventeen and twenty-two. For Polycrates, fever and the phenomena of the sweat were such as have been described. He had drastic purging below after a drug, and the phenomena of the fever were so mild as to be imperceptible, except at the temples. He had the sweats toward evening around the head, neck, chest, then again on to the whole body cavity, and again he became fevered. Around the twelfth and fourteenth days his fever increased. Few feces. He took gruel after the purge. About the fifteenth, pains in the belly by the spleen and left abdomen. Applications of heat helped less than cold. His pain was relieved when he used a soft clyster. (The same thing helped Cleocydes too for a similar pain,

---

2 σφόδρα η (sic) M
3 recc.: τὰ MV
4 ψυχρὸν M

ὀδύνην ὁμοίως ἔχουσαν καὶ ἐν πυρετῷ. περὶ δὲ ἑκκαι-
δεκάτην ἠπιώτεραι αἱ θέρμαι ἐδόκεον εἶναι· καὶ ὑποχωρήσιες ἀκρήτου χολῆς, καὶ ἡ διάνοια θρασυτέρη·
πνεῦμα μέτριον, ἔστι δ' ὅτε καὶ ἅλες ἑλκύσας πάλιν
ἀθρόον ἐξέπνει, ὥσπερ ὑπ' ἀψυχίης, ἢ ὡς ἂν διὰ
366 πνίγους πορευθεὶς ἐν σκιῇ καθεζόμενος ὥς τις⁵ | ἀνα-
πνεύσειε.

Τῇ δὲ οὖν ἑπτακαιδεκάτῃ ἑσπέρης, ἀνακαθεζόμενος
ἐς δίφρον ἠψύχησε, καὶ ἄφωνος πολὺν χρόνον καὶ
ἀναίσθητος ἔκειτο· μελικρήτου μόγις κατεδέξατο ξυν-
τείνων τὰς ἐν τῷ τραχήλῳ ἶνας ὡς κατεξηρασμένης
τῆς φάρυγγος καὶ τῆς πάσης ἀδυναμίης παρεούσης·
μόγις δ' οὖν ἐντὸς ἑωυτοῦ ἐγένετο, καὶ αἱ θέρμαι
ἠπιώτεραι μετὰ ταῦτα· ἐπαύσατο δευτέρῃ καὶ εἰκοστῇ.

2. Πυθοδώρῳ περὶ τὸν αὐτὸν χρόνον πυρετὸς ξυνεχής. ὀγδοαίῳ ἱδρὼς ἐγένετο, καὶ πάλιν ἐπεθέρμηνε.
δεκάτῃ πάλιν ἱδρώς. δωδεκάτῃ ἐρρύφησε χυλοῦ· καὶ
μέχρι τεσσαρεσκαιδεκάτης ἄδηλος ἦν, ἐν κροτάφοισι
δὲ ἐνῆν· ἄδιψος δέ, καὶ αὐτὸς ἑωυτῷ ὑγιὴς ἐδόκει εἶναι·
ἱδρῶτες ἐγίνοντο ἑκάστης ἡμέρης. πεντεκαιδεκάτῃ ζω-
μίον νεοσσοῦ ῥυφῶν ἤμεσε χολήν, καὶ κοιλίη κάτω
ἐξεταράχθη· καὶ ὁ πυρετὸς παρωξύνθη, πάλιν ἔληξεν·
καὶ ἱδρῶτες ἐγένοντο πολλοί, καὶ τὸ σῶμα πᾶν περι-
εψυγμένον⁶ πλὴν κροτάφων· σφυγμὸς οὐκ ἔλιπεν,⁷
ἐδόκει δὲ κοπιάσαι⁸ ὀλίγον χρόνον ὡς δόξαι ἐπιθερ-
μαίνειν. τετάρτῃ καὶ εἰκοστῇ, ἤδη ἀπογευομένῳ σιτί-

⁵ ὥς τις] ὅστις M   ⁶ περιψυγμένων M

and also in a fever.)[1] And about the sixteenth day the fever seemed milder. There were bowel movements of pure bile, and his alertness increased. Breathing moderate; sometimes he would take a large breath and exhale again in great volume, like one in a faint or as one who has walked in stifling heat would sit in the shade and catch his breath.

Then on the seventeenth, in the evening, as he was getting up to sit on a chair, he fainted. He lay speechless for a long time, perceiving nothing. He took melicrat with difficulty, straining the cords in his neck, as though his throat was dry and he was quite powerless. He came to himself with difficulty. After that the fevers became milder. His illness ceased on the twenty-second day.[2]

2. At about the same time Pythodorus had a continuous fever. On the eighth day there was sweat, and again he grew hot. On the tenth, sweat again. On the twelfth he had barley broth. Till the fourteenth day it was not apparent, but was there at the temples. But he was not thirsty, and he seemed to himself to be healthy. There were sweats every day. On the fifteenth day he drank broth made from a nestling bird, and vomited bile and was disturbed in the lower intestines. The fever intensified and again abated. There was much sweat, and his whole body was chilled, save the temples. The pulse did not abate, but seemed to labor for a short time so that it seemed that he would become fevered. On the twenty-fourth day when he had already been eating

---

[1] This comparison with the case of Cleocydes seems intrusive in the sequence of the case. One might suspect that it is a note made by a later professional reader.   [2] Cf. *Epid.* 5.73.

ων συχνὰς ἡμέρας, καὶ ἀριστήσαντι πολὺς ἦν ὁ πυρετός, καὶ πρὸς τὴν ἑσπέρην· παραλήρησις προσιόντι
ἅμα τῷ ὕπνῳ· εἶχε δ' οὖν ἤδη ξυνεχὴς καὶ ἰσχυρός·
καὶ ὁτὲ μὲν μίαν, ὁτὲ δὲ δύο νύκτας ἄγρυπνος, τὸν δὲ
λοιπὸν ἅπαντα χρόνον κατακορὴς ὕπνος, ἐγείρειν⁹
ἔργον, καὶ παραλήρησις ἐν τῷ ὕπνῳ, καὶ εἴ ποτε ἐξ
ὕπνου ἐγερθείη, μόγις ἐντὸς ἑωυτοῦ· ἄδιψος· καὶ τὸ
πνεῦμα μέτριον, τοιοῦτον δὲ ἐνίοτε οἷον Πολυκράτει·
γλῶσσα οὐκ ἄχρως. μετὰ δὲ ἑβδόμην τῆς ὑποστροφῆς χυλοὶ προσεφέροντο, μετὰ δὲ τεσσαρεσκαιδεκάτην σιτίον· περὶ τὰς πρώτας ἑπτά, ἐρυγμοί· καὶ ἀπή-
368 μει ἔστιν ὅτε μετὰ τοῦ | ποτοῦ ὑπόχολον ἄνευ ἄσης
ἕως κάτω ἐλύθη ἡ κοιλίη. ἱδρῶτες ἐξ οὗ ὑπέστρεψεν
ἐξέλιπον, εἰ μή τις ὅσον ὡς σημεῖον περὶ μέτωπον·
γλῶσσα ἐκ τοῦ ὕπνου, εἰ μὴ διακλύσαιτο, ὑπότραυλος
ὑπὸ ξηρότητος, καὶ ῥήγματα περὶ αὐτὴν ἑλκέων, καὶ
ἐν χείλει τῷ κάτω, καὶ παρ' ὀδόντας· σμικραὶ ὑποχω-
ρήσιες· περὶ πεντεκαιδεκάτην ὡς ὑπέστρεψε πυκνότε-
ραι καὶ γλοιώδεις· τὸ ἀπὸ τῶν σιδίων ἔπαυσεν αὐτάς·
οὖρα, οἷα τὰ πολυχρόνια. ὑπὸ δὲ τὸν τελευταῖον χρόνον, ἄλγημα στήθεος καταπίνοντι τὸ ποτόν, καὶ ἡ
χεὶρ ἐπὶ τῷ στήθει, κύμινον καὶ ὠὸν ῥυφαίνοντι¹⁰
κατέστη· ἡ γλῶσσα δέ, τὸ μαννῶδες ξυνήνεγκεν.
πεντηκοστῇ ἀπὸ τῆς πρώτης, περὶ ἀρκτοῦρον, ἱδρώτιον κατ' ὀσφῦν καὶ στήθεα βραχύ· καὶ τὸ σῶμα
περιεψύχετο¹¹ πλὴν κροτάφων ὀλίγον χρόνον πάνυ.
πρώτη καὶ πεντηκοστῇ ἐχάλασε, καὶ δευτέρη οὐκ ἔτι
ἔλαβεν.

food several days, he had a high fever after his afternoon meal and into the evening. Delirious talk as he was going to sleep. Then the fever persisted, strong and continuous. He was sleepless sometimes one night, sometimes two, but all the rest of the time he was a glutton for sleep. It was great effort to wake him, he babbled in his sleep, and if ever he was roused he had trouble coming to himself. No thirst. Breathing measured, and like that of Polycrates at times. Tongue not pale. After the seventh day of relapse he was given broth, and after the fourteenth, food. About the first seven days, belching, and at times, after drinking, he vomited bilious material without nausea, up until the time when the lower bowel was opened. After the time of relapse, the sweats left him, unless some appeared on the forehead, like a sign. After sleep his tongue lisped from the dryness unless his mouth was rinsed, and there were fissures from sores around it, and on the lower lip as well, and around the teeth. Small bowel movements; on the fifteenth day of relapse they were more frequent and gluey; pomegranate extract stopped them. Urine as in protracted illnesses. Towards the end he had chest pain when he swallowed liquid, or even had a hand on his chest. When he drank cumin and egg in broth the pain subsided. His tongue: manna helped it. On the fiftieth day after the first, at the rising of Arcturus, brief sweating around the loins and chest. The body, except the temples, was chilled intensely for a short time. On the fifty-first day it remitted. On the fifty-second he was free from fever.

---

9 Foës: ἐγείρει mss. Asul.
10 ῥυμφαίνοντι V: ῥυμφάνοντι M: corr. Foës
11 περιέψυχε V

3. Ὁ Ἐρατολάου περὶ φθινοπωρινὴν ἰσημερίην δυσεντερικὸς ἐγένετο, καὶ πυρετὸς εἶχε. τὰ ὑποχωρήματα ἦν χολώδεα, λεπτά, πολλά, καὶ ὕφαιμα μετρίως, ἡ δὲ ὀδύνη τῆς γαστρὸς σφοδρή. ὀροποτήσαντι δὲ καὶ γάλα πεπυρωμένον πιόντι, μετριώτερα ἐγένετο τὰ ἀλγήματα καὶ τὰ ὕφαιμα· τὰ δὲ χολώδεα παρηκολούθει, καὶ ἀναστάσιες πυκναί, ἀπονώτεραι. τὸ πυρέτιον ἐδόκει ἀρρωστέοντι καὶ τοῖσι πολλοῖσιν δὲ ὅλως οὐκ ἔχειν διὰ παντὸς τοῦ χρόνου μετὰ τὰς πρώτας πέντε ἢ ἓξ ἡμέρας, οὕτως ἄδηλος ἦν· κατὰ κροτάφους δὲ σφυγμὸς ἦν, καὶ γλῶσσα ὑπὸ ξηρότητος ὑπότραυλος, καὶ διψώδης μετρίως καὶ ἄγρυπνος· ῥυφήμασι δὲ ἤδη ἐχρῆτο καὶ οἴνοισιν. περὶ τεσσαρεσκαίδεκα ἐόντος ἡμέρας[12] ἤδη, ἐγένετο τὰ παρ᾽ οὖς ἐν μέρει παρ᾽ ἑκάτερον σκληρά, καὶ ἄπεπτα πάντα ἐμωλύνθη, ὀδυνώδεα[13] δὲ μετρίως. οὐκ ἀποληγόντων δὲ τῶν ὑποχωρημάτων, καταχόλων τε διὰ παντὸς ἐόντων, τὴν ἐν τῷ ἀλεύρῳ βοτάνην ῥυφέοντι μετριώτερα τὰ χολώδεα καὶ τὰ ἀλγήματα ἐγένετο μέχρι τινὸς χρόνου, ὑγρὰ δὲ πολλὰ καὶ[14] πολλάκις· καὶ ἀπόσιτος σφόδρα, καὶ μετὰ πάσης ἀνάγκης προσδεχόμενος· τὰ δὲ τῆς θέρμης καὶ γλώσσης καὶ δίψης τοιαῦτα παρηκολούθει, οἷα εἴρηται· καὶ ἱδρῶτες, οὐδέν. λήθη δέ τις τοιαύτη· ἐρωτήσας ὅτι πύθοιτο, σμικρὸν καὶ διαλιπών, πάλιν εἰρώτα,[15] καὶ ἔλεγεν αὖτις ὡς οὐκ εἴη εἰρηκώς· καθεζόμενός τε ἐπελανθάνετο, εἰ μή τις ὑπομιμνήσκοι[16] αὐτόν· καὶ αὐτὸς ἑωυτῷ ξυνῄδει τὸ πάθος, οὐδ᾽ ἠγνόει. πνεῦμα, ὁποῖον ὑγιαίνοντι. ὑπὲρ δὲ τὰς

3. Near the time of the fall equinox, the son of Erato-
laus got dysentery, and fever came on him: his bowel movements were bilious, thin, copious, moderately bloody; violent pain in the belly. When he drank whey and boiled milk
his pains and the bloodiness moderated, but the bilious excrement persisted and he had to get up often, but it was
less painful. The feverishness seemed to the patient and to
people in general not to be there at all in the whole time after the first five or six days, so obscure was it, but there was
pulsation at the temples. His tongue lisped from dryness,
and he was moderately thirsty and wakeful. But he was already taking broth and wine. When he was already at the
fourteenth day the areas by his ears grew hard, one after
the other; without coction they softened again entirely, but
were moderately painful. The bowel movements did not
cease, and were bilious throughout; when he took the herb
in gruel the biliousness and pains moderated for a time,
but movements remained large, frequent, and watery. He
had an intense aversion to food, and had to be forced to eat
anything. Symptoms of fever, of the tongue, and of thirst
persisted such as I have described. No sweats. Loss of
memory of this sort: he would ask something he wanted
to know, subside awhile and ask it again, and repeat it
as though he had not spoken. He would forget that he was
sitting at stool unless someone were to remind him of
it. He himself noticed the ailment. He was not unaware.
Breathing like that of a healthy person. From the thirtieth

---

12 ἡμέρας ἐόντος V   13 ὀδυνώδεες V
14 om. V   15 Smith: ἐρωτᾷ MV: ἠρώτα recc.
16 HIR: ὑπομιμνήσκει MV

τριήκοντα μέχρι τῶν τεσσαράκοντα ἥ τε ὀδύνη πολὺ
ἐπεδίδου τῆς γαστρός· ὕπτιός τε κατέκειτο, καὶ ἐπι-
στρέφεσθαί τε οὐδ' ἐπὶ ποσὸν ἐδύνατο, τὸ δὲ μὴ τὸ
ἄλγημα δεινόν· ψωμίζειν ἄλλον δεῖ. ἡ δὲ ὑποχώρησις
πολλή, καὶ διακεκριμένη, λεπτή, καὶ χρώματα οἰνωπά,
ὅσαπερ προσδέχοιτο, καὶ ὕφαιμα ἔστιν ὅτε· καὶ ὁ
τόνος τοῦ σώματος, ἔκτηξις ἐσχάτη[17] καὶ ἀδυναμίη,
οὐδ' ἀνίστασθαι ἄλλου ἐπαίροντος ἔτι δυνατὸς ἦν.
μέσον δὲ ὀμφαλοῦ καὶ χόνδρου[18] κατὰ ταύτην τὴν
καταγωγήν, ἁπτομένῳ τῇ χειρὶ τοιοῦτος παλμὸς ἦν
οἷος οὐδὲ ὑπὸ δρόμου οὔτε ὑπὸ δείματος[19] περὶ καρδίην ἂν γενηθείη. πίνοντι ἄπεφθον ὡς[20] ἐννέα ἀττικὰς[21] ὄνειον[22] ἐπὶ δύο ἡμέρας, χολώδης ἐγένετο σφό-
δρη κάθαρσις, καὶ ἔληξαν αἱ ὀδύναι, καὶ σιτίων
372 ἐπιθυμίη ἐγένετο. μετὰ δὲ | ταῦτα, βοείου γάλακτος ὡς
τέσσαρας κοτύλας ἀττικὰς ὁμοῦ ἔπινεν, κατὰ δύο
κυάθους δι' ἡμέρης, τὸ πρῶτον ὕδατος τὸ ἕκτον
μίσγων καὶ μέλανος οἴνου καὶ αὐστηροῦ σμικρόν.
ἐμονοσίτει δὲ ἑσπέρην· ἄρτος ὡς ἡμιχοίνικος ἐγκρυ-
φίης καὶ ἰχθύδιον πετραῖον ἁπλοῦν, ἢ κρεάδιον αἰγὸς
ἢ προβάτου· ἡ δὲ πόσις τοῦ γάλακτος, ἐπὶ τεσσαράκοντα ἡμέρας ἄνευ ὕδατος μετὰ τὰς πρώτας δέκα
ἡμέρας, οἴνου δὲ ὡς σμικρὸν μέλανος μέρος. ἱδρώτιον
δὲ ἐγένετο μετὰ τὰς ἑβδομήκοντα ἤδη ἀπὸ τῆς πρώτης
μετὰ λουτρὸν ἐς νύκτα. ὀλιγοποσίη δὲ ἐχρῆτο, καὶ
μετὰ τὸ σιτίον τῷ ποτῷ αὐστηροτέρῳ, τοῖσιν ἄλλοισιν οὔ.[23]

to the fortieth day the belly pain increased considerably. He lay on his back and could not turn even a little way without the pain being dreadful. Someone else had to feed him. There was much fecal matter, thin and separated, its colors wine dark, the same quantity as what he ate, and sometimes bloody. The tone of his body: extreme emaciation and weakness; he could no longer stand when someone supported him. And to the touch in the area between the navel and the bottom of the breastbone there was such palpitation as occurs not even at the heart from running or from fright. When he drank boiled ass's milk to the amount of nine Attic measures in two days, he had a violent bilious purge, the pains ceased, and he got an appetite for food. After this he drank about four Attic cotyls of cow's milk all told, at the rate of two cyathi daily.[3] First it was mixed with a sixth part of water and a little red astringent wine. He ate one meal a day, in the evening, bread up to a half-choenix,[4] ash-baked, with plain rockfish, or goat's or sheep's flesh. He drank the milk, after the first ten days and up to the fortieth, without water, and with a small portion of red wine. Sweating occurred after the seventieth day from the beginning, after the bath towards night. He took little wine; after food a rather astringent one, and no others.

[3] Four Attic cotyls is ca. 1 quart. Two cyathi are ca. one-fourth pint.   [4] Ca. two cups.

---

[17] om. V   [18] χοννάρου M: χωνάρου V: corr. Foës
[19] δειγματος M: δίγματος V: corr. I
[20] HIR: ἀπεφρόνως MV   [21] ἀττικὰς κοτύλας HIR
[22] HIR: ὤνιον V: ὀνιον M
[23] ἢ αὐστηροτέρου τῆς ἁλωσίμου MV: corr. Li.

ΕΠΙΔΗΜΙΑΙ

4. Κτησικράτει δὲ τὸ ἐν τῷ ἀλεύρῳ μᾶλλον τοῦ αἰγείου[24] ὀροῦ ξυνήνεγκεν, ὀδύνης ἐούσης περὶ ὅλην τὴν κοιλίην καὶ πόνων καὶ ἀναστάσιος πολλῆς καὶ ὑφαίμου, καὶ ἐπάρματος περὶ τοὺς πόδας σχεδὸν ἤδη περὶ πέντε καὶ εἴκοσιν ἡμέρας ἐόντι. καὶ Ἀγριάνῳ[25] ταῦτά·[26] τῷ δὲ Καινίου τὸ ὄνειον ἐφθόν.

5. Τῷ Κύδιος περὶ χειμερινὰς ἡλίου τροπὰς ῥῖγος καὶ πυρετός, καὶ ὠτὸς δεξιοῦ ἄλγημα, καὶ κεφαλῆς ὀδύνη· τὸ δὲ τοιοῦτον[27] ἄλγημα εὐθὺς ἐκ σμικροῦ παιδίου παρηκολούθει ῥευματῶδες καὶ συριγγῶδες καὶ ἔνοδμον, ἔχον δὲ οὕτω τὰ πολλὰ ἀνώδυνον ἦν· τότε δὲ ἡ ὀδύνη ἦν δεινὴ καὶ ἡ κεφαλαλγίη. δευτεραίῳ ἢ τριταίῳ ἐόντι χολῆς ἔμετος· ἀνακαθιζομένῳ ἐγένετο ὑπόχολον, γλίσχρον, ὡς ἐξ ᾠοῦ, ὕπωχρον. τετάρτῃ ἐς νύκτα καὶ πέμπτῃ ὑποπαρελήρει· καὶ ἡ ὀδύνη τῆς κεφαλῆς καὶ τοῦ ὠτὸς δεινή, καὶ ὁ πυρετός. ἕκτῃ ὑποχώρησις[28] ἀπὸ λινοζώστιος, καὶ ἡ θέρμη λῆξαι ἐδόκει καὶ ἡ ὀδύνη. τῇ ἑβδόμῃ ὡς ὑγιής· | κροτάφῳ δὲ σφυγμὸς οὐκ ἔλιπεν· ἱδρῶτος οὐδὲν ἐγένετο. ὀγδόῃ χυλοῦ ἐρρύφησεν, ἐς δὲ τὴν ἑσπέρην σεύτλου· καὶ τὴν νύκτα ὕπνος καὶ σφόδρα ἀνώδυνος, καὶ τὴν ἐνάτην ἔς τε τὸ πρὸς ἡλίου δυσμάς· ἐς δὲ νύκτα, τῆς κεφαλῆς δεινὴ ἡ ὀδύνη καὶ τοῦ ὠτός· ξυνέβαινε δὲ καὶ πυορροεῖν τὸ οὖς περὶ τοῦτον τὸν καιρὸν ὁπότε μάλιστα πονοίη, εὐθὺς ἀπ' ἀρχῆς. ὅλην δὲ τὴν νύκτα τὴν ἐνάτην καὶ τὴν ἐπιοῦσαν ἡμέρην καὶ τῆς νυκτὸς τὸ πλεῖον οὐκ ἐπεγίνωσκεν οὐδένα, στένων δὲ διετέλει· ἡμέρην δὲ ἐντὸς ἑωυτοῦ ἐγένετο, καὶ αἱ ὀδύναι ἔληξαν

374

286

4. But Ctesicrates was helped more by milk in gruel than by goat's whey when he had pain all through the intestines, distress, many bowel movements and bloody ones, and swelling about the feet beginning around the twenty-fifth day. Agrianus also. But Caenias' son was helped by boiled ass's milk.

5. Cydis' son, about the winter solstice, had shivering and fever, distress in the right ear, and headache. But that kind of distress had persisted from babyhood, with draining, a fistular sore and a bad smell, but still was not for the most part painful. But at that time the pain and the headache were terrible. On the second or third day he vomited bile. When he sat at stool it was bilious, gluey, egglike, pale yellow. Towards night on the fourth day and the fifth, some delirious talk. The pain in the head and the ear was dreadful, and the fever. On the sixth, bowel movement from *linozostis*; the heat and the pain appeared to abate. On the seventh he seemed well, but the throbbing in the temples did not abate. No sweating occurred. On the eighth day he had barley broth, and beet broth towards evening; sleep at night, totally without pain, and on the ninth, until the period towards sundown. But towards night terrible pain in the head and the ear. Flows of pus from the ear coincided with his times of greatest discomfort, right from the start. The whole ninth night, and the next day, and most of the night, he did not recognize anyone, and groaned continuously. During the day he came to himself and the pains

---

καὶ τὰ τῆς θέρμης μετριώτερα. ῥυφήσαντι δὲ λινόζω-
στιν τῇ ἑνδεκάτῃ ὑπεχώρησε φλεγματώδεα, μυξώδεα,
κάκοδμα. δυοκαιδεκάτῃ καὶ τρισκαιδεκάτῃ, μετρίως.[29]
τεσσαρεσκαιδεκάτῃ, ἀρξάμενος ἀφ᾽ ἑωθινοῦ μέχρι ἐς
μέσον ἡμέρης ἵδρου ὅλον τὸ σῶμα, μετὰ ὕπνου καὶ
κώματος πολλοῦ· ἐγεῖραι ἔργον ἦν· πρὸς ἑσπέρην δὲ
διηγέρθη, καὶ τὸ μὲν σῶμα μετρίως κατεψύχθη, ἐν
κροτάφοισι δὲ σφυγμὸς διετέλει. πεντεκαιδεκάτῃ καὶ
ἑκκαιδεκάτῃ χυλοῖσιν ἐχρήσατο. ἑπτακαιδεκάτῃ ἧκε
πάλιν ἐς νύκτα ὀδύνη τῶν αὐτῶν, καὶ παραλήρησις,
καὶ ἐπυρρόει. ὀκτωκαιδεκάτῃ, ἐννεακαιδεκάτῃ καὶ εἰκοστῇ μανικῶς· ἦν δὲ κεκραγὼς ἐπαίρειν ἑωυτὸν εἰ
πειρώμενος, οὐ δυνάμενος δὲ κρατεῖν τῆς κεφαλῆς,
τῇσι χερσὶν ἐπορεγόμενος καὶ αἰεί τι διακενῆς θη-
ρεύων. πρώτῃ καὶ εἰκοστῇ ἱδρώτιον περὶ πλευρὸν δεξιὸν καὶ στήθεα καὶ κεφαλήν, δευτέρῃ καὶ εἰκοστῇ
περὶ τὸ πρόσωπον πλεῖστον ἦν· τῇ δὲ φωνῇ κατὰ τὸν
χρόνον τοῦτον, εἰ μὲν σφόδρα ἀπεβιάσατο,[30] εἶπεν ἂν
τελέως ἃ ἠβούλετο, εἰ δὲ προχείρως, ἡμιτελέα· καὶ
στόμα λελυμένον, καὶ αἱ γένυες καὶ χείλεα αἰεὶ ἐν
κινήσει, ὥς τι θέλοντος λέγειν· καὶ τῶν ὀφθαλμῶν
376 πυκνὴ κίνησις καὶ ἔμβλεψις, καὶ χρῶμα | ἐπ᾽ ὀφθαλ-
μοῦ δεξιοῦ, οἷον εἴρηται τὸ ὕφαιμον, καὶ βλέφαρον τὸ
ἐπάνω ἐπῴδησε, καὶ κατὰ γνάθον ἔρευθος ἐπὶ τελευτῆς, καὶ φλέβες πᾶσαι ἐν τῷ προσώπῳ φανεραί, οὕτω
ξυντετρυμέναι·[31] καὶ τοῖσιν ὀφθαλμοῖσιν οὐκ ἔτι ξυμ-
μύων, ἀλλ᾽ ἀτενὲς ἐνορῶν, καὶ διαίρων τὰ βλέφαρα ἐς

slackened and signs of fever were more moderate. When he took *linozostis* on the eleventh day, he passed bilious movements with mucus, foul-smelling. On the twelfth and the thirteenth day, moderately better. On the fourteenth, beginning at dawn and up to midday, sweat over the whole body, with sleep and much coma. It was hard to rouse him. But he wakened towards evening and his body cooled off moderately, but the throbbing in the temples persisted. On the fifteenth and sixteenth day he had barley broth. And on the seventeenth pain in the same places came on again towards night, and delirious talk and the flow of pus. On the eighteenth, nineteenth and twentieth, delirium. He cried out if he tried to raise himself, and could not control his head; he reached out with his hands continually, looking for something in the empty air. On the twenty-first day, sweating about the right side of the ribs and the chest and head. On the twenty-second there was much sweat about the face. With his voice in this period, if he was very forceful he succeeded in saying what he wished, but if he was casual, it was imperfect. His mouth was slack, the jaws and lips always moving as though he wanted to say something; and there was was much moving of the eyes and glancing about, and color at the right eye such as we call bloodshot; and the upper eyelid swelled, and there was redness in the cheek as he was approaching the end, and all the blood vessels in his forehead were obvious, so great was their distress. He no longer closed his eyes, but he had a fixed stare, raising his eyelids upward as when some-

---

29 η μετρίως M (the descendants of M omit η)

30 ἀπεβιάσαιτο M: corr. Lind.

31 Smith: ξυντετραμμέναι M

τὸ ἄνω μέρος, ὡς ἐπήν τι ἐμπέσῃ ἐς τὸ ὄμμα· καὶ ὁπότε πίοι, κατιόντος ἐς τὰ στήθεα καὶ τὴν κοιλίην ψόφος,
οἷος καὶ Χαρτάδει. πνεῦμα δὲ ἐπιεικῶς διὰ παντὸς μέ-
τριον· γλῶσσα οἵη ἐστὶ τοῖσι περιπλευμονικοῖσιν,
ὠχρόλευκος· ἀπ᾽ ἀρχῆς καὶ διὰ παντὸς κεφαλαλγίη·
τράχηλος διὰ παντὸς ἀκίνητος· ξυμπεριάγειν τῇ κεφαλῇ ἐδεῖτο· καὶ τὸ κατὰ ῥάχιν ἐκ τραχήλου ἰθὺ καὶ
ἄκαμπτον· καὶ κλίσιες,[32] ὁποῖαι εἴρηνται, καὶ οὐκ αἰεὶ
ὕπτιος· τὸ δὲ πύον εἴδει ὀρῶδες, λευκόν, πολύ, ἔργον
ἀποσπογγίσαι, ὀδμὴ ὑπερβεβλημένη· κατὰ τὸν τελευταῖον χρόνον ποδῶν ἄψιος οὐ πάνυ καταισθανό-
μενος.

6. Τῇ Ἁρπαλίδεω ἀδελφῇ περὶ τέταρτον μῆνα ἢ
πέμπτον κυεύσῃ, οἰδήματα περὶ τοὺς πόδας ὑδατώδεα[33] ἃ ἐγένετο, καὶ τὰ κύκλα τῶν ὀφθαλμῶν ἐπῴδει,
καὶ ἅπαν τὸ χρῶμα μετέωρον, οἷον τοῖσι φλεγματί-
ῃσιν· βὴξ ξηρή· ὀρθοπνοίη δὲ καὶ ἆσθμα τοιοῦτον καὶ
πνιγμοὶ ἔστιν ὅτε ὑπὸ τοῦ πνεύματος, ὥστε καθημένη
διετέλει, κατακλῖναι[34] δὲ οὐχ οἵη τε[35] ἦν, ἀλλ᾽ εἴ τις
καὶ ὕπνου δόξα γένοιτο, καθημένη ἦν· ἄπειρος δὲ
ἐπιεικῶς· καὶ τὸ κύημα ἐπὶ πολὺν[36] χρόνον ἀκίνητον
ἦν, ὡς διεφθαρμένον. καὶ μετέπιπτεν, παρηκολούθησε
δὲ[37] τὸ ἆσθμα σχεδὸν δύο μῆνας. κυάμοισι δὲ χρω-
μένη μελι|χροῖσι καὶ μέλιτος ἐκλείξει, καὶ τοῦ αἰθιο-
πικοῦ κυμίνου πιοῦσα ἐν οἴνῳ, ἐρρήισεν.[38] μετὰ ταῦτα
ἀνῆγε βήσσουσα πολλά, πέπονα, φλεγματώδεα, λευ-

378

---

[32] κλισίαις M: corr. recc.   [33] οἰδατώδεα M: corr. recc.

thing falls into the eye. And when he drank there was a gurgling as it went into the chest and the intestine, like that in
the case of Chartades. The breathing was properly moderate throughout. The tongue yellowish-white like that of
those with pneumonia. The headache was there throughout from the beginning, the neck immovable throughout.
He had to turn it about with his head. The spine at the base
of the neck was straight and unbending. He lay as described, not always on his back. The pus was serous to the
view, white, copious, hard to sponge up, its smell overwhelming. In the last period he could hardly perceive a
touch on his feet.

6. Harpalides' sister, pregnant four or five months, had
swellings about the feet which became watery. The circles
of her eyes swelled, and her whole skin was puffed as it is
in those with anasarca. Dry cough. Orthopnea, and such
breathing trouble and choking from time to time that she
stayed sitting in her bed. She could not lie prone, but if she
ever decided to sleep, it was while she was sitting up. She
was virtually without sensation, and the fetus was immo-
bile for a long time, as though it had died. She underwent a
change, and the breathing trouble persisted for almost two
months, but she took beans with honey and a honey linctus
and drank Ethiopian cumin in wine, and it grew easier.
Following that she coughed and brought up much con-

---

34 Smith: κατὰ κλίνην αἱ M

35 οἷοί τε M: corr. Corn.

36 Smith: πουλὺ M: πλεῖστον recc.

37 recc.: τε M

38 ἐρηϊς (sic) M: ἐρ(ρ)άϊσε recc.: corr. Lind.

291

κά, καὶ τὸ πνεῦμα ἔληξεν· ἔτεκε γόνον θῆλυν.[39]

7. Τῇ Πολυκράτεος, θέρεος περὶ τὸ ἄστρον, πυρε-
τός· πνεῦμα τὸ ἑωθινὸν ἧσσον, ἀπὸ μέσου ἡμέρης
μᾶλλον τὸ πνεῦμα, πυκνότερον σμικρῷ· βὴξ καὶ ἀπόχρεμψις εὐθὺς ἀπ᾽ ἀρχῆς ὁμοίη πυώδεσιν· ἔσω περὶ
ἀρτηρίην καὶ φάρυγγα ὑπεσύριζε κερχναλέον· πρόσωπον εὔχροον ἐπὶ γνάθοισιν ἐρύθημα, οὐ κατακορές,
ἀλλ᾽ ἐπιεικῶς ἀνθηρόν. προϊόντος δὲ τοῦ χρόνου, καὶ
ἡ φωνὴ βραγχώδης, καὶ τοῦ σώματος ξύντηξις, καὶ
περὶ ὀσφῦν ἐκρήγματα, καὶ ἡ κοιλίη κατὰ τὸν τελευταῖον χρόνον ὑγροτέρη. ἑβδομηκοστῇ ὁ πυρετὸς ἔξωθεν σφόδρα περιέψυχε, καὶ ἐν κροτάφοισιν ἡσυχίη· τὸ
δὲ πνεῦμα πυκνότερον ἐγίνετο. μετὰ δὲ τὴν παῦσιν
ταύτην τὸ πνεῦμα πυκνότερον οὕτως ὥστε καθημένη
διετέλεσεν. ὥστε ἀπέθανεν.

Ἐν δὲ τῇ ἀρτηρίῃ ψόφος πολὺς ἐνῆν, καὶ ἱδρῶτες
πονηροί, καὶ ἐμβλέψιες ἐμφρονώδεις σφόδρα μέχρι
τοῦ ἐσχάτου χρόνου. ἡμέραι δ᾽ ἐπεγένοντο ἤδη κατ-
εψυγμένῃ πλείους ἢ πέντε· μετὰ δὲ τὰς πρώτας διετέλει πυῶδες ἀποχρεμπτομένη.

8. Τῇ ἄνω τῶν πυλέων οἰκεούσῃ, ἤδη πρεσβυτέρῃ,
πυρέτιον ἐγένετο, καί, ἤδη ἀπολήγοντος, τοῦ τραχήλου ἄλγημα μέχρις ἐς ῥάχιν καὶ ὀσφῦν· καὶ οὐ πάνυ
ἐγκρατὴς τούτων· γέννες δὲ ξυνηγμέναι καὶ ἑωυτοὺς
ὀδόντας πλέον ἢ μήλην παρεῖναι οὐκ ἦν· ἥ τε φωνὴ
ψελλὴ διὰ τὸ παραλελυμένον καὶ ἀκίνητον καὶ ἀσθε-
νὲς εἶναι τὸ σῶμα· ἔμφρων δέ. χλιάσμασι καὶ μελικρήτῳ χλιερῷ ἐχάλασε τριταίη σχεδόν, καὶ μετὰ

380

292

cocted, phlegmatic white matter, and the breathing improved. She bore a female child.

7. Polycrates' wife, in summer near the Dogstar, had fever. Her breathing in the morning was less; after the mid-part of the day, her breathing was greater, slightly more rapid. A productive cough right from the beginning, as in people with purulence. Inside, by the pharynx and trachea, a hoarse whistling. Face of good color, redness on the cheeks, not intense but properly fresh. As time passed, the voice hoarse, and wasting of the body, eruptions on the lower back, bowels loose towards the final period. On the seventieth day the external fever much cooler, and quietness at the temples. But the breathing grew more rapid, and after this respite the breathing was so rapid that she had to remain sitting. And so she died.

In her trachea there was much noise, and there were unfavorable sweats. And very alert looks until the final moments. She passed more than five days without fever. She kept coughing up pus after the first days.

8. The woman who lived over the gate, already rather old, got a fever and, after it had abated, a pain in the spine from the neck to the lower back. She had not much strength there. Her jaws were clenched and the teeth could not relax themselves more than the width of a probe. Her speech was unintelligible because her body was paralyzed, immovable and weak. But she retained her reason. With fomentations and warm hydromel she relaxed on about the third day and afterwards became healthy with

---

39 γόνυ θῆλυ M: corr. R

ταῦτα χυλοῖσι καὶ ζωμοῖσιν ὑγιὴς ἐγένετο. ξυνέβη δὲ τελευτῶντος τοῦ μετοπωρινοῦ.

9. Ὁ παρὰ Ἁρπαλίδῃ ἀλείπτης,[40] ἀκρατέστερος σκελέων καὶ χειρῶν περὶ φθινόπωρον γενόμενος, ἔπιεν ἐκεῖ φάρμακον ἄνω καὶ κάτω· ἐκ δὲ τῆς καθάρσιος πυρετός· καὶ ἐς τὴν ἀρτηρίην κατερρύη τοιοῦτον οἷον ἐπισχεῖν διαλεγόμενον καὶ ἀσθμαίνειν ἐν τῷ διαλέγεσθαι ὁμοίως κυναγχικῷ βραγχώδει· πνιγμὸς καταπίνοντι, καὶ ἄλλα κυναγχικά, οἴδημα δὲ οὐκ ἦν. ὁ δὲ πυρετὸς ἐπέτεινε, καὶ ἡ βήξ, καὶ ἀπόχρεμψις ὑγροῦ καὶ πολλοῦ φλέγματος. προϊόντος δὲ καὶ ὀδύνη κατὰ στῆθος καὶ μαζὸν ἀριστερόν. ὁπότε δὲ ἐξαναισταίη ἢ μετακινηθείη ἆσθμα πολὺ καὶ ἱδρὼς ἀπὸ μετώπου καὶ κεφαλῆς· καὶ τὰ περὶ τὴν φάρυγγα κατεῖχε, μαλακώτερον δέ, ἐς τὸ στῆθος τῆς ὀδύνης ἀπελθούσης. ἀπ᾽ ἀρχῆς μὲν οὖν κυάμοισι μελιχροῖσι θερμοῖσιν ἐχρῆτο· ἐπεὶ δὲ οἱ πυρετοὶ ἐπεῖχον, μᾶλλον ὀξυμέλιτι θερμῷ καὶ μέλιτος ἐκλίξει[41] πολλῇ. παρελθουσέων δὲ τεσσάρων καὶ δέκα ἡμερέων ἅπαντα ἔληξε, καὶ τῶν περὶ χεῖρας καὶ σκέλεα οὐ πολὺ ὕστερον ἐγκρατὴς ἐγένετο.

10. Χαρτάδει, πυρετὸς καῦσος, ἔμετος χολῆς πολλῆς, καὶ κάτω ὑποχώρησις· ἄγρυπνος· καὶ κατὰ σπλῆνα ἔπαρμα στρογγύλον. ἐνάτῃ πρωῒ ἐξανέστη, ψόφου περὶ τὴν κοιλίην ἄνευ ὀδύνης γενομένου· ὡς ἀφοδεύων, ὑπῆλθεν αἵματος πλέον ἢ χοεὺς προσφάτου, καὶ σμικρὸν ἐπισχόντι, καὶ τρίτον· πεπηγότες θρόμβοι. ἄση δὲ περὶ τὴν καρδίην, καὶ ἱδρώτιον

294

barley gruel and soup. This occurred at the end of autumn.

9. The masseur at the house of Harpalides, having grown rather weak in arms and legs towards autumn, drank at that time a drug that purged upward and downward. After the purge, fever. There flowed into the trachea such material as to stop his speech and cause him to choke when talking, like one whose throat is sore with quinsy. He gagged when he was drinking, and had other symptoms of quinsy, but there was no swelling. The fever grew intense, and the cough, and there was bringing up of much watery phlegm. As time went on there was pain in the chest and left breast. Whenever he stood up or moved he had much breathing difficulty and sweat on the forehead and head. Symptoms in the pharynx persisted, but more gently as the pain moved down to the chest. From the outset he ate warm honeyed beans. But when the fever came on he took instead oxymel, warm, and much honey as a linctus. When fourteen days had passed all symptoms ceased, and shortly later he regained his strength in arms and legs.

10. When Chartades had burning fever he vomited much bile and passed much in stools. He was sleepless. He had a round swelling by the spleen. On the ninth morning early, he got up with noise in his intestines, without pain. But as he sat at stool there came forth more than a choeus[5] of fresh blood, and after he waited a brief time a third of a choeus; and there were blood clots. He had distress in the

---

5 An impossible amount, ca. six pints.

---

40 ἀλήπτης M: corr. R
41 ἕλιξει M: corr. Lind.

σχεδὸν καθ' ὅλον τὸ σῶμα· καὶ τὸ πυρέτιον[42] καταψύ-
χειν[43] ἐδόκει, καὶ ἔμφρων τὸ πρῶτον, προϊούσης δὲ
382 τῆς | ἡμέρης ἥ τε ἄση πλείων,[44] καὶ ἀλυσμός, καὶ
πνεῦμα σμικρῷ πυκνότερον· καὶ θρασύτερον καὶ φιλο-
φρονώτερον τοῦ καιροῦ προσηγόρευε καὶ ἐδεξιοῦτο.
καί τινες λειποψυχίαι ἐδόκεον ἐπιγίνεσθαι· προσφε-
ρόντων δέ τινων χυλοὺς καὶ τὸ ἀπὸ κρίμνων ὕδωρ, οὐκ
ἔληγεν· ἀλλὰ τὸ πνεῦμα πρὸς τὴν ἑσπέρην ὑπέρπολυ
ἦν, καὶ ῥιπτασμὸς πολύς, καὶ ἐπὶ τὰ δεξιὰ καὶ τὰ
ἀριστερὰ μεταρρίπτων ἑωυτὸν οὐδένα χρόνον[45] ἀτρε-
μίζειν δυνατὸς ἦν· πόδες ψυχροί· ἐν κροτάφοισι καὶ
κεφαλῇ θέρμη μᾶλλον ὑπογύου τῆς τελευτῆς· καὶ
ἱδρώτια πονηρά· καὶ πίνοντι τὰ τοῦ ψόφου περί τε
στήθεα καὶ κοιλίην κατιόντος τοῦ πόματος. οἷον δὲ
κάκιστον. φάμενος δὲ θέλειν τι ἑωυτῷ ὑπελθεῖν, καὶ
ἀτενίσας τοῖσιν ὄμμασιν, οὐ πολὺ ἐπισχών, ἐτελεύτησεν.

11. Τῇ Ἑρμοπτολέμου χειμῶνος πυρετὸς καὶ κεφαλῆς ἀλγήματα· καὶ ὁπότε πίοι ὡς χαλεπῶς κατα-
πιοῦσα ἐξανέστη, καὶ τὴν καρδίην οἱ γυιοῦσθαι ἔφη·
γλῶσσα πελιδνὴ ἀπ' ἀρχῆς. ἡ δὲ πρόφασις[46] ἐδόκει
ἐκ φρίκης μετὰ λουτρὸν γενέσθαι. ἄγρυπνος καὶ
νύκτα καὶ ἡμέρην. μετὰ τὰς πρώτας ἡμέρας ἐρωτωμέ-
νη οὐκ ἔτι κεφαλὴν ἀλλ' ὅλον τὸ σῶμα πονεῖν ἔφη.
δίψα ὁτὲ μὲν κατακορής, ὁτὲ δὲ μετρίη. πέμπτη καὶ
ἕκτη καὶ μέχρι τῆς ἐνάτης σχεδὸν παραλήρησις, καὶ
αὖτις πρὸς ἑωυτὴν ἐλάλει μετὰ κώματος ἡμιτελέα· καὶ
τῇ χειρὶ ἔστιν ὅτε ἐπωρέγετο πρὸς τὸ κονίημα καὶ

area of the heart and sweat over virtually the whole body. The fever seemed to cool. At first he was rational, but as the day went on the distress increased, there was delirium and slightly more rapid breathing. He spoke more aggressively and greeted people more warmly than the occasion warranted; he appeared to have lapses of consciousness. When people offered broth and barley water there was no improvement. Towards evening his breathing was very heavy and there was much tossing about. He threw himself from the right side to the left and back again, and was not able to hold still for any time. His feet grew cold, there was heat more at the temples and the head as the end was imminent. Bad sweats. When he drank, signs of noise in the chest and intestines as the drink went down. All signs were bad. He said he wanted something under him, stared fixedly, resisted a brief time, and died.

11. The wife of Hermoptolemus, in the winter, had fever and headache. Whenever she drank she sat upright because of difficulty swallowing. She said that her heart had been damaged. Tongue livid from the outset. The cause seemed to be a chill after a bath. She was sleepless, night and day. After the first days, when asked, she no longer said that her head hurt, but that her whole body hurt. Thirst, sometimes insatiable, sometimes moderate. On the fifth and sixth days and up to the ninth, almost constantly delirious. Later, she babbled to herself half-intelligible things in the midst of coma. She would reach out with her hand from time to time towards the plaster wall, and to the cold

---

⁴² πῦρ αἴτιον M here and often   ⁴³ κατὰ ψυχὴν M: corr. Lind.   ⁴⁴ πλείω M: corr. R   ⁴⁵ οὐδὲν ἄχροον M: corr. R ⁴⁶ πρόφασι M: corr. H   V resumes with ἐδόκει

προσκεφάλαιόν τι ψυχρὸν ἐνεὸν τῇ κεφαλῇ· καὶ τοῖσι
στήθεσι προσεῖχε, καὶ τὸ ἱμάτιον ἔστιν ὅτε ἀπερ-
ρίπτει· καὶ ἐπ' ὀφθαλμοῦ τοῦ δεξιοῦ τὸ ὕφαιμον ἦν,
καὶ δάκρυον ἦν· οὖρον δὲ ὃ⁴⁷ τοῖσι παισὶ πονηρόν
ἐστιν αἰεί. ὑπεχώρει δὲ ἀπ' ἀρχῆς μὲν ὑπόκιρρα,
ὕστερον δὲ ὑδατώδεα σφόδρα καὶ τοιουτόχροα. ἐν-
384 δεκάτη, ἐδόκει μετριωτέρη εἶναι | θερμή,⁴⁸ καὶ ἄδιψος
ἔστιν ὅτε ἐγένετο οὕτως ὡς εἰ μή τις διδοίη⁴⁹ οὐκ
ᾔτει.⁵⁰ ὕπνοι μετὰ τὸν πρῶτον χρόνον ἐγένοντο ἐπιεικῶς ἡμέρης, ἐς δὲ νύκτα ἄγρυπνος, καὶ ἐπόνει μᾶλλον ἐς νύκτα. ἐνάτῃ ἡ γαστὴρ ἐξεταράχθη ὑδατώδεα,
καὶ δεκάτῃ· καὶ τὰς ἐπιούσας ἐπιεικῶς πολλὴ ἡ δια-
νάστασις καὶ τοιαύτη. ἦσαν δὲ ἐν τῇσι πρόσθεν
ἡμέρῃσιν ἀκρηχολίαι καὶ κλαυθμοὶ οἷον παιδαρίου,
καὶ βοή, καὶ δείματα,⁵¹ καὶ περιβλέψιες ὁπότε ἐκ τοῦ
κώματος ἐγείροιτο. τῇ δὲ⁵² τεσσαρεσκαιδεκάτῃ, ἔργον
κατέχειν ἦν ἀναπηδῶσαν καὶ βοῶσαν ἐξαίφνης καὶ
ξυντόνως, ὥσπερ ἂν ἐκ πληγῆς καὶ δεινῆς ὀδύνης καὶ
φόβου, ὡς καταλαβών τις αὐτὴν κατάσχοι χρόνον
ὀλίγον· εἶτα πάλιν ἡσυχίην τε εἶχε κεκωματισμένη,⁵³
καὶ ὑπνώσσουσα διετέλει⁵⁴ οὐχ ὁρῶσα, ἔστι δ' ὅτε
οὐδὲ ἀκούουσα·⁵⁵ μετέβαλλε δὲ ἐς ἀμφότερα θόρυβόν
τε καὶ ἡσυχίην πυκνὰ σχεδὸν ὅλην τὴν ἡμέρην ταύ-
την· ἐς νύκτα δὲ τὴν ἐπομένην ὑπῆλθέ τι ὕφαιμον οἷον
μυξῶδες, καὶ πάλιν οἷον ἰλυῶδες, μετὰ δὲ πρασοειδέα
σφόδρα καὶ μέλανα. τῇ δὲ πεντεκαιδεκάτῃ ὀξεῖς

---

⁴⁷ recc.: om. MV   ⁴⁸ θερμὴ εἶναι M

pack on her head. She would put her hands on her chest
and sometimes throw off the cover. There was a bloodshot
area in her right eye and a tendency to weep. The urine of
the sort that is always bad in children. The stool from the
outset yellowish, later very watery and the same color. On
the eleventh day the heat seemed more moderate, and
she was without thirst sometimes to the extent that if one
did not give something to her she did not ask. Sleep, after
the initial period, was generally in the day, but towards
night she was sleepless and towards night her suffering was
worse. On the ninth day her intestines were upset, passing
watery excrement, and on the tenth. Through the subsequent days the movements were generally frequent and
similar. She had bursts of temper in the previous days, and
childish weeping, crying out, frights, and glancing about,
whenever she roused from coma. On the fourteenth day it
was a task to restrain her as she leaped up and shouted suddenly and intensely, as though from a blow or a dreadful
pain or fright, whenever anyone took hold of her and held
her down briefly. Then again she lapsed into coma and was
silent and persisted in drowsiness, seeing nothing, and
sometimes not hearing. And she alternated often between
uproar and quiet virtually that whole day. Towards the
following night she passed a bloody movement, like mucus, and again muddy, and later very greenish and dark. On
the fifteenth day sharp tossing about. The frights and the

---

49 δ᾿ οἰδείη M   50 ἦν τι MV: corr. Asul.
51 δήματα M   52 τῇ δὲ om. V
53 καὶ κωματισμένη M   54 διετέλοι M
55 οὐδ᾿ ἄκουσα M

ῥιπτασμοί· καὶ οἱ[56] φόβοι καὶ ἡ βοὴ ἐγίνετο ἤπιος,
παρηκολούθει δὲ τὸ ἀγριοῦσθαι καὶ τὸ θυμαίνειν καὶ
κλαίειν εἰ μὴ οἱ ταχέως ὅτι βούλοιτο ῥεχθείη· καὶ
ἐπεγίνωσκεν μὲν πάντας[57] καὶ πάντα ἤδη μετὰ πρώτας εὐθὺς ἡμέρας·[58] καὶ τὸ τοῦ ὀφθαλμοῦ κατέστη· ἡ
δὲ μανίη καὶ τὸ παρακαιρὸν καὶ ἡ βοὴ καὶ ἡ μεταβολὴ[59] ἡ εἰρημένη παρηκολούθει ἐς τὸ κῶμα· ἤκουεν
ἀνωμάλως, τὰ μὲν σφόδρα καὶ εἰ σμικρόν τις[60] λέγοι,
ἐπ᾽ ἐνίων δὲ μέζον ἔδει διαλέγεσθαι· πόδες αἰεὶ ὁμαλῶς τῷ ἄλλῳ σώματι θερμοὶ ἔς τε τὰς τελευταίας
386   ἡμέρας, τῇ δὲ ἑκκαιδε|κάτῃ,[61] ἧσσον. ἑπτακαιδεκάτῃ
μετριώτερον τῶν ἄλλων ἡμερῶν ἔχουσα, ἐς νύκτα ὡς
φρίκης αὐτῇ γενομένης ξυνάγουσα ἐπυρέτηνε μᾶλ-
λον· καὶ ἡ δίψα πολλὴ ἦν, τὰ δ᾽ ἄλλα ὅμοια παρηκο-
λούθει· τρόμοι δὲ περὶ τὰς χεῖρας ἐγένοντο, καὶ κεφαλὴν ὑπέσειεν· ὑπώπια[62] καὶ ἐμβλέψιες τῶν ὀφθαλμῶν
πονηραί·[63] καὶ ἡ δίψα ἰσχυρή· πιέουσα[64] πάλιν ᾔτει,
καὶ ἥρπαζε καὶ λάβρως ἔπινεν. ἀποσπάσαι οὐκ ἐδύναντο. γλῶσσα ξηρή, ἐρυθρὴ σφόδρα, καὶ τὸ στόμα
ὅλον καὶ χείλεα καθηλκωμένα,[65] ξηρά· καὶ τὰς χεῖρας
ἀμφοτέρας ἐπὶ τὸ στόμα ἀφαιροῦσα ἐμασᾶτο, τρομώδης ἐοῦσα, καὶ εἴ τι προσενέγκαι τις μασήσασθαι ἢ
ῥυφήσασθαι, λάβρως καὶ μανικῶς κατέπινε καὶ ἐρρύ-
φανεν· καὶ τὰ περὶ τὴν ὄψιν πονηρά. ἡμέρῃσι δὲ
τρισὶν[66] ἢ τέσσαρσι πρὸ τῆς τελευτῆς, φρῖκαί τε αὐτῇ

56 οἱ om. V
57 ἐπεγίνωσκον μὲν πάντες MV: corr. Corn.

shouts became gentle, but she persisted in her wildness, her anger and tantrums if what she wanted was not done for her quickly. She recognized everyone and all objects straightway after the first days. The eye condition settled down. But the mania, inappropriate behavior, crying out, and the alternations I described, persisted until the coma. She heard irregularly: some things very clearly, even if one said them quietly, but for some it was necessary to talk louder. Her feet were throughout as warm as the rest of her body until the final days, but on the sixteenth day less so. On the seventeenth day she was more comfortable than on the other days, but towards night she drew herself together as though a chill was coming on, and her fever increased. There was much thirst. Everything else continued the same. But trembling developed in her hands and she kept tossing her head. The area under her eyes, and the looks of her eyes were bad. Her thirst was powerful. When she had drunk she would ask for more, snatch it and drink violently. They could not take it away from her. Her tongue was dry, quite red, and her whole mouth and lips ulcerated, parched. She kept moving both hands to her mouth and chewing them, trembling, and if anyone gave her something to chew or sip, she drank and sipped it violently, madly. The area around her eyes was bad. On the three or four days before the end shivers came on her at

---

58 ἡμέρας εὐθὺς V   59 καὶ ἡ μεταβολὴ om. V
60 τι M   61 ἐξκαιδεκάτῃ MV
62 ὁπόπιοι M   63 πονηρή M
64 ποιέουσα MV: corr. recc.
65 καθελκώμενα M
66 Asul.: τρίτῃσιν MV

ἔστιν ὅτε ἐνέπιπτον, ὥστε ξυνάγειν τὸ σῶμα, καὶ
ξυγκαλύπτειν,[67] καὶ πνευστιᾶν· τέτανοί τε τὰ σκέλεα,
καὶ ψύξις ποδῶν· ἡ δίψα δὲ ὁμοίη, καὶ τὰ περὶ τὴν
διάνοιαν ὅμοια· καὶ ἐξαναστάσιες ἢ διὰ κενῆς ἢ σμικρὰ καὶ λεπτὰ μετά τινος βραχέος τόνου. τῇ δὲ τελευ-
ταίῃ, τῇ τρίτῃ καὶ εἰκοστῇ, τὸ ὄμμα μέγα πρωὶ ἦν, καὶ
περίβλεψις βραχείη· καὶ ἡσυχίην ἔσχεν ἔστιν ὅτε καὶ
ἄνευ τοῦ ξυγκεκαλύφθαι τε καὶ κεκωματίσθαι.[68] πρὸς
δὲ τὴν ἑσπέρην τοῦ δεξιοῦ ὄμματος κίνησις οἵη ὀρε-
ούσης ἤ τινος βουλήσιος, ἐκ τοῦ ἔξω κανθοῦ πρὸς
ῥῖνα· καὶ ἐπεγίνωσκε καὶ πρὸς τὸ ἐρωτώμενον ὑπεκρί-
νετο· φωνὴ μετὰ πολλὰ ὑπότραυλος, καὶ ὑπὸ τῆς
βοῆς[69] ἀπερρωγυῖα καὶ βραγχώδης. |

388   12. Τῷ Ἀμφιφράδεος θέρεος πλευροῦ ἀριστεροῦ
ὀδύνη, καὶ βήξ, καὶ ὑποχωρήματα πολλὰ ὑδατώδεα
καὶ ὑπόχολα. ὁ πυρετὸς ἐδόκει λῆξαι περὶ ἑβδόμην· ἡ
βὴξ ἐνῆν· χρῶμα ὑπόλευκον καὶ ὕπωχρον. περὶ δὲ τὰς
δώδεκα ὑπόχλωρον ἔπτυεν· τὸ πνεῦμα προϊόντος τοῦ
νοσήματος αἰεὶ πυκνότερον, καὶ[70] ὑποκαρχάλεον περὶ
στήθεα καὶ ἀρτηρίην. ῥυφήμασιν ἐχρῆτο· ἔμφρων
ἅπαντα τὸν χρόνον. περὶ δὲ εἰκοστὴν καὶ ὀγδόην
ἐτελεύτησεν· ἱδρῶτες ἔστιν ὅτε ἐγένοντο.

13. Ὁ[71] ἔξω κάπηλος ὁ περιπλευμονικός, κοιλίη
εὐθὺς ὑπῄει. περὶ τετάρτην, ἱδρὼς πολύς· ἐδόκει λῆξαι
τὸ πυρέτιον· βηχίον οὐδὲν ὡς εἰπεῖν. πέμπτῃ καὶ
ἕκτῃ καὶ ἑβδόμῃ ἐπεῖχεν ὁ πυρετός. ἱδρὼς ὀγδόῃ.
ἐνάτῃ, ἀπεχρέμψατο ὠχρόν. δεκάτῃ ἅλες[72] οὐ πολ-

times so that she would draw her body together, cover up, and breathe hard. Cramps in the legs, cold feet. The thirst as before, and mental affection similar. Bowel movements either nothing or small and thin with brief straining. On the last day, the twenty-third, her eye was large in the morning and her vision short. She was quiet at times, without huddling under covers or coma. Towards evening there was movement of the right eye, as though seeing or seeking something, from the outer corner towards the nose. She showed recognition and answered what was asked. Her voice lisping after much talking, and broken and hoarse from the shouting.

12. Amphiphrades' son, in summer, had a pain in the left side of the rib cage, cough, much watery bilious excrement. His fever seemed to stop about the seventh day. The cough persisted. Complexion whitish, yellowish. About day twelve he spat up greenish-yellow matter. His breathing, as the illness progressed, was ever more rapid; there was rattling in the chest and trachea. He took gruel. He was conscious the whole time. He died about the twenty-eighth day. There were sweats sometimes.

13. The salesman from outside who had peripneumonia: his bowels immediately moved. On the fourth day much sweat. The fever seemed to stop. No cough to speak of. On the fifth, sixth, and seventh days the fever persisted. Sweat on the eighth. On the ninth he coughed up yellowish matter. On the tenth abundant bowel movements, not

---

67 ξυγκαλύπτεσθαί τε M   68 κωματίσθαι M
69 βοῆς om. V   70 καὶ ἆσθμα καὶ M   71 om.
M (which attaches ἔξω to the preceding sentence: "external sweats")   72 Smith: ἄλες V: εἰαλες M: εἴη ἄλες recc.

λάκις. περὶ ἑνδεκάτην, ἠπιώτερος. ἐν τῇ τεσσαρεσκαιδεκάτῃ, ὑγιής.

14. Ἑρμοπτολέμῳ μετὰ Πληϊάδος δύσιν πυρετός,
βὴξ οὐ πάνυ· γλῶσσα δὲ περιπλευμονική. ἐνάτῃ ἐξίδρωσεν ὅλος κατεψύχθη τε ὡς[73] ἐδόκει· προσήνεγκαν
αὐτῷ χυλόν· περὶ μέσον ἡμέρης ἐθερμαίνετο. ἑνδεκάτῃ ἱδρώς, καὶ κοιλίη ἐξυγραίνετο· κατάχολα ὑπο-
χωρήματα· τὸ βηχίον ἐπεγένετο. τεσσαρεσκαιδεκάτῃ
ὠχρὸν ἀπεχρέμψατο,[74] καὶ ῥεγχώδης ἦν. καὶ πεντεκαιδεκάτῃ, ἔμφρων δὲ πάντα τὸν χρόνον ἐών, ἐτελεύτη-
σεν. |

390   15. Ἕτερός τις ἐπὶ τοῦ ὑπερῴου ῥεγχώδης, γλῶσσα
ξηρή, περιπλευμονική, ἔμφρων ἐτελεύτησεν.

16. Καὶ Ποσειδώνιος ἔτι τοῦ θέρεος κατὰ στῆθος
καὶ ὑποχόνδρια καὶ πλευρὸν ἐπόνει χρόνον πολὺν[75]
ἄνευ πυρετῶν· πολλοῖσι δὲ ἔτεσιν ἔμπροσθεν ἔμπυος
ἐγένετο. τοῦ δὲ χειμῶνος φρίξαντι ἐπέτεινεν ἡ ὀδύνη,
καὶ τὸ πυρέτιον[76] λεπτόν, καὶ ἀπόχρεμψις πυώδης,
βὴξ κερχαλέη περὶ φάρυγγα, καὶ ῥεγχώδης, ἔμφρων
δὲ σφόδρα ἐὼν τεταρταῖος ἐτελεύτησεν.

17. Ὁ δὲ Βαλοῖος[77] ἐκ τοῦ ὄρεος πάντα ἡμαρτηκώς,
ἐννεακαιδεκάτῃ γλῶσσα πονηρή, ὑπέρυθρος, καὶ κατὰ
φωνὴν ἦν ἐν τῇ ῥέμβῃ·[78] ὀφθαλμοὶ κεχρωσμένοι,
πλέοντες ὥσπερ τῶν νυσταζόντων· χρῶμα καὶ τοῦ
ἄλλου σώματος οὐκ ἰκτεριῶδες[79] σφόδρα, ἀλλ᾽ ὑπ-

[73] τε ὡς] τελέως Μ   [74] ἀπεχρήμψατο Μ
[75] πολὺν χρόνον V

often. Around the eleventh day the symptoms were less severe. On the fourteenth, he was healthy.

14. Hermoptolemus had fever with some coughing after the setting of the Pleiades; his tongue peripneumonic. On the ninth, his whole body sweated, and he was chilled, so it appeared. They gave him broth. About midday he grew hot. On the eleventh, sweat; intestines grew moist; bilious feces; the cough supervened. On the fourteenth he coughed up yellowish matter and began to wheeze. On the fifteenth day, having been conscious throughout, he died.

15. Another man, in the upper town. Wheezing. Tongue dry and peripneumonic. Died conscious.[6]

16. Posidonius, still in the summer, suffered for a considerable time in chest, hypochondria, and ribs, without fever. Many years earlier he had been empyemic. In the winter after shivering the pain intensified, the fever was light. He coughed up purulent matter, had a rattling cough in his throat, wheezing. He remained quite conscious and died on the fourth day.

17. The man from Baloea who lived on the hill had been very careless in his way of life; on the nineteenth day his tongue was bad, quite red, and he was imprecise in speech. Eyes colored, swimming like those of very tired people. Color, even of the rest of the body, not very like

6 Cf. *Epid.* 5.105.

---

76 πῦρ αἴτιον M
77 Gal. *Gloss.*: Βάλλεος MV
78 Gal. *Gloss.*: ῥεμβίη MV
79 ἰκτεριώδεες M

ωχρον, πελιδνόν· φωνὴ πονηρή, ἀσαφής· γλῶσσα
περιπλευμονική· οὐκ ἔμφρων· πνεῦμα πρὸς χεῖρα πονηρόν, οὐ πυκνὸν οὐδὲ μέγα· πόδες ψυχροί, λιθώδεις.
περὶ εἰκοστὴν[80] ἐτελεύτησεν.

18. Κυναγχικὴ ἡ παρὰ Μέτρωνι χεῖρα δεξιήν, σκέλος ἤλγησε, πυρέτιον ἐπεῖχε, βηχίον, πνίγμα. τρίτῃ,
ἐχάλασεν. ἑβδόμῃ σπασμώδης, ἄφωνος, ῥέγχος,
ὀδόντων ξυνέρεισις, γνάθων ἔρευθος πλέον· ἐτελεύτησε. πέμπτῃ, ἕκτῃ σημεῖον περὶ χεῖρα ὑποπέλιον.

19. Βίων ἐξ ὑδρωπικοῦ πολυχρονίου ἀπόσιτος ἐγέ-
392 νετο πολλὰς | ἡμέρας καὶ στραγγουριώδης· ἐπὶ γού-
νατος ἀριστεροῦ ἀπόστημα ἐγένετο, ἐξεπύησεν, ἐτελεύτησεν.

20. Κτησιφῶν ὑδρωπικὸς ἐκ καύσου πολλοῦ, καὶ
πρότερον ὑδρωπικὸς καὶ σπληνώδης, σφόδρα ξυν-
επληρώθη καὶ ὄσχεον[81] καὶ σκέλεα καὶ περιτόναια.
ἐπὶ τῇ τελευτῇ βὴξ καὶ πνιγμοὶ ἐγένοντο, ἐς νύκτα
μᾶλλον, ἀπὸ τοῦ πλεύμονος, ὥσπερ τοῖσι πλευμονώ-
δεσιν. πρὸ δὲ τῆς τελευτῆς ἡμέρῃσι[82] τρισὶν ἢ τέσσαρσι ῥῖγος, πυρετός· κατὰ μηρὸν δεξιόν, ἔσω κατὰ
φλέβα μέσην τὴν ἀπὸ τοῦ βουβῶνος, ὡς πυρὸς ἀγρί-
ου σύστρεμμα ὑποπέλιον ἔχον ἔρευθος· ἐς νύκτα,
καρδίης ἄλγος, καὶ οὐ πολὺ ὕστερον ἀφωνίη, πνιγμὸς
μετὰ ῥέγχους, καὶ ἐτελεύτησεν.

21. Καὶ ὁ ἐν Ὀλύνθῳ ὑδρωπικὸς ἐξαίφνης ἄφωνος,
ἔκφρων νύκτα καὶ ἡμέρην, ἐτελεύτησεν.

---

[80] ἐνάτην MV: corr. Li. ex ms. Q´

jaundice, but pale yellow, livid. Speech bad, unclear. Tongue peripneumonic. He was not conscious. Breathing, to feel, was bad, not frequent or large. Feet cold, stony. He died around the twentieth day.

18. The woman at Metron's house with quinsy had pain in the right arm and leg. Fever held her, cough, choking. On the third day it relaxed. On the seventh she had spasms, was voiceless; there was wheezing, clenching of teeth, increased redness of the cheeks; she died. On the fifth or sixth day, a sign: lividness around the hand.[7]

19. Bion, after a long hydropic illness, could not eat for many days, and had strangury. An abscess appeared on his left knee; it festered. He died.

20. Ctesiphon was hydropic after a lengthy burning fever, having also been previously hydropic and having had spleen problems. He was quite filled up in the scrotum, legs and peritoneum. At the end, coughing and choking occurred, more towards night, from the lungs, as in pneumoniacs. In the three or four days before the end, shivering, fever. On the right thigh, inside along the central vein from the groin, there was a livid gathering, as of a fierce fire, containing a redness. Towards night, pain at the heart, and not much later voicelessness, choking and wheezing. And he died.

21. And the hydropic man at Olynthus. Suddenly voiceless. Delirious a night and a day. He died.[8]

---

[7] Cf. *Epid.* 5.104.   [8] Cf. *Epid.* 5.106.

---

[81] ἰσχίον V
[82] πρὸ τῆς δὲ τελευτῆς ἡμέρης M

22. Ἡ δὲ Προδρόμου, θέρεος, ὑπότραυλος, καυσώδης, γλῶσσα ὑπόξηρος, ἀσαφής· κάτω πολλὴ ἄφοδος· περιεγένετο.

23. Λεωφορβείδῃ, πυρετὸς ὀξὺς μετὰ χειμερινὰς τροπάς, ὑποχονδρίων καὶ κατὰ κοιλίην ἄλγημα· ὑποχωρήματα ὑγρά, χολώδεα πολλά· καὶ μεθ᾿ ἡμέρην κωματώδης· γλῶσσα περιπλευμονική· βὴξ οὐκ ἐνῆν. δωδεκάτῃ μέλανα σμικρὰ καὶ πρασοειδέα ὑπεχώρησεν. τεσσαρεσκαιδεκάτῃ λῆξαι πυρέτιον ἐδόκει, μετὰ δὲ ῥυφήμασιν ἐχρῆτο. ἑκκαιδεκάτῃ στόμα σφόδρα[83] ἁλμῶδες, ξηρὸν ἐγένετο· ἀκρέσπερον δὲ φρίκη, πυρετός. μιῇ καὶ εἰκοστῇ περὶ μέσον ἡμέρης ῥῖγος καὶ ἱδρώς· τὸ πυρέτιον ἔληξε, θέρμη δὲ ὑπῆν λεπτή· ἐς 394 νύκτα πάλιν ἱδρώς· | καὶ δευτέρῃ καὶ εἰκοστῇ ἐς νύκτα ἱδρώς·[84] καὶ ἡ θέρμη ἐχάλασεν. τῇσι δὲ ἔμπροσθεν πάσῃσιν ἀνίδρωτος ἦν, ἡ δὲ κοιλίη ὑγράνθη καὶ ἐν τῇ ὕστερον δοκεύσῃ ὑποστροφῇ.

24. Ἡ ἄνω οἰκοῦσα ἡ Θεοκλῆ προσήκουσα, ὑπὸ Πληϊάδα πυρετὸς ὀξύς. ἕκτῃ ἐδόκει λῆξαι· ἐλούσατο ὡς πεπαυμένη. ἑβδόμῃ πρωὶ γνάθος σφόδρα ἐρυθρή, ὁποτέρη οὐ μέμνημαι· ἐς τὴν ἑσπέρην ὁ πυρετὸς ἧκε πολύς, καὶ λειποψυχίη, καὶ ἀφωνίη ἦν· οὐ πολὺ δὲ ὕστερον ἱδρώς, καὶ παῦσις ἑβδομαίη.

25. Καὶ ἡ Θεοδώρου σφόδρα ἐν πυρετῷ, αἱμορραγίης γενομένης, χειμῶνος· λήξαντος δὲ τοῦ πυρετοῦ περὶ δευτέρην,[85] οὐ πολὺ ὕστερον πλευροῦ δεξιοῦ ὡς ἀπὸ ὑστερέων βάρος· καὶ πρῶτον[86] ἐγεγόνει, καὶ τὰς

22. The wife of Prodromus, in summer, was lisping, had burning fever. Tongue dry, unintelligible. Much excrement from bowels. She survived.

23. Leophorbides had acute fever after the winter solstice, pain in the hypochondria and in the intestines. Watery, large, bilious movements. Even in the daytime, comatose. Tongue peripneumonic. There was no cough. On the twelfth day he passed small dark green bowel movements. On the fourteenth day the fever seemed to relent, and afterwards he had gruel. On the sixteenth day his mouth became very salty and dry. In the evening shivering, fever. On the twenty-first in the middle of the day, chill and sweat. The fever relented but there was moderate heat. At night again sweat. And on the twenty-second day towards night, sweat. And the heat abated. On all previous days he was without sweat, but the bowels were moist, even at what seemed in retrospect the relapse.

24. The woman who lived above, the relative of Theocles, acute fever towards the setting of the Pleiades. On the sixth day it seemed to abate. She bathed as though she was cured. Early on the seventh day her cheek was suddenly red, I don't recall which. Towards the evening the fever came on severely, and fainting, and she was voiceless. Not much later, sweat, and cessation of the sickness on the seventh day.

25. The wife of Theodorus, greatly in fever, hemorrhage having occurred, in winter. The fever abated the second day; shortly afterwards she had heaviness of the right side, as from the womb. This was the first time it had hap-

---

83 om. V   84 καὶ δευτέρῃ . . . ἱδρώς om. V
85 ἐνάτην mss.: corr. Foës   86 πρῶτον δὲ M

ἐχομένας ἡ ὀδύνη κατὰ στῆθος δεινή· καὶ πλευρὸν
δεξιὸν πυριωμένη ἐχάλασεν. τεταρταίη τὰ ἀλγήματα,
τὸ πνεῦμα πυκνότερον· ἀρτηρίη[87] μόγις ἀναπνεούσῃ
ὑπεσύριζεν· κλισίη[88] ὑπτίη, ἐπιστρέφεσθαι χαλεπῶς·
ἐς νύκτα ὀξύτερος ὁ πυρετός, καὶ λῆρος βραχὺς ἐγένετο. πέμπτῃ πρωῒ ἐδόκει ἠπιώτερος εἶναι· ἱδρώτιον ἀπὸ
μετώπου κατεχύθη ὀλίγον πρῶτον, ἔπειτα πολὺν χρόνον ἐς ὅλον τὸ σῶμα καὶ πόδας· μετὰ δὲ ταῦτα ἐδόκει
αὐτῇ κεχαλακέναι τὸ πῦρ· ἦν δὲ πρὸς χεῖρα ψυχρότερον τὸ σῶμα· αἱ δὲ ἐν κροτάφοισι καὶ μᾶλλον ἐπήδων,
καὶ πνεῦμα πυκνότερον, καὶ ὑπελήρει ἄλλοτε καὶ
ἄλλοτε, καὶ πάντα ἐπὶ τὸ χεῖρον. γλῶσσα διὰ τέλεος
λευκὴ λίην· ἥ τε βὴξ οὐκ ἐνῆν ὅτι μὴ τριταίη καὶ
πεμπταίη ὀλίγον χρόνον· δίψα οὐκ ἐνῆν, πτυσμὸς ἦν·
ὑποχόνδριον δεξιὸν σφόδρα ἐπήρθη[89] περὶ τὴν[90] πέμ-
396   πτην, μετὰ δὲ μαλα|κώτερον· ὑποχώρησις τριταίη
ἀπὸ βαλάνου[91] κόπρου ὀλίγης, πέμπτῃ πάλιν ὑγρὸν
ὀλίγον· κοιλίη δὲ λαπαρή· οὖρα στρυφνά, ὁποειδέα·
ὄμματα ὡς κοπιώσης, χαλεπῶς ἀνέβλεπε καὶ περι-
έφερεν. πέμπτῃ, ἐς νύκτα χαλεπῶς, καὶ λῆρος. ἕκτῃ
πάλιν, τὴν αὐτὴν ὥρην περὶ πλήθουσαν ἀγορήν,
ἱδρὼς πολὺς κατεχεῖτο,[92] ἀπὸ μετώπου ἤρχετο ἐς ὅλον
τὸ σῶμα πολὺν χρόνον· ἐμφρόνως[93] διετίθετο τὰ
ἑωυτῆς· πρὸς μέσον δὲ ἡμέρης σφόδρα ἐλήρει, καὶ τὰ
τῆς καταψύξιος ὅμοια, βαρύτερα δὲ τὰ κατὰ τὸν
χρῶτα πάντα· πρὸς δὲ τὴν ἑσπέρην ἡ κνήμη αὐτῆς ἐκ

[87] ἀρτηρίου V   [88] καὶ σιη M   [89] ἐπήρθαι M

pened. On succeeding days the pain in the chest was terrible. With fomentations on the right side she improved. On the fourth day, the pains; the breathing quicker. Her trachea whistled as she breathed with difficulty. She lay on her back; difficulty turning over. Towards night the fever grew more acute. There was, briefly, delirious talk. Early on the fifth day the fever seemed more mild. Sweat poured from her forehead, briefly at first, then for a long period over her whole body and feet. After this she thought the burning had grown less, and her body was cooler to the touch, but the vessels at the temples jumped more, her breath was more rapid, she talked deliriously from time to time, and all signs changed for the worse. Throughout, the tongue was extremely white; there was no cough except on the third and fifth day for a short time. There was no thirst, but expectoration. The right hypochondrium was much swollen on the fifth day but afterwards softer. Some solid excrement on the third day after a suppository. On the fifth again, a little liquid. The belly soft. Urine astringent, like fig juice. The eyes as of one who is weary. She had difficulty seeing things and looking about. On the fifth day towards night, difficulty and delirious talk. On the sixth, again, at the same hour, that of the filling of the marketplace, much sweat poured from her. It went from the forehead to the whole body for a long time. She conducted herself rationally. Towards midday, however, she talked very deliriously, the chilling was similar but she had greater heaviness all over the body. Towards evening her lower leg

---

⁹⁰ om. V   ⁹¹ βαλανείου MV (here and frequently)
⁹² κατείχετο mss.: corr. Foës: V adds οὐ before κατείχετο
above the line   ⁹³ ἀφρόνως V

τῆς κλίνης κατερρύη, καὶ τῷ παιδὶ παραλόγως ἠπεί-
λησέ τε[94] καὶ πάλιν ἐσιώπησε, καὶ ἐς ἡσυχίην μετ-
έβαλεν· περὶ δὲ τὸν πρῶτον ὕπνον, δίψα, πολλὴ μανίη,
καὶ ἀνεκάθιζε, καὶ τοῖσι παροῦσιν ἐλοιδορεῖτο, καὶ
πάλιν ἀπεσιώπησε καὶ ἐν ἡσυχίῃ ἦν· καὶ ἐδόκει τὴν
ἐπίλοιπον νύκτα κεκωματίσθαι·[95] τοὺς δὲ ὀφθαλμοὺς
οὐ ξυνῆγεν. ἐπὶ δ' ἡμέρην ὑπεκρίνετο τὰ πλεῖστα
νεύμασιν, ἀτρεμίζουσα τὸ σῶμα, καὶ κατανοοῦσα ἐπι-
εικῶς· πάλιν δὲ ἱδρὼς τὴν αὐτὴν ὥρην· ὁμοίως οἱ
ὀφθαλμοὶ κατηφεῖς, ἐς τὸ κάτω βλέφαρον μᾶλλον
ἐγκείμενοι, ἀτενίζοντες, κεκαρωμένοι,[96] τὰ λευκὰ τῶν
ὀφθαλμῶν ὠχρὰ καὶ νεκρώδεα, καὶ τὸ πᾶν χρῶμα
ὠχρὸν καὶ μέλαν[97] ἐόν. τῇ χειρὶ τὰ πολλὰ πρὸς τοῖχον
ἢ πρὸς ἱμάτιον· οἱ ψόφοι δὲ πινούσῃ ἐγίνοντο, ἀπ-
επύτιζε καὶ ἄνω ἐς τὴν ῥῖνα ἐφόρει,[98] καὶ ἐκροκυδο-
λόγει, καὶ ξυνεκαλύπτετο πρόσωπον· μετὰ δὲ τὸν
ἱδρῶτα, χεῖρες ὥσπερ κρυστάλλιναι· ὁ ἱδρὼς παρηκολούθει ψυχρός· σῶμα πρὸς χεῖρα ψυχρόν· ἀνεπήδα,
ἐκεκράγει, ἐμαίνετο· πνεῦμα πολύ· τρομώδης χεῖρας
ἐγίνετο, ὑπὸ δὲ τὸν θάνατον σπασμώ ιδης. ἑβδομαίη
ἐτελεύτησεν. ἐνούρησε τῇ ἕκτῃ ἐν νυκτὶ ὀλίγον· τὸ
οὐρούμενον τῷ κάρφει εἵλκετο, γλίσχρον, γονοειδές·
ἄγρυπνος ἁπάσας· μετὰ τὴν ἕκτην οὖρον ὕφαιμον.

26. Τῷ Ἀντιφάνους χειμῶνος ἄλγημα πλευροῦ
δεξιοῦ, βήξ, πυρετός· ἤσθιεν, ἐπορεύετο ὑποπυρεταί-
νων, ἐδόκει ῥηγματώδης εἶναι. ἐνάτῳ ἀφῖκτο ὁ πυρετός, οὐκ ἔλιπεν· ἡ βὴξ πολλή, καὶ παχέα ἀφρώδεα· τὸ
πλευρὸν ἐπόνει. περὶ τὴν τεσσαρεσκαιδεκάτην καὶ

slipped out of bed, she threatened her child irrationally and again fell silent, changed into quiescence. About the first sleep, thirst. Much delirium: she sat up and rebuked those who were there. Again she fell silent and remained quiescent. She seemed to be in a coma the rest of the night. She did not close her eyes. Towards day she answered mostly with nods, her body unmoving, reasonably alert. Again the sweat at the same hour. The eyes similarly downcast, leaning more on the lower lid, staring, torpid, the whites of the eyes yellowish and corpselike. Her whole color yellowish and dark. Mostly reaching with her hand towards the wall or the bedclothes. The gurgling occurred when she drank, and she spurted it out and brought it up through her nose. She plucked at the blankets, and kept her face covered up. After the sweats her hands were like ice. Cold sweat persisted. Body cold to the touch. She jumped up, cried out, raved. Breathing very rapid. She developed trembling in the hands. At the point of death she twitched. She died on the seventh day. She urinated a little in her bed on the sixth day in the night. The urine was picked up on a twig, sticky, like semen. Sleepless all nights. After the sixth day bloody urine.

26. Antiphanes' son, in the winter, had pain in the right side, cough, fever. He kept eating, went outdoors with a slight fever. He seemed to have some fissuring. On the ninth day the fever came on and did not again leave him. The cough extensive; thick, foamy matter. Pains in the thorax. On the fourteenth and again on the twentieth day

---

94 ταῖ M: corr. M(02   95 καὶ κωματίσθαι M
96 κεκαρωμένῳ MV: corr. Lind.   97 τὸ μέλαν V
98 Smith: ῥιναφορεῖ M: ῥῖνα V

πάλιν τὴν εἰκοστὴν ἐδόκεον ἀπολήγειν οἱ πυρετοί, καὶ
πάλιν ἐπελάμβανον· ἦν δὲ λεπτὴ θέρμη,[99] βραχύ τι
ἐξέλιπεν· ἡ[100] δὲ βὴξ ὀτὲ μὲν ἐξέλιπεν,[101] ὀτὲ δὲ κατακορὴς μετὰ πνίγματος πολλοῦ, τότε δὲ ἐχάλα· καὶ
ἀπόχρεμψις μετὰ ταῦτα[102] πολλὴ μετὰ πνιγμώδους
βηχός, καὶ πυώδεα ζέοντα ὑπὲρ τοῦ ἀγγείου καὶ ἀφρέ-
οντα· καὶ ἐν τῇ φάρυγγι τὰ πολλὰ κερχαλέα ὑπεσύρι-
ζεν· ἆσθμα αἰεὶ κατεῖχε, καὶ πνεῦμα πυκνότερον, ὀλιγάκις εὔπνοος. ὑπὲρ δὲ τὰς τεσσαράκοντα, ἐγγὺς
οἶμαι τῶν ἑξήκοντα, ὀφθαλμὸς ἀριστερὸς ἐτυφλώθη
μετὰ οἰδήματος ἄνευ ὀδύνης, οὐ πολὺ δ' ὕστερον καὶ ὁ
δεξιός· καὶ σφόδρα αἱ κόραι λευκαὶ καὶ ξηραὶ ἐγένον-
το· ἐτελεύτησε μετὰ τὴν τύφλωσιν οὐ πολὺ ὑπὲρ ἑπτὰ
ἡμέρας μετὰ ῥέγχου καὶ ληρήσιος.

27. Ὅμοια δὲ καὶ ἐξ ὁμοίων τὴν ὥρην τὴν αὐτὴν
ξυνέβη Θεσσαλίωνι, τὰ ζέοντα καὶ ἀφρέοντα καὶ
πυώδεα, καὶ βῆχες, καὶ οἱ κερχμοί. |

400   28. Τῇ Πολεμάρχου χειμῶνος κυναγχικῇ, οἴδημα
ὑπὸ τὸν βρόγχον,[103] πολὺς πυρετός· φλέβα ἐτμήθη·
ἔληξεν ὁ πνιγμὸς ἐκ τῆς φάρυγγος· ὁ πυρετὸς[104]
παρείπετο.[105] περὶ πέμπτην γούνατος ἄλγημα καὶ οἴδημα ἀριστεροῦ· καὶ κατὰ τὴν καρδίην ἔφη δοκεῖν τι
ξυνάγεσθαι ἑωυτῇ, καὶ ἀνέπνει οἷον ἐκ τοῦ βεβαπτίσθαι ἀναπνεούσῃ, καὶ ἐκ τοῦ στήθεος ὑπεψόφει· ὡς[106]
ἐγγαστρίμυθοι λεγόμεναι, τοιοῦτόν τι ξυνέβαινεν.
περὶ δὲ τὴν ὀγδόην ἢ ἐνάτην ἐς νύκτα κοιλίη κατερ-

99 καὶ θέρμη M   100 recc.: om. M

the fever appeared to be stopping, and again seized him,
but it was a mild heat and ceased for a brief time. The
cough sometimes left him, sometimes was overwhelming
with much choking, but at that time (day twenty) it dimin-
ished. Afterwards much material was brought up with a
choking cough, purulent matter seething and foaming
over the basin. In his pharynx there was generally a hoarse
whistling. Asthma possessed him continuously, and quite
rapid breathing; occasionally he breathed easily. After the
fortieth day, approaching the sixtieth, I think, his left eye
went blind with a swelling and without pain. Not much
later, the right also. The pupils became very white and dry.
He died after the blindness, not much beyond seven days,
with wheezing and delirium.

27. Thessalion had similar phenomena in the same season, starting from similar symptoms; he had the seething,
foaming, and purulent matter, the cough and the hoarse
whistling.

28. Polemarchus' wife had quinsy in the winter, swelling by the windpipe, much fever. A vein was opened. The
choking in her pharynx was relieved, the fever persisted.
About the fifth day, pain and swelling in the left knee. She
said she felt as though there was a gathering around her
heart; she breathed as though she was catching her breath
after being submerged. She made a noise from her chest.
It was something like that which so-called ventriloquists
make. Towards the eighth or ninth day her intestines broke

---

101 ἡ δὲ . . . ἐξέλιπεν om. V
102 ταύτας τὰς ἡμέρας M   103 βρόχον MV: corr. recc.
104 φλέβα . . . πυρετὸς om. V   105 παρήπετο V
106 V adds αἱ above the line after ὡς

ΕΠΙΔΗΜΙΑΙ

ράγη· ὑγρά, πολλὰ καὶ ἀλέα καὶ κάκοδμα· ἀφωνίη
ἔσχεν· ἐτελεύτησεν.

29. Ἀρίστιππος ἐς τὴν κοιλίην ἐτοξεύθη ἄνω βίῃ
χαλεπῶς· ἄλγος κοιλίης δεινόν· καὶ ἐπίμπρατο[107] τα-
χέως· κάτω δὲ οὐδὲν διεχώρειν· ἀσώδης ἦν· ἤμει
χολώδεα[108] κατακορέα· ὁπότε[109] δὴ ἀπεμέσειεν, ἐδόκει
ῥῆϊον εἶναι· μετ᾽ ὀλίγον δὲ τὰ ἀλγήματα πάλιν δεινά·
καὶ ἡ κοιλίη ὁμοίως ἐπίμπρατο[107] ὡς ἐν εἰλεοῖσιν·
θέρμαι, δίψαι· ἐν ἑπτὰ ἡμέρῃσιν ἐτελεύτησεν.

30. Ὁ δὲ Νεόπολις, πληγεὶς ὁμοίως, ταῦτα ἔπα-
σχεν· κλυσθέντι δὲ δριμεῖ ἡ κοιλίη κατερράγη· χρῶμα
κατεχύθη λεπτόν, ὠχρόν, μελανόν·[110] ὄμματα αὐχμη-
ρά, καρώδεα, ἐνδεδινημένα, ἀτενίζοντα.

31. Τῷ δὲ καθ᾽ ἧπαρ ἐγγὺς πληγέντι ἀκοντίῳ εὐθὺ
τὸ χρῶμα κατεχύθη νεκρῶδες· τὰ ὄμματα κοῖλα·
ἀλυσμός· δυσφορίη· ἀπέθανε πρὶν ἀγορὴν λυθῆναι,
ἅμ᾽ ἡμέρῃ πληγείς.

32. Ἐπὶ τὸν Μακεδονικόν· ὁ τὴν κεφαλὴν ἀπὸ
Μακεδόνος λίθῳ πληγείς, ὑπὲρ κροτάφου ἀριστεροῦ
ὅσην ἀμυχὴν διεκόπη· ἐσκοτώθη πληγείς, καὶ ἔπεσεν.
τριταῖος ἄφωνος ἦν· ἀλυσμός· πυρετὸς οὐ πάνυ σφο-
δρός· σφυγμὸς ἐν κροτάφοισιν ὡς λεπτῆς θέρμης·
ἤκουεν οὐδέν, οὐδὲ ἐφρόνειν, οὐδ᾽ ἠτρέμιζεν· νοτὶς[111]
περὶ μέτωπον καὶ ὑπὸ ῥῖνα καὶ ἄχρι ἀνθερεῶνος·
πεμπταῖος ἐτελεύτησεν.

[107] ἐμπίπρατο M   [108] ἡμιχολώδεα M
[109] ποτὲ M   [110] Smith: μέλαν ἐὸν MV: μελανέον Li.

402

316

loose towards night: much wet, gushing, foul-smelling excrement. Voicelessness possessed her. She died.[9]

29. Aristippus was struck a powerful blow by an arrow in the upper belly. Terrible pain in the intestine. It was quickly inflamed, but no excrement passed below. He was nauseous. He vomited very bilious matter; whenever he vomited he seemed easier, but shortly later the pains were again terrible. And the intestine became inflamed as in intestinal obstructions. Fever, thirst. He died in seven days.[10]

30. Neopolis, wounded similarly, had the same affections. But with an acrid clyster his bowel broke loose. A light color was diffused over him, yellow and black. His eyes were dry, stuprous, rolling inwards, staring.[11]

31. The man struck near the liver with a javelin: immediately he was suffused with a corpselike color. His eyes were hollow; tossing about; distress; he died before the marketplace closed, having been wounded at daybreak.[12]

32. The Macedonian. The man struck in the head with a stone thrown by the Macedonian: he was struck above the left temple, a superficial wound. When hit he blacked out and fell. On the third day he was voiceless. Tossing about. Fever, not very intense. Throbbing in temples as of moderate heat. He heard nothing, nor was he conscious, nor was he still. Moisture around the forehead and beneath the nose to the chin. He died on the fifth day.[13]

---

[9] Cf. *Epid.* 5.63. [10] Cf. *Epid.* 5.98.
[11] Cf. *Epid.* 5.99. [12] Cf. *Epid.* 5.62.
[13] Cf. *Epid.* 5.60.

---

111 ὅστις MV (cf. *Epid.* 5.60)

33. Ὁ Αἰνιήτης[112] ἐν Δήλῳ ἀκοντίῳ πληγεὶς ἐς τοὔπισθεν τοῦ πλευροῦ κατὰ τὸ ἀριστερὸν μέρος· τὸ μὲν ἕλκος ἄπονον, τριταίῳ δὲ γαστρὸς ὀδύνη σμικρή· οὐχ ὑπεχώρει, κλυσθέντι δὲ κόπρος ἐς νύκτα ἦν· ὁ πόνος διαλιπών. ἕδρη ἔξω ἐς τοὺς ὄρχιας τεταρταίῳ, καὶ ἐς ἥβην καὶ κοιλίην ὅλην δεινὸς ὁ πόνος καταιγίζων· ἀτρεμεῖν οὐκ ἐδύνατο· χολώδεα ἤμεσε κατακορέα· χλοώδεις οἱ[113] ὀφθαλμοί, καὶ οἷοι[114] τῶν λειποθυμεόντων. μετὰ πέμπτην ἐτελεύτησεν· θέρμη λεπτή τις ἐνῆν.

34. Αὐδέλλῳ πληγέντι ἐς τὸν νῶτον, πνεῦμα πολὺ κατὰ τὸ τρῶμα μετὰ ψόφου[115] ἐχώρει, καὶ ἡμορράγει· τῷ ἐναίμῳ καταδεθεὶς ὑγιής. ξυνέβη δὲ καὶ τῷ Δυσχύτᾳ.

35. Τῷ Φιλίας παιδίῳ, ψιλώματος ἐν μετώπῳ[116] γενομένου, ἐναταίῳ πυρετός· ἐπελιάνθη τὸ ὀστέον· ἐτελεύτησεν. καὶ τῷ Φανίου καὶ τῷ Εὐέργου· πελαινομένων δὲ τῶν ὀστέων καὶ πυρεταινόντων, ἀφίστατο τὸ δέρμα ἀπὸ τοῦ ὀστέου, καὶ πύον ὑπεμένετο· τούτοισι τρυπωμένοισιν ἐξ αὐτοῦ τοῦ ὀστέου ἀνήρχετο ἰχὼρ λεπτός, ὀρώδης, ὕπωχρος, κάκοδμος, θανάσιμος. ξυμβαίνει δὲ τοῖσι τοιούτοισι, καὶ ἐμέτους ἐπιγίνεσθαι καὶ τὰ σπασμώδεα ἐπὶ τελευτῇ, καὶ ἐνίους κλαγγώδεας εἶναι, καὶ ἀκρατέας ἐνίους, καὶ ἦν μὲν ἐν τοῖσι

112 γενιήτης Μ: γενειήτης V (cf. Epid. 5.61)
113 om. Μ
114 οἱ Μ

33. The man from Aenus, wounded at Delos, by a javelin in the upper back on the left. The wound was painless, but on the third day he had a small pain in the belly.
He had no bowel movements, but when he had a clyster
there were solid feces towards night, and the pain departed. In the seat, and outwards towards the testicles on
the fourth day and into the pubis and whole intestinal cavity, there was terrible pain which came like a thunderclap.
He could not keep still. He vomited very bilious matter.
The eyes were greenish-yellow, like those of people who
have fainted. He died after the fifth day. He had a slight
fever.[14]

34. Audellus, struck in the back. Much wind escaped
from the wound noisily, and he hemorrhaged. Bandaged
with drugs for stopping blood, he recovered. The same
happened to Dyschytas.[15]

35. The child of Philia, whose skull was laid bare on his
forehead, had fever on the ninth day. The bone became
livid. The child died. Also the children of Phanias and
Euergus: when their bones became livid and they became
feverish, the skin stood away from the bone and pus gathered below. When they were trephined a thin serum came
out of the bone itself, like fig juice, slightly yellow, foul-
smelling, deathly. And it occurs in such cases that there
are vomiting and convulsions at the end, and some cry

[14] Cf. *Epid.* 5.61.
[15] Cf. *Epid.* 5.96.

---

[115] φόβου MV (cf. *Epid.* 5.96)
[116] ἐν μετώπου V
[117] ἐν V

404 δεξιοῖσι τὸ τρῶμα ᾖ, τὰ ἀριστερά· ἦν[117] δὲ | τοῖσιν
ἀριστεροῖσι, τὰ δεξιά. τῷ Θεοδώρου ἐναταίῳ ἡλιω-
θέντι πυρετὸς δεκαταίῳ ἐκ ψιλώματος, οὐδενὸς ἄξιος
εἰπεῖν, κατὰ τὸ ὀστέον· ἐν δὲ τῷ πυρετῷ ἐμελάνθη,
ἀπέστη τὸ δέρμα· ἐπὶ πολὺ κλαγγώδης· δευτέρῃ καὶ
εἰκοστῇ, ἡ γαστὴρ ἐπήρθη, μάλιστα δὲ κατὰ τὰ ὑπο-
χόνδρια· τρίτῃ καὶ εἰκοστῇ ἐτελεύτησεν.[118] οἷσι δ' ἂν
ὀστέα καταγῇ, τούτοισιν ἑβδομαίοισιν οἱ πυρετοί· ἦν
δὲ θερμοτέρη ἡ ὥρη, καὶ θᾶσσον· ἦν δὲ μᾶλλον
κλασθῇ, καὶ παραχρῆμα. καὶ ὁ Ἐξαρμόδου παιδίσκος
παραπλησίως, καὶ ἄλγημα ἐς μηρὸν οὐ κατ' ἴξιν τοῦ
τρώματος, καὶ κλαγγώδης, καὶ τραχήλου ὀδύνη. καὶ ὁ
Ποσειδοκρέων, τρίτῃ σπασμός· θέρμη οὐκ ἔλιπεν·
ἐτελεύτησεν ὀκτωκαιδεκαταῖος. ὁ τοῦ Ἰσαγόρα ὄπισθεν ἐπλήγη τῆς κεφαλῆς, φλασθέντος τοῦ ὀστέου
καὶ μελανθέντος πεμπταίῳ,[119] περιεγένετο,[120] ὀστέον
δὲ[121] οὐκ ἀπέστη.

36. Τῷ ἐκ τοῦ μεγάλου πλοίου διόπῳ, ᾧ[122] ἡ ἄγκυρα
τὸν λιχανὸν δάκτυλον καὶ τὸ ὀστέον ξυνέφλασε[123] τῆς
δεξιῆς χειρός, φλεγμασίη ἐπεγένετο, καὶ σφάκελος
καὶ πυρετός. ὑπεκαθάρθη πέμπτῃ μετρίως, αἱ θέρμαι
ἀνῆκαν καὶ αἱ ὀδύναι, τοῦ δακτύλου τι ἀπέπεσεν. μετὰ
δὲ τὴν ἑβδόμην ἐξήει ἰχὼρ ἐπιεικῶς· μετὰ ταῦτα τῇ
γλώσσῃ οὐ πάντα ἔφη δύνασθαι ἑρμηνεύειν· πρόρρη-

118 ἐτέλευσε Μ
119 πεμπταῖος V
120 περιγένετο V

320

shrilly; some are paralyzed: if the wound is on the right side paralysis on the left, if on the left, the right. Theodorus' child, having been exposed to the sun on the ninth day, got an insignificant fever on the tenth day from the exposed bone. But in the fever it turned black, the skin fell away. Much shrill crying. On the twenty-second day the child's belly swelled up, especially at the hypochondria. On the twenty-third death. Those whose bones are broken have fever on the seventh day. If the weather is warm, even earlier. The more the bones are shattered the quicker it comes. Exarmodus' young slave had it immediately, and the pain in the thigh, not on the side of the wound, and he cried shrilly, and had pain in the neck. Poseidocreon had convulsions on the third day. The fever did not leave him; he died on the eighteenth day. Isagoras' son was struck behind the head, the bone was shattered and grew black on the fifth day. He survived, and the bone did not come away.[16]

36. The commander of the large ship, whose right forefinger and its bone the anchor crushed: inflammation developed, gangrene, and fever. He was purged moderately on the fifth day. The fever relaxed and the pain; part of the finger fell away. After the seventh day a fair amount of serum ran out. After that he said that he could not articulate everything with his tongue. The prediction: *opis-*

---

[16] Cf. *Epid.* 5.97.

σις, ὀπισθότονος· ξυνεφέροντο[124] αἱ γνάθοι ξυνερειδόμεναι, ἔπειτα ἐς τράχηλον, τριταῖος ὅλος ἐσπᾶτο ἐς
τοὐπίσω ξὺν ἱδρῶτι· ἑκταῖος ἀπὸ τῆς προρρήσιος
ἀπέθανεν.

37. Ὁ δὲ[125] ἐκ τῆς Ἁρπάλου ἀπελευθέρης Τηλεφάνης τύμμα κάτωθεν μεγάλου δακτύλου ἔλαβεν· ἐπ-
406 εφλέγμηνε, καὶ σφόδρα | ἐπώδυνος ἦν, καί, ἐπεὶ ἀνῆ-
κεν, ᾤχετο ἐς ἀγρόν. ὀσφῦν ἤλγησεν, ἐλούσατο, αἱ
γέννες ξυνήγοντο ἐς νύκτα·[126] καὶ ὀπισθότονος παρῆν·
τὸ σίαλον ἀφρῶδες, μόγις ἔξω διὰ τῶν ὀδόντων διῄει·
τριταῖος ἀπέθανεν.

38. Ῥίνων ὁ τοῦ Δάμωνος, περὶ κνήμην καὶ σφυρὸν
ἕλκος κατὰ νεῦρον, ἤδη καθαρόν· τούτῳ δηχθέντι ὑπὸ
φαρμάκου ξυνέβη ὀπισθοτόνῳ ἀποθανεῖν.

39. Δείνωνι[127] ὑπ᾽ Ἀρκτοῦρον, καὶ πρότερον ἐκ πυρετοῦ θερινοῦ καὶ διαρροίης ἀσθενέως[128] διατεθέντι,
ἐκ πορείης κοπιάσαντι, καὶ πλευροῦ ὀδύνη ἀριστεροῦ·
καὶ βήξ, ἔχουσα μὲν ἐκ καταρρόου καὶ πρότερον,[129]
τότε δὲ ἦν κατακορής· καὶ ἄγρυπνος, καὶ δυσφόρως
φέρων τὸν πυρετὸν εὐθὺς ἀπ᾽ ἀρχῆς. καὶ ἀνακαθίζων
τριταῖος ἔπτυσεν ὠχρόν· ἀρτηρίη ὑπεσύριζε ῥεγχῶ-
δες. περὶ τὴν πέμπτην, πνεῦμα ἐπιεικῶς πυκνόν· πόδες, καὶ αἱ κνῆμαι, ἄκρεα τὰ πλεῖστα ψυχρά, καὶ ἔξω
τοῦ ἱματίου ὑποχώρησις ἀπ᾽ ἀρχῆς ἐπεγένετο χολώ-

[124] τι πρόρρησις ὀπισθότονος ηεξει ξυνεφ. M
[125] M writes ὁ δὲ twice   [126] ἐς νύκτα om. V
[127] Δείνωνι R: Δεινόνη V: δεινὸν εἰ M: Δίνωνι HI
[128] ἀσθενῶς M   [129] καὶ οὐ πρότερον MV: corr. Asul.

thotonos.[17] His jaws became fixed together, then it went to the neck, on the third day he was entirely convulsed backward, with sweating. On the sixth day after the prediction he died.

37. Telephanes, son of Harpalus' freedwoman, received a blow at the base of the thumb. It became inflamed and was extremely painful, and when it desisted he went into the fields. He had pain in the lower back. He bathed. His jaws were drawn together towards night and *opisthotonos* was present. His saliva was frothy; it passed out through the teeth with difficulty. He died on the third day.

38. Rhinon, Damon's son, had a wound in the tendon at the lower leg and the ankle; it had already become clean. But, eroded by a drug, he died from *opisthotonos*.[18]

39. Deinon, towards the rising of Arcturus, having previously been weakly disposed because of summer fever and diarrhea, was fatigued from a journey and got a pain in his left thorax. And the cough, which he had had earlier from a catarrh, became at that time severe. He was sleepless and bore the fever badly from the beginning. And sitting up on the third day he spat up yellowish matter. His trachea whistled and wheezed. About the fifth day, breathing somewhat rapid. His feet and calves, most of his extremities, were cold and outside the covers. The excrement was bilious from the beginning, not excessively

[17] The kind of tetanus that draws the patient backwards into a bow shape.

[18] It looks as though the writer blames a purgative drug for the tetanus which killed the patient.

δης, οὔτε λίην ὀλίγη, οὔτε πολλή. ἑβδομαῖος καὶ
ὀγδοαῖος καὶ ἐναταῖος ῥᾷον ἐδόκει φέρειν, καί τινες
ὕπνοι ἐγένοντο, καὶ τὰ ἀποχρεμπτόμενα πεπτότερα.¹³⁰
δεκάτῃ καὶ μέχρι τρισκαιδεκάτης, σφόδρα λευκὰ καὶ
καθαρά· καὶ ὑποχόνδριον λαπαρώτερον ἐγένετο, ἀριστερὸν ξυντεταμένον· καὶ εὐπνούστερος·¹³¹ πρὸς βάλανον ὑπῆλθε μετρίως.¹³² τρισκαιδεκάτῃ πάλιν ἔπτυσεν ὠχρόν, τεσσαρεσκαιδεκάτῃ δὲ μᾶλλον, πεντεκαιδεκάτῃ δὲ πρασοειδές· κοιλίη δὲ κακώδεα, χολώδεα,
ὑγρά, συχνὰ ὑπῆλθεν· ἀριστερὸν ὑποχόνδριον ἐπήρε-
408   το, ἑκκαιδε|κάτῃ δὲ καὶ σφόδρα ἐπώδησε· καὶ τὸ πνεῦ-
μα ἤδη ῥεγχῶδες· ἱδρὼς περὶ μέτωπον καὶ αὐχένα,
ὀλιγάκις ἐπὶ στῆθος· ἄκρεα καὶ μέτωπον ἐπιεικῶς
διετέλει ψυχρά· πηδηθμὸς δὲ φλεβῶν περὶ κροτάφους
κατεῖχεν· ὕπνοι κωματώδεις καὶ ἡμέρην καὶ νύκτα
τοὺς τελευταίους χρόνους· οὖρον ἀπ᾽ ἀρχῆς ὠμόν,
σποδοειδές· περὶ δεκάτην μέχρι τρισκαιδεκάτης λεπτὰ καὶ οὐκ ἄχροα, ἀπὸ δὲ τῆς τρίτης καὶ δεκάτης
οἷάπερ ἀπ᾽ ἀρχῆς.

40. Κλεόχῳ πλευροῦ ἄλγημα· ἀνῆκεν ὁ πυρετός·
ἵδρωσε τὸ σῶμα ὅλον· ἐν τῷ οὔρῳ πολλὰ τὰ ἐμφερό-
μενα ἦν, ἐθορυβήθη μετὰ ταῦτα.

41. Μετὰ Πληϊάδων δύσιν τὴν Ὀλυμπιάδεω, ὀκτά-
μηνον ἔχουσαν, ἐκ πτώματος πυρετὸς ὀξὺς ἔλαβε·
γλῶσσα καυσώδης, ξηρή, τρηχείη, ὠχρή· ὀφθαλμοὶ
ὠχροί, καὶ τὸ χρῶμα νεκρῶδες. διέφθειρε πεμπταίη·
ῥηϊδίως ἀπήλλαξε, καὶ ὕπνος, ὡς ἐδόκει, κωματώ-

little, not a lot. On the seventh, eighth, and ninth days he seemed to be better and had some sleep; the material coughed up was more ripened. On the tenth day and up to the thirteenth it was quite white and clean. And his hypochondrium became softer; left side tight. He breathed more easily. A suppository produced a moderate movement. On the thirteenth day he again spat up yellowish matter, more on the fourteenth, and on the fifteenth leek-colored matter. Bowels produced large, foul-smelling, bilious, damp movements. His left hypochondrium became elevated, and on the sixteenth day was very swollen. By now his breath was wheezing. Sweat about forehead and neck, sometimes on to the chest. Extremities and forehead stayed rather cold. There was persistent leaping of the blood-vessels in the temples. Sleep was comatose night and day in the last period. The urine from the beginning raw, full of particles. About the tenth to the thirteenth day thin and not colorless, but after the thirteenth as from the beginning.

40. Cleochus had pain in the thorax. The fever remitted. He sweated over the whole body. In the urine there were many particles suspended; later it was turbid.

41. After the setting of the Pleiades an acute fever seized Olympiades' wife after a fall in her eighth month of pregnancy. Her tongue was burnt, dry, rough, pale yellowish. Her eyes were yellowish and her skin corpselike. She aborted on the fifth day. The abortion was easy, and her sleep, as it appeared, was comatose. When they tried to

---

130 Smith: ἀπεπτότερα MV: ἀμεμπτότερα Li.

131 εὐπνούστερον MV: corr. Li.

132 μετθίως (sic) V

δης·[133] δείλης διεγειρόντων οὐκ ᾐσθάνετο, πταρμικῷ
ὑπήκουσε, πόμα κατεδέξατο καὶ χυλοῦ, ὑπέβησσε
καταπίνουσα[134] τὸ πόμα· ἡ φωνὴ οὐκ ἐλύετο, οὐδὲ
αὐτή τι ἀνέφερεν· τὰ ὄμματα κατηφέα· πνεῦμα μετέωρον, κατὰ ῥῖνα σπώμενον· χρῶμα πονηρόν· ἱδρὼς περὶ
τοὺς πόδας καὶ σκέλεα τελευτώσης.

42. Τῇ Νικολάου ἐκ καύσου τὰ παρ' οὖς ἐγένετο ἐπ'
ἀμφότερα, ὀλίγῳ ὕστερον τὸ ἕτερον, ἤδη δοκέοντος
χαλᾶν τοῦ πυρετοῦ, ὡς οἶμαι, περὶ τεσσαρεσκαι-
δεκάτην· μεγάλα ἀσήμως κατέστη· ὑπέστρεψεν· χρῶ-
μα νεκρῶδες, γλῶσσα τρηχείη, δασέη σφόδρα, | ὑπό-
λευκος, διψώδης· ὑποχώρησις κάτω, πολλή, ὑγρή,
κακώδης παρὰ πάντα τὸν χρόνον, πρὸ τῆς τελευτῆς
ἐφθάρη τὸ σῶμα τῷ πλήθει· ἐτελεύτησεν ὑπὲρ τὰς
εἴκοσιν.

43. Ἀνδρέαν πρὸ Πληϊάδος, φρίκη, πυρετός, ἔμε-
τος· ἀπ' ἀρχῆς ἡμιτριταῖος ἐφαίνετο. τριταῖος δ' οὖν
ἐὼν πάλιν ἀγοράζων ἔφριξε· πυρετὸς ὀξύς· ἔμετος
χολῆς ἀκρήτου· παραλήρησις ἐς νύκτα· ῥᾴων[135] πάλιν, πέμπτῃ, χαλεπῶς. ἕκτῃ, ἀπὸ λινοζώστιος εὖ[136]
ὑπῆλθεν. ἑβδόμῃ, χαλεπώτερον· καὶ τὰς ἐφεξῆς ξυν-
εχέστερος ἤδη, καὶ ἀνίδρωτος ἀπ' ἀρχῆς καὶ διψώδης·
μάλιστα δὲ τὸ στόμα ἀπεξηραίνετο, καὶ πόμα οὐδὲν
ἡδέως προσεδέχετο, ἀηδίης πολλῆς ἐούσης περὶ τὸ
στόμα· γλῶσσα ξηρή, ἄκροπις,[137] τρηχύτης ἐπήνθει
ὠχρόλευκος· ἄγρυπνος, ἀσώδης, ἐκλελυμένος, κεκλα-
σμένος, γλῶσσα ὑπὸ ξηρότητος ἐνίοτε ὑπότραυλος,

rouse her in the evening she did not notice; after a sternutatory she could hear. She took a drink and some broth;
she coughed while drinking the drink. Her voice was not
released, nor did she bring up anything. Her eyes were
dim. Breath shallow, drawn through the nose. Color bad.
Sweat about the feet and legs as she was dying.

42. Nicolaus' wife, as a result of a burning fever, developed swellings by both ears, one shortly after the other,
when it seemed that the fever was already stopping, I believe about the fourteenth day. They remained large, without signs. She relapsed. Color corpselike, tongue rough,
very fuzzy, whitish; thirsty. Bowel movements numerous,
wet, foul-smelling the whole time. Her body was wasted in
its bulk before the end. She died after the twentieth day.

43. Before the rising of the Pleiades, Andreas had shivering, fever, vomiting. From the outset it looked like a
semitertian. On the third day he again had a fit of shivering
in the marketplace. Acute fever. Vomitus of pure bile. Delirious towards night; easy again. On the fifth day, bad. On
the sixth, a good bowel movement from *linozostis*. On the
seventh, worse. On the following days it became more continuous, and was without sweat from the beginning, and
accompanied by thirst. Especially his mouth was parched
and there was no drink he could take with pleasure because of the great discomfort of the mouth. The tongue
was dry, inarticulate. A yellow-white hardness bloomed on
it. He was sleepless, nauseous, uncoordinated, helpless.
The tongue, from dryness, sometimes lisped until he wet-

---

133 καυματώδης MV: corr. ms. rec. D (Li.)
134 καταπίσα M: corr. M2   135 ῥέων MV: corr. M2 recc.
136 καλῶς V   137 M: κόπρις V: ἀκροαπὶς Gal. *Gloss.*

ἕως διαβρέξειεν· χυλὸν μάλιστα προσεδέχετο. ἐναταίῳ ἢ δεκαταίῳ παρ' οὓς ἀριστερὸν καὶ παρὰ τὸ
ἕτερον ἐπάρματα σμικρά· ἀσήμως ἠφανίσθη· οὖρα
διὰ παντὸς οὐκ ἄχροα, ἄνευ δὲ ὑποστάσιος. τεσσα-
ρεσκαιδεκαταίῳ ἱδρώτιον περὶ τὰ ἄνω· οὐ πολὺ μετριώτερον ἡ θέρμη, περὶ τὰς ἑπτακαίδεκα ἐμωλύνθη.
κοιλίη μετὰ τὰς δέκα ξηρή, οὐκ ἄνευ βαλάνων ὑπο-
χωρέουσα. περὶ δὲ τὰς πέντε καὶ εἴκοσιν, ἐξανθήματα[138] ὀλίγου κνησμώδεα, θερμά, ὥσπερ πυρίκαυ-
στα. ὀδύνη δὲ ἦν περὶ μασχάλας καὶ πλευρά· ἐς
σκέλεα διῆλθεν ἀσήμως, καὶ ἔληξεν. λουτρὸν ὠφέλει
καὶ χρῖμα[139] τὸ ἐν τῷ ὄξει. μηνὶ δὲ[140] δευτέρῳ ἴσως ἢ
τρίτῳ, ἐς νεφροὺς ἡ ὀδύνη, καὶ πρότερόν ποτε γενο-
μένη, κατέστη. |

412   44. Ἀριστοκράτει περὶ ἡλίου τροπὰς χειμερινὰς[141]
κόπος καὶ φρίκη καὶ θέρμη· μετὰ δὲ ἤρξατο τριταίῳ
πλευροῦ ὀδύνη καὶ ὀσφύος, καὶ οἴδημα ἐκ τῆς μασχάλης ἀρξάμενον παρ' ὅλον τὸ δεξιὸν πλευρόν, σκληρόν·
κατὰ δ' αὐτὴν τὴν πλευρὴν ἐκ μασχάλης ἀρξάμενον
ἐρυθρὸν καὶ πελιδνὸν ὡς ὑπὸ πυρὸς θαλφθὲν καὶ
ἐκκεκαυμένον.[142] ἀσώδης, δυσφόρως ἔχων, σφόδρα
διψώδης, γλῶσσα ὑπόλευκος, οὖρα οὐκ ἐχώρει, σκέλεα ὑπόψυχρα· ὑποχώρησις ἀπὸ λινοζώστιος ὀλίγη,
ὑγρά, ὑπόλευκα, ἀφρώδεα. ἐς νύκτα πνεῦμα ἐμετεω-
ρίζετο· ἱδρώτιον περὶ μέτωπον· τὰ κάτω ψυχρά· ἀσώ-
δης· τράχηλος ἐνεφυσᾶτο· βὴξ οὐκ ἐνῆν· ἐτελεύτησεν
ἔμφρων.

ted it. He took barley broth for the most part. On the ninth or tenth day there were small swellings by the left ear and then beside the other one. They disappeared without a trace. Throughout, the urine was not of bad color but without sediment. On the fourteenth day, sweat around the upper parts. The fever continued not much moderated; it went down on the seventeenth day. Bowel dry after the tenth day, passing nothing without a suppository. About the twenty-fifth day, an eruption appeared, almost itching, hot, like burns. There was pain around the armpits and the thorax. It went down into the legs without signs, and ceased. Bathing helped, and ointment made with vinegar. In the second month, perhaps, or the third, the pain went to the kidneys, having appeared there sometime before. He recovered.

44. Aristocrates, around the winter solstice, had fatigue, shivering and fever. Later, on the third day, a pain developed in his ribs and lower back, and a swelling, commencing from the armpit, down the whole right side; it was hard. On the ribs themselves, starting from the armpit, it was red and livid as though scalded and burned by fire. He was nauseous, uncomfortable, very thirsty, his tongue whitish; no urine passed; legs chilly. A bowel movement from *linozostis*, small, wet, whitish, foamy. Towards night his breathing became shallow; sweat on the forehead; lower limbs cold. Nauseous. Neck inflated. There was no cough. He died while conscious.

---

138 recc.: ἐξάνθημα MV
139 Smith: χρῆμα MV: χρίσμα recc., edd.   140 om. V
141 Corn.: τροπέας χ. V: τροπέων χειμερινέων M
142 ἐκκαυμένον (sic) M

45. Μνησιάνακτι περὶ φθινόπωρον ὀφθαλμίη, μετὰ
δὲ τεταρταῖος πυρετός· ἀρχόμενος τοῦ τεταρταίου
σφόδρα ἀπόσιτος, προσιόντος δὲ ἡδέως πρὸς σιτίον·
καὶ Πολυχάρει[143] δὲ ἐν τεταρταίῳ ὅμοια τὰ περὶ τὴν
σίτισιν. ξυνέβη δὲ τῷ Μνησιάνακτι ὑποχώρησις ἔμπροσθεν τοῦ πυρετοῦ, καὶ μετὰ ἐπὶ πολὺν χρόνον
παρηκολούθει πολλῶν, λευκῶν, μυξοποιῶν, καὶ ἔστιν
ὅτε σμικρὸν αἷμα ἄνευ τόνων καὶ ὀδύνης· ψόφοι δὲ ἐν
γαστρί. μετὰ τὸν πυρετόν, ἀπέστη παρὰ τὴν ἕδρην
φῦμα σκληρόν, πολὺν χρόνον παρηκολούθει ἄπεπτον,
ἐρράγη ἐς τὸ ἔντερον, καὶ ἔξω συριγγῶδες ἐγένετο.
περιπατοῦντι δ᾽ αὐτῷ ἐν τῇ ἀγορῇ, μαρμαρυγαὶ πρὸ
τῶν ὀφθαλμῶν, καὶ τὸν ἥλιον οὐ πάνυ καθεώρα· ἀπο-
χωρήσας δὲ σμικρὸν ἐξ ἑωυτοῦ ἦν καὶ τράχηλον
σπασμώδης. ἐπεὶ[144] δὲ ἐκομίσθη ἐς οἶκον, μόγις ἀν-
έβλεψε, καὶ αὐτὸς ἑωυτοῦ μόγις ἐγένετο· τὸ πρῶτον δὲ
περιέβλεπε τοὺς περιεστῶτας, καὶ τὸ σῶμα | κατ-
414 εψύχθη, μόγις δὲ ἀνεθερμάνθη ἀσκίοισι καὶ πυρίῃ
ὑπὸ τῇ κλίνῃ.[145] ἐπεὶ δ᾽ ἐντὸς ἑωυτοῦ ἦν, καὶ ἐξανίστατο, οὐκ ἐξιέναι ἤθελεν, ἀλλὰ δεδιέναι ἔφη· εἴ τε
τις περὶ νοσημάτων χαλεπῶν διαλέγοιτο, ὑπεξῄειν
φόβῳ·[146] ἔστι δ᾽ ὅτε προσπίπτειν αὐτῷ πρὸς τὰ ὑποχόνδρια θερμασίην ἔφη, καὶ τῶν ὀφθαλμῶν μαρμαρυ-
γὰς παρακολουθεῖν. καὶ ἡ ὑποχώρησις πολλὴ καὶ
πολλάκις, καὶ ὁμοίη χειμῶνος ξυνέβη. φλεβοτομίη·
ἐλλέβοροι· γαλακτοποσίη βοείου, πρότερον δὲ ὀνείου,
ξυνήνεγκε, καὶ τὰς ὑποχωρήσιας ἔπαυσεν· ὑδροποσίη
ἀπ᾽ ἀρχῆς, καὶ περίπατοί τε καὶ κεφαλῆς καθάρσιες.

330

45. Mnesianax had eye trouble towards fall. Later, quartan fever. When he began his quartan he was without appetite, but as it progressed he enjoyed food. (Polychares, in a quartan, had similar responses to meals.) It occurred with Mnesianax that before the fever and for a long time afterwards his bowel movements continued to be of much white, mucous matter, and sometimes there was a little blood without stretching and pain. And noise in the belly. After the fever there rose by his anus a hard swelling. It persisted unripened a long time, broke through into the gut, and came outward as a fistula. As he was walking about in the marketplace there were sparks before his eyes and he did not see the sunlight well. He withdrew a short distance. He was beside himself, and began to have spasms in the neck. When he was taken home he could hardly see and had difficulty coming to himself. At first he peered around at the bystanders and his body was cold, and was warmed only with difficulty, with bags of hot water and with steam beneath his bed. When he had come to himself and stood up he did not want to go out, but said he was afraid. And if anyone spoke of severe illnesses he would withdraw in fright. Sometimes he said heat fell on his hypochondria and that the sparks before his eyes continued. Bowel movements large and frequent, and they were like those of winter. Phlebotomy; hellebore. Drinking cow's milk, but earlier, ass's, helped, and stopped the bowel movements. Drinking of water from the beginning and walks, along with purges of the head.

---

143 Πολύχαρι MV: corr. Li. ex ms. rec. K   144 ἐπὶ M

145 κλίνη (sic) V: καινῇ M

146 Smith: ὑπεξέειν φόβον V: ὑπέξη φόβῳ M

46. Τῷ Ἀνεχέτου ταῦτα·[147] χειμῶνος ἐν βαλανείῳ πρὸς πυρὶ χριόμενος ἐθερμάνθη, καὶ παραχρῆμα ἐπι-ληπτικοῖσι σπασμοῖς·[148] ἐπεὶ δ᾽ ἀνῆκαν οἱ[149] σπα-σμοί, περιέβλεπεν, οὐ παρὰ[150] ἑωυτῷ ἦν. ἐπεὶ δὲ ἐντὸς ἐγένετο ἑωυτοῦ, πάλιν τῇ ὑστεραίῃ πρωῒ ἐλήφθη· σπασμώδης· ἀφρὸς δὲ οὐ πάνυ· καὶ τρίτῃ ἄκροπις· καὶ τετάρτῃ ἐπεσήμανεν αὐτῇ τῇ γλώσσῃ, ἔπταιεν, οὐχ οἷός τε ἦν λέγειν, ἀλλ᾽ ἴσχετο ἐν τῇσιν ἀρχῇσι τῶν ὀνομάτων. καὶ τῇ πέμπτῃ[151] γλῶσσα σφοδρά, καὶ ὁ σπασμὸς ἐπεγένετο, καὶ ἐξ ἑωυτοῦ ἐγένετο· ὅτε δὲ ἠνίει[152] ταῦτα, ἡ γλῶσσα μόγις ἀποκαθίστατο εἰς τωυτό. ἑκταίῳ ἀποσχομένῳ πάντων, καὶ ῥυφήματος καὶ ποτοῦ, οὐκ ἔτι ἐλάμβανεν.

47. Κλεόχῳ ἐκ κόπων καὶ γυμνασίων, μέλιτι τὰς ἡμέρας διαχρωμένῳ, οἴδημα ἐς γόνυ δεξιόν, μᾶλλον δὲ ἐς τὸ κάτω περὶ τοὺς τένον|τας τοὺς ὑπὸ τῷ γούνατι· περιήει ὑποχωλαίνων·[153] καὶ ἡ γαστροκνημίη ᾤδει, καὶ σκληρὴ ἦν καὶ ἐς τὸν πόδα καὶ ἐς τὸ σφυρὸν τὸ[154] δεξιόν· καὶ ἐς τὰ οὖλα παρ᾽ ὀδόντας μεγάλα, ὡς ῥᾶγες, πελιδνά, μελαινόμενα, ἀνώδυνα ὁπότε μὴ ἐσθίοι, καὶ τὰ σκέλεα, εἰ μὴ ἐξανίσταιτο·[155] ἦλθε γὰρ καὶ ἐς τὸ ἀριστερὸν τὸ οἴδημα, ἧσσον δέ· καὶ ἀπελειαίνετο[156] ἐν τοῖσιν οἰδήμασι τοῖσι περὶ γούνατα καὶ πόδας, ὥσπερ ὑπώπια.[157] τέλος δὲ οὐχ οἷός τε ἦν ἵστασθαι, οὐδ᾽ ἐπὶ τὰς πτέρνας ἐπιβαίνειν, ἀλλὰ κλινοπετής. θέρμαι δή-

416

147 ταῦτα V   148 σπάσμασιν V
149 ἐπεὶ δ᾽ ἂν ἱκανοὶ MV: corr. Li.   150 περὶ M

46. Anechetus' son had these symptoms: in winter, in the bath, he was heated by the fire as he was being rubbed with oil. Suddenly he had epileptic convulsions. When the convulsions stopped he gazed about; he was not with himself. And when he came to himself he was again seized on the next day in the morning. Convulsions, some foam. On the third day inarticulate. On the fourth day he gave a sign with his tongue, fell, was unable to speak, but hung up on the beginnings of words. On the fifth day, tongue severely affected; the convulsion came on and he was beside himself. When these things ceased, his tongue with difficulty returned to its own condition. On the sixth day, as he abstained from everything, both gruel and drink, there were no further seizures.

47. Cleochus, after exercise and fatigue, and after having used honey for some days, had a swelling on the right knee, especially below it on the tendons under the knee. He walked about lamed. His calf swelled up. It was hard, clear down to the foot and the right ankle. On the gums by the teeth developed large grapelike swellings, livid, blackish, painless except when he was eating, as were the legs if he did not stand up, for the swelling had gone to his left side too, though less. There was a smoothness in the swelling about the knees and feet, like that of a black eye. Finally he was not able to stand or to put weight on his heels, but was confined to bed. Fever sometimes evident.

---

151 πέμπτῃ τε M   152 ἦν εἴη V

153 ὑποχολαίνων M   154 τὸν σφυρὸν τὸν V

155 ἐξανίστατο V

156 ἀπολιαίνετο M: ἐπεμελαίνετο Langholf

157 ὑπόπια M

λοι ἐνίοτε· ἀπόσιτος, οὐ πάνυ διψώδης· οὐδὲ ἐπὶ θῶκον[158] ἀνίστατο, ἀσώδης, καὶ ἔστιν ὅτε ὀλιγοψυχίη.
ἐλλέβοροι, κεφαλῆς καθάρσιες· πρὸς τὸ στόμα μαννῶδες[159] ξὺν τοῖσιν μισγομένοισι ξυνήνεγκεν· πρὸς τὰ
ἐν τῷ στόματι ἕλκεα, ῥύφημα φακῆ ἐπιτήδειον. περὶ
δὲ[160] ἑξηκοστὴν κατέστη τὰ οἰδήματα πρὸς τοῦ δευ-
τέρου ἐλλεβόρου μοῦνον· ὀδύναι ἐς τὰ γούνατα κατα-
κειμένῳ ἦλθον, ὑγρὸν δὲ καὶ χολὴ[161] ἀπέστη ἐς γούνατα, καὶ πλείους ἡμέρας πρὸ[162] τοῦ ἐλλεβόρου.

48. Πεισιστράτῳ ὤμου ἄλγημα, καὶ βάρος πολυ-
χρόνιον περιπατοῦντι[163] καὶ τἆλλα ὑγιαίνοντι· χειμῶνος δὲ ἐπέπεσε πλευροῦ ὀδύνη, καὶ θέρμη, καὶ βήξ,
ἀπόχρεμψίς τε αἵματος ἀφρώδεος· τούτῳ[164] καὶ ῥεγ-
χῶδες ἐν τῇ φάρυγγι· εὔφορος δὲ καὶ παρ' ἑωυτῷ. καὶ
ἡ θέρμη ἐχάλασεν, ἅμα δὲ καὶ ἡ ἀπόχρεμψις, καὶ τὸ
κέρχνον· καὶ περὶ τετάρτην ἢ πέμπτην ἡμέρην ὑγιής. |

49. Τῇ Σίμου ἐν τόκῳ σεισθείσῃ ἄλγημα περὶ
στῆθος καὶ πλευρόν· βήξ, πυρετοί, ἀποχρέμψιες ὑπο-
πυώδεις. ἐς φθίσεας[165] καὶ κατέστη· καὶ ἐξ μῆνας οἱ
πυρετοί, καὶ διάρροια αἰεί· ἐπὶ τέλει παῦσις πυρετοῦ·
κοιλίη ἔστη μετὰ τὴν παῦσιν· ἡμέρας μεθ' ἑπτὰ ἐτελεύτησεν.

50. Καὶ ἡ Εὐξένου· ἐκ πυριήσιος ἐδόκει· θέρμαι οὐκ
ἔλειπον οὐδένα χρόνον, μᾶλλον πρὸς ἑσπέρην ἐπέτει-
νον· ἱδρῶτες ἐγίνοντο ἐς ὅλον τὸ σῶμα· μέλλοντος

158 Smith: θάκον MV   159 μανῶδες M: μανιώδεες V
160 om. M   161 πολλὴ χολὴ V

418

334

No appetite, very little thirst. He did not get up for the toilet, was nauseous, and periodically had faintness. Hellebore, purging of the head. *Manna* in a mixture was of benefit for his mouth. For the ulcers on the mouth lentil soup as a porridge was a help. About the sixtieth day, the swellings went down only because of the second hellebore. Pains came on his knees as he was lying down, and water and bile settled on the knees even for many days before the hellebore.

48. Peisistratus had pain in the shoulder and heaviness for a long time, though he walked about and was otherwise healthy. In the winter there came on much pain in the thorax, fever, cough, coughing up of frothy blood. And with this there was wheezing in the pharynx. But he was generally sound and rational. The fever receded, and with it the expectoration and the hoarseness. On the fourth or fifth day he was healthy.

49. The wife of Simus, shaken in childbirth, had a pain in the chest and ribs. Cough, fever, bringing up of somewhat purulent matter. She went into consumption and for six months had fever and constant diarrhea. At the end, cessation of fever. Her bowels stabilized after the cessation. After seven days she died.[19]

50. Also Euxenes' wife. It seemed to come after a steam bath. The fever did not depart for any period, but intensified more towards evening. Sweat occurred over her whole

19 Cf. *Epid.* 5.103.

---

162 πρὸς M   163 περικρατοῦντι V   164 τοῦτο M
165 Smith: πυώδεες φθίσις M: ὑποπυώδεες ἐς φθίσιες V

ἐπιτείνειν τοῦ πυρετοῦ ψύξιες τῶν ποδῶν, ὁτὲ δὲ καὶ
κνημέων καὶ γουνάτων, ἐγίνοντο· βηχίον[166] ξηρὸν
ὀλίγον χρόνον ἀρχομένου τοῦ πυρετοῦ παροξύνεσθαι,
εἶτ᾽[167] ἔλεγε, διὰ δὲ χρόνου πολλοῦ καὶ ὅλου[168] τοῦ
σώματος ῥῖγος ἐγένετο· ἄδιψος διὰ παντός. φάρμακον
πιοῦσα καὶ ὀρὸν ἐβλάβη μᾶλλον. ἀπ᾽ ἀρχῆς πάντων
ἀνώδυνος καὶ εὔπνοος· μεσοῦντος δὲ τοῦ χρόνου πλευροῦ δεξιοῦ ἐγένετο ἄλγημα καὶ ἡ βὴξ ἐκινήθη, καὶ
ἆσθμα, καὶ ἀποχρέμψιες σμικραί, λευκαί, ὑπόλεπτοι·
καὶ ἡ φρίκη, οὐκ ἔτι ἐκ ποδῶν ἀλλὰ ἀπὸ τραχήλου καὶ
νώτου· κοιλίη ὑγροτέρη. ἐχάλασεν ὁ πυρετὸς μετὰ
πολλοῦ ἱδρῶτος, καὶ κατεψύχθη· ἆσθμα δὲ ἦν ποι-
κίλον· ἐτελεύτησε μετὰ τὴν ἄφεσιν ἑβδόμῃ ἔμφρων.

51. Καὶ ἡ Πολεμάρχου θέρεος ἤρξατο πυρεταίνειν·
ἀφῆκε δεκαταίην·[169] μετὰ δὲ ὑπεφέρετο· ἐς νύκτα θέρ-
μαι·[170] πάλιν δὲ διαλιπὼν ἔλαβεν ὁ πυρετός, καὶ οὐκ
ἀνῆκε σχεδὸν τριῶν μηνῶν. βὴξ πολλή· ἀπόχρεμψις
φλέγματος· ἐπεὶ περὶ[171] τὰς εἴκοσιν ἐγένετο, | πνεῦμα
αἰεὶ πυκνόν· ἐν τῷ στήθει ψόφοι· ἱδρώδης τὰ πολλά·
πρωὶ ἠπιώτερος ὁ πυρετός· καὶ φρῖκαι ἔστιν ὅτε ἐλάμ-
βανον· ὕπνοι ἐγίνοντο· καὶ κοιλίη ἔστιν ὅτε καθυγραί-
νετο, καὶ πάλιν ξυνίστατο· ἐγένετο ἐπιεικῶς. μεσοῦντος δὲ τοῦ χρόνου ἐς γούνατα καὶ κνήμας ἄλγημα· καὶ
ξυγκάμπτειν καὶ ἐκτείνειν ἄλλον ἔδει· διετέλει τὰ τῶν
σκελέων μέχρι τελευτῆς· ὑπόγυον δέ· καὶ ἐπῴδησαν[172]

420

166 βήξιον M 167 εἶτ᾽ M 168 ὀλίγου V
169 δεκαταίη V 170 θέρμη V

body. When the fever was going to intensify, her feet were cold, and sometimes her calves and knees. Dry cough for a short time as the fever was beginning to intensify, then it stopped and for a long time there was shivering of her whole body. No thirst throughout. She drank a purgative drug and whey and was further harmed. From the beginning she was without all pain and breathed easily. About the middle of the time, pain developed in her right thorax, the cough was aroused, and asthma, and small expectorations, white, thinnish. And the shivering, no longer from the feet but from the neck and back. Bowels rather watery. The fever ceased with much sweat, and she grew chilled. The asthma was variegated. She died conscious on the seventh day after the withdrawal of fever.

51. Also Polemarchus' wife, in summer, started to be feverish. It left her on the tenth day. But afterwards she relapsed. Fever towards night. Again the fever left and then seized her and did not leave her for nearly three months. Much coughing, bringing up of phlegm. When she came to about twenty days her breathing became continuously rapid. There were noises in her chest. She was sweaty for the most part. The fever was milder early in the day, and shivering sometimes seized her. She slept. Her bowels were watery sometimes and again stabilized. She had a fairly good appetite. But in the middle of the time she developed pain in the knees and lower legs, and it was necessary for someone else to bend and extend them. The symptoms in the legs persisted to the end, and that was

---

171 ἐπείπερ M
172 ὑπῴδησαν V

οἱ πόδες μέχρι κνημέων, καὶ ἐφαπτομένων ἤλγει· καὶ
οἱ ἱδρῶτες ἔληξαν καὶ τὰ ῥίγεα· ὁ δὲ πυρετὸς αἰεὶ
ἐπέτεινεν. πρὸ δὲ τῆς τελευτῆς κοιλίη κατερράγη·
ἔμφρων διετέλει· πρὸ τριῶν ῥεγχώδης ἐν φάρυγγι, καὶ
πάλιν ἐπανίετο.

52. Ἡγησιπόλιος παιδίον σχεδὸν τέσσαρας μῆνας
ἄλγημα περὶ ὀμφαλόν, βρωτικῶς[173] εἶχεν· προϊόντος
ἐπέτεινεν ἡ ὀδύνη, ἔκοπτε τὴν γαστέρα, ἐτίλλετο.
θέρμαι ἐπελάμβανον· ἐτήκετο· ὀστέα ἐλείφθη· τὰ πό-
δια ἐπῴδει, ὄρχιες· γαστρὸς τὰ περὶ ὀμφαλὸν πεφυ-
σιγγωμένον[174] ἀραιὸν οἶσι μέλλουσι κοιλίαι ὑπο-
ταράσσεσθαι· ἀπόσιτος ἐγένετο, γάλα προσεδέχετο·
ὑπόγυον, καὶ ἡ κοιλίη καθυγράνθη, καὶ ὕφαιμος ἰχὼρ
ὑπῄει κάκοδμος· κοιλίη ἐπίμπρατο.[175] ἐτελεύτησεν
ἐμέσας σμικρόν, βραχύ, φλεγματῶδες, ὥστε δόξαι
οἷον γονὴν τῆς πλατείης. τελευτήσαντι δ᾽ ἡ ῥαφὴ[176]
τῆς κεφαλῆς σφόδρα ἐκοιλάνθη· ἀρρωστῶν δ᾽ αἰεὶ τῇ
χειρὶ κατῆγε κατὰ τοῦ βρέγματος, μάλιστα δ᾽ ὑπόγυον, οὐκ ἤλγει δὲ τὴν κεφαλήν· καὶ ἐν μηρῷ ἀρι-
στερῷ ὑπὸ | βουβῶνα τὸν[177] κάτω, πελιδνόν· ἴσως τῇ
προτεραίῃ ὄρχιες[178] κατισχνάνθησαν. ὅμοια[179] δὲ καὶ
τῷ Ἡγητορίδεω παιδίῳ. ἀπέθανε. πλὴν ὅτι ἔμετοι
προσεγένοντο ὑπὸ τὴν τελευτὴν πλείους.

422

---

173 βρωτικὸς M   174 πεφυσηγγωμένον M
175 ἐμπίπρατο M   176 γόνη M
177 τῷ M   178 ὄρχιες δὲ M   179 om. V

at hand. Her feet swelled up as far as the calves, and were painful when touched. The sweats and the shivering abated, but the fever continually intensified. And before the end the bowel broke loose. She remained conscious. Three days before, she had wheezing in the pharynx, and again that stopped.

52. The infant son of Hegesipolis had, for nearly four months, a gnawing pain by the navel. As time passed, the pain intensified, he beat on his belly, plucked at it. Fever seized him. He wasted away. The bones were seized. His feet swelled; testicles. The parts of the belly around the navel formed a loose-textured stalk of material that was going to make a disturbance in the bowel.[20] He did not want food. He would take milk. The end was near. The bowel became watery, and bloody serous matter came out, foul-smelling. The intestine was inflamed. He died after vomiting a small short phlegmy object that seemed like the embryo of the flatworm. On his death the suture of his skull became very hollow. While he was sick he kept drawing down with the hand from the front of his head, especially as the end was imminent, but he had no pain in the head. And on the left thigh underneath the lower gland, a livid area. The testicles lost their swelling, perhaps on the previous day. Similar symptoms occurred for the child of Hegetorides, which died. Except that more vomiting occurred towards the end.

[20] The grammar and meaning here are not very clear, but the comparison seems to be drawn between the swelling of the area of navel and the growth of a garlic stalk which "puffs" as it grows.

53. Ἡ[180] Ἱππίου[181] ἀδελφεή, χειμῶνος, φρενιτική, ἁμαρτάνουσα, τῇσι χερσὶ πραγματευομένη, ἀμύσσουσα ἑωυτὴν πέμπτη. ἕκτη, ἐς νύκτα ἄφωνος, κωματώδης, ἐμφυσῶσα ἐς γνάθους καὶ χείλεα, ὡς οἱ καθεύδοντες· ἐτελεύτησε περὶ ἑβδόμην.

54. Ἄσανδρος[182] φρίξας· πλευροῦ ἄλγος, ἐς γούνατα καὶ μηρὸν ὀδύνη. φαγὼν παρεφρόνει, ἐτελεύτησε ταχέως.

55. Τῷ Κλεοτίμου σκυτεῖ, κοιλίης ὑγρανθείσης πολὺν χρόνον, καὶ θέρμης γενομένης, καθ᾽ ἧπαρ ἔπαρμα φυματῶδες ἐς ὑπογάστριον κατέβη· καὶ κοιλίη ὑγραίνετο· καὶ ἕτερον αὐτῷ καθ᾽ ἧπαρ ἄνω πρὸς ὑποχόνδριον φῦμα· ἐτελεύτησεν.

56. Οἷσι κεφαλῆς ὀδύνη δεινὴ ξὺν θέρμῃ, οἷσι μὲν ἐς τὸ ἥμισυ τῆς κεφαλῆς, καὶ κατὰ ῥῖνάς τι ὑγρὸν ἀποχωρεῖ λεπτὸν ἢ πέπον, ἢ ἐς ὦτα, ἐς φάρυγγα ἐκ κεφαλῆς, ἀσφαλέστερον· οἷσι δὲ ξηρὰ ταῦτα, ὁ δὲ σφάκελος δεινός, ἐπικίνδυνα· ἢν δὲ προσῇ ἀσώδης, ἢ χολώδης[183] ἔμετος, ἢ κατάπληξις ὀμμάτων, ἢ ἀφωνίη, καὶ σπάνιόν τι φθέγγηται, ἢ λῆρός τις, θανάσιμα καὶ σπασμώδεα. ὁπόσοι δ᾽ ἂν ἐκ κατάρρου τὸ ἥμισυ τῆς κεφαλῆς πονέοντες, καὶ κατὰ ῥῖνας ὑγροῦ ὑποχωρέοντος, ἐπιπυρετήνωσιν, ἐπιεικῶς ἐν τῇ πέμπτῃ ἢ ἕκτῃ περιψύχονται.

---

[180] om. V   [181] Ἵππιος MV: corr. recc.
[182] ὁσάνδιος MV: ὅσανδρος recc.: corr. Asul.
[183] χυλώδης M

53. The sister of Hippias, in the winter, was phrenitic; regimen bad; very busy with her hands, lacerating herself on the fifth day. On the sixth day, towards night, speechless, comatose, puffing into her jaws and lips like sleeping people. She died towards the seventh day.

54. Asandrus had been shivering; pain in thorax, painfulness into knees and thighs. He ate, became delirious, died swiftly.

55. For Cleotimus' shoemaker, after his bowels had been watery for a long time and a fever had come on, a tumorous swelling developed by the liver and descended to the lower abdomen. The bowels were watery. Another tumor developed above the liver towards the hypochondrium. He died.

56. People who have dreadful pain in the head with fever: if they have it in half the head, and their nose exudes something watery, or thin, or concocted, or if it comes to the ears or into the pharynx from the head, there is less danger. But when these things are dry, and the lesions are frightful, there is much danger. And if nausea be present, or bilious vomiting, or fixity of the eyes, or voicelessness and infrequent speech, or delirious talk; these signs are fatal and convulsive. Those who suffer from flux in half their head, and develop fever while there is a watery discharge by the nose, generally lose the fever on the fifth or sixth day.[21]

[21] Cf. *Epid.* 5.102.

57. Ἐχεκράτει τῷ τυφλῷ, κεφαλῆς ὀδύνη δεινή, μᾶλλον ἐς τοὔπισθεν, καὶ τραχήλου, ᾗ ἡ πρόσφυσις, 424 καὶ ἐς κορυφὴν ἐχώρει, προϊόντος δὲ καὶ ἐς οὖς ἀριστερόν· καὶ τὸ ἥμισυ τῆς κεφαλῆς ἐπώδυνον· μυξώδεα αἰεὶ ἐχώρει ἐπιεικῶς ξυγκεκαυμένα· καὶ θέρμη παρηκολούθει λεπτή· καὶ ἀπόσιτος· τὴν ἡμέρην ῥάων, ἐς νύκτα δὲ ὀδύνη. ἐπεὶ δὲ τὸ κατὰ τὸ οὖς ἐρράγη πύον, ἔληξε πάντα· ἐρράγη δὲ περὶ χειμῶνα. ἦρά γε ἐν[184] πᾶσι τοῖσιν ἐμπυήμασιν, καὶ τοῖσι περὶ ὀφθαλμόν, ἐς νύκτα οἱ πόνοι;

58. Οἷσι βῆχες χειμῶνος, μάλιστα δὲ ἐν[185] νότοισι, παχέα καὶ πολλὰ χρεμπτομένοισι, πυρετοὶ ἐπιγίνονται, ἐπιεικῶς δὲ πεμπταῖοι παύονται· αἱ βῆχες δὲ περὶ τὰς τεσσαράκοντα, οἷον Ἡγησιπόλει. οἷσι θέρμαι λεπταὶ ἔστιν ὅτε παυομένων, οὐχ ὅλον τὸ σῶμα, ἀλλ᾽ ἢ περὶ αὐχένα καὶ ὑπὸ μασχάλας, ἢ κεφαλὴν ἀφιδρώσαντες,[186] παύονται.

59. Χάρητι,[187] χειμῶνος, ἐκ βηχίου ἐπιδημίου προσγενόμενος πυρετὸς ἐπέλαβεν ὀξύς· τὰ ἱμάτια ἀπέβαλλε· κῶμα μετὰ πόνου ἐγένετο· οὖρα[188] ἐρυθρά, οἷον ὀρόβων πλύμα· ὑπόστασις εὐθὺς ἀπ᾽ ἀρχῆς πολλὴ λευκή, ὕστερον δὲ καὶ ὑπέρυθρος. ἑβδόμῃ, ἀπὸ βαλάνου σμικρὰ ὑπῆλθεν· τὸ κῶμα κατεῖχεν ἄλυπον· νοτὶς ἐπὶ μετώπῳ· ὕπνος ἐς νύκτα, καὶ θέρμη ἠπι-

---

184 ἦρά γε ἐν] ἠράγεε V: ἠραγεεν M (cf. Epid. 5.77)
185 om. M (cf. Epid. 5.78)
186 ἀφιδρώεντες M

342

57. Echecrates, who was blind, had a dreadful pain in the head, more to the back and in the neck at the top of the spine, and it reached to the crown of the head and as time passed also to the left ear. Half his head was painful. There was continuous mucous discharge, rather burnt. A light fever persisted, and he did not want food. In the daytime he was easier, but he had pain towards night. When the pus broke out by the ear all symptoms abated. It broke out towards winter. Is it true that in all suppurations, including those around the eye, the distress comes towards night?[22]

58. Those who have coughs in winter, especially in south winds, and who cough up much thick matter, develop fevers, but the fevers generally cease on the fifth day, while the coughs last until around the fortieth day, as with Hegesipolis. If people have light fevers of the sort that stop periodically, when they have sweats that are not over the whole of the body but either about the neck and down to the armpits or about the head, they are recovering.[23]

59. Chares, in winter. After an epidemic cough an acute fever adding to the cough seized him. He kept throwing off the bedcovers. Coma with distress developed. Urine red, like the water that vetch seeds have soaked in. There was much white sediment immediately from the beginning, later reddish. On the seventh day, from a suppository, a small movement. The coma persisted, without pain. Moisture on the forehead. Sleep towards night and more gentle

[22] Cf. *Epid.* 5.77.
[23] Cf. *Epid.* 5.78.

---

187 χάριτι M
188 ἐγένετονετοόυρα M

ωτέρη.[189] ὀγδόη, χυλὸν προσεδέξατο· κωματώδης διετέλει μέχρις ἑνδεκάτης. ταύτῃ δὲ καὶ ἡ θέρμη μάλιστα ἔληξεν· ὑπῆν δὲ βήξ,[190] ἀπόχρεμψις διετέλει αἰεὶ
πολλή, ῥηϊδίως, πρῶτον γλίσχρη, λευκή, παχέη, ἐπεὶ
δὲ ξυνεπεπαίνετο, ὁμοίη πυώδεσιν· οὖρα ἀπὸ ἑνδεκά-
της[191] καθαρώτερα, ὑπόστασις τρηχέη. τρισ|και-
δεκάτῃ ἄλγημα ἐπὶ δεξιὰ μέχρι κενεῶνος ἐς ὑπο-
γάστριον· οὖρον ἔσχετο· πόμα τὸ ἀπὸ τοῦ καλλιφύλ-
λου ξυνήνεγκεν. πεντεκαιδεκάτῃ πάλιν τὸ ἄλγημα.
ἑκκαιδεκάτῃ ἐς νύκτα μᾶλλον τὸ ἄλγημα[192] ὑποχονδρίου ἤρχετο ἐς κοιλίην· λινοζώστιος ὕδωρ ὑπήγαγεν.
ἡ θέρμη ἐντὸς τῶν εἴκοσιν ἐμωλύνθη, καὶ[193] ἀποχρέμψιες παρηκολούθουν παχέαι, ῥηϊδίως, ἐς τεσσαράκοντα.

60. Ὑποκαθαίρειν τὰς κοιλίας ἐν τοῖσι νοσήμασιν,
ἐπὴν πέπονα ἢ πεμπταῖα, τὰς μὲν κάτω ἐπὴν ἱδρυμένα
ἴδῃς· σημεῖον, ἢν μὴ ἀσώδεις μηδὲ καρηβαρικοὶ ἔωσι,
καὶ ὅτε αἱ θέρμαι πρηΰταται, καὶ ὅταν λήγωσι[194] μετὰ
τοὺς παροξυσμούς· τὰς δὲ ἄνω ἐν τοῖσι παροξυσμοῖσι, τότε γὰρ καὶ αὗται μετεωρίζονται, ἐπὴν ἀσώδεις
καὶ βαρεῖς τὰ ἄνω ἔωσιν. διὰ τοῦτο δὲ μὴ ἐν ἀρχῇσι
καθαίρειν, ὅτι ἀπὸ τοῦ αὐτομάτου ἐν τοῖσι χρόνοισι[195]
τούτοισιν[196] ἡ χρονίη ἐπικίνδυνος.[197]

189 ἠπιώτερον MV: corr. recc.
190 ὑπῆν δὲ ἦν βὴξ M
191 ἀπὸ ἑνδεκάτην M
192 ἑκκαιδεκάτη . . . ἄλγημα om. V
193 om. V

fever. On the eighth day he took broth. He stayed comatose until the eleventh day. On that day the fever, too,
abated for the most part. But the cough was there; much
matter continued to be brought up, easily, sticky at first,
white, thick, and, when it was concocted, like purulent discharges. Urine from the eleventh day clearer, the deposit
coarse. On the thirteenth day, pain on the right side and
stretching to the flank, towards the lower belly. The urine
was retained. The drink made from *kalliphyllon* helped.
On the fifteenth day, again the pain. On the sixteenth,
more towards night, the pain of the hypochondrium pro-
ceeded towards the lower belly. Water of *linozostis* produced a bowel movement. The fever was relieved before
the twentieth day, and the expectorations continued thick,
easily produced, to the fortieth day.

60. In diseases purge the intestines when the diseases
are ripe (concocted), or in their fifth day; the lower intestines, when you see the diseases are settled. Indication: if
the patients are not nauseous or heavy-headed, and when
the fever is mildest, and when it abates after the exacerbations. Purge the upper intestines during the exacerbations,
because the intestines are then elevated when patients are
nauseous and heavy in the upper parts. But here is the reason not to purge at the beginning: because chronic illness
is automatically a danger in this period.[24]

[24] Cf. *Epid.* 5.64.

---

[194] ληρῶσι MV (cf. *Epid.* 5.64)
[195] χρονίοισι MV: corr. Li.
[196] om. V
[197] ἐπικίνδυνα M

61. Ἐπὶ ὠλεκράνου[198] ἐκ τρώματος τρωθέντος
πῆχυς ἐπισφακελίσας πυοῦται·[199] πεπαινομένου δὲ
γλίσχρος ἰχὼρ καὶ κολλώδης ἐκθλίβεται· ταχὺ προσ-
ίσταται, ὡς καὶ Κλεογενίσκῳ καὶ Δημάρχῳ τῷ Ἀγλα-
οτέλεος· ὁμοίως δὲ καὶ πάνυ ἐκ τῶν αὐτῶν πύον οὐδέν,
οἷον τῷ Αἰσχύλου παιδὶ ξυνέβη· πυομένου δὲ τοῖσι
πλείστοισι φρῖκαι καὶ πυρετοὶ ἐπιγίνονται.

62. Ἀλκμᾶνι ἐκ νεφριτικῶν[200] ἀνακομιζομένῳ, κάτω
καὶ αἵματος ἀφαιρεθέντος ἄνω καθ᾽ ἧπαρ ἐτράφη καὶ
πρὸς καρδίην. ἄλγος δεινόν, καὶ τὸ πνεῦμα ὑπὸ τοῦ
πόνου κατείχετο· καὶ ἡ κοιλίη χαλεπῶς | ὑπεχώρει
428 σμικρὰ σπυραθώδεα·[201] ἄση οὐκ ἐνῆν· ῥῖγος δ᾽ ἔστιν
ὅτε καὶ πυρετὸς ἐπελάμβανε, καὶ ἱδρώς, καὶ ἔμετος. ἐν
τῷ ἀλγήματι οὐ ξυνέφερεν ὑποκλύσαι θαλάσσῃ, ξυνήνεγκεν ἀπὸ πιτύρων. ἡσίτησεν ἡμέρας ἑπτά· μελίκρητον ἀκρητέστερον. μετὰ δὲ χυλὸς φακῶν, ὁτὲ δὲ
λεπτὸν τὸ ἔτνος,[202] ἐπέπινεν ὕδωρ. μετὰ δὲ σκυλακίου
ἐφθοῦ, μάζης μικρὸν ὅτι μάλιστα πάλαι ξυγκειμένης·
προϊόντος δὲ ἢ βόεια τραχήλια ἢ κωλῆνας ὑείων
κρεῶν[203] ἐφθῶν. τῇ προτεραίῃ ὑδροποσίῃ, ἡσυχίη,
σκέπη· πρὸς τὸ[204] νεφριτικόν, ἐκ τῆς σικύης κλυσμός.

63. Τῷ Παρμενίσκου παιδί, κώφωσις· ξυνήνεγκε μὴ
κλύζειν, διακαθαίρειν δὲ εἰρίῳ· μοῦνον ἐγχεῖν ἔλαιον ἢ

198 ὠλεκράνου MV (proper name in V): corr. Asul.
199 recc.: πυοῦταί τε MV
200 φρενιτικῶν V
201 σπυθαρώδεα V
202 Asul.: ἔθνος M: ἔλνος V

61. The one pierced by a wound in the tip of the elbow: his forearm grew purulent from mortification. But when it ripened a sticky, gluey serum was expressed. It quickly stopped, as it did also for Cleogeniscus and for Demarchus, son of Aglaoteles. Similarly, it happens that there is no pus in very much the same condition, as was the case with Aeschylus' son. But where there is purulence, most develop shivering and fever.[25]

62. Alcman was recovering from nephritic affections and, when blood was removed below, the disease was diverted up along the liver and towards the heart. The pain was terrible and his breathing was checked by the suffering. And the bowels, with difficulty, produced small pellets. There was no nausea, but shivering at times and fever seized him, and sweat and vomiting. And in the midst of the pain it did not help to give seawater clysters; there was help from bran-husk clysters. He abstained from food for seven days, drank melicrat in a strong mixture. Afterwards bean broth and sometimes thin pea soup; he drank water; later some boiled young dog, and a little barley cake which had been made as long before as possible. As time went on either beef neck-meat or pork leg-bones, boiled. The previous day, drinking water, quiet, keeping covered. For the nephritic condition, a clyster of wild cucumber.

63. Parmeniscus' child was afflicted with deafness. Not to wash it (the ear?) out was helpful, but to clean it with

[25] Cf. *Epid.* 5.65.

---

203 ὑιου ἀκρέων M
204 τὸν MV: corr. Asul.

νέτωπον·[205] περιπατεῖν, ἐγείρεσθαι πρωΐ, οἶνον πίνεν λευκόν, λαχάνων[206] ἀπέχεσθαι. ἄρτον, ἰχθῦς[207] πετραίους.

64. Τῇ Ἀσπασίου ὀδόντος δεινὸν ἄλγημα καὶ γνάθου· καστόριον καὶ πέπερι διακλυζομένη καὶ κατέχουσα ἐν τῷ στόματι ἀνῆκε, καὶ στραγγουρικὸν αὐτὴ ἀνῆκε. προσθερμαίνει τὸ ἄλευρον τὸ ξὺν τῷ ῥοδίνῳ. τὰς ἀφ᾽ ὑστερέων κεφαλαλγίας καστόριον παύει. τὰ πλεῖστα τῶν ὑστερικῶν, αἱ φῦσαι· σημεῖον, οἱ ἐρευγμοὶ καὶ οἱ περὶ γαστέρα ψόφοι, καὶ ἐπάρματα ὀσφύος, καὶ περὶ νεφροὺς ἀλγήματα καὶ ἰσχία· καὶ ἐκγεγλευκισμένος[208] μέλας, ἢ ἀρωμάτων τρίτον μέρος, ἀλεύρου δύο, ἐν οἴνῳ εὐώδει ἑψῆσαι λευκῷ, ἐπ᾽ ὀθόνιον ἐπι|χέοντα, ἐπαλείψαντα,[209] καταπλάσσειν, ἐπὶ τῆς γαστρὸς ἀλγήματα ὑστερικά.

430

65. Τῷ Καλλιμέδοντος ξυνήνεγκε πρὸς τὸ φῦμα τὸ ἐν τῷ τραχήλῳ, σκληρὸν ἐὸν καὶ μέγα καὶ ἄπεπτον καὶ ἐπώδυνον, καὶ ἀπόσχασις βραχίονος, λίνον καταπλάσσειν πεφυρημένον.[210] οἴνῳ λευκῷ καὶ ἐλαίῳ δεύοντα ἐπιδεῖν μὴ θερμὸν μηδ᾽ ἑφθὸν ἄγαν, ἢ ξὺν μελικρήτῳ ἑψεῖν ἄλευρον τήλιος, ἢ κριθῶν, ἢ πυρῶν.

66. Μελησάνδρῳ τοῦ οὔλου[211] ἐπιβαλόντος, καὶ ὄντος ἐπωδύνου, καὶ σφόδρα οἰδέοντος, ἀπόσχασις βραχίονος, στυπτηρίη αἰγυπτίη, ἐν ἀκμῇ παραστέλλειν.

---

[205] μέτοπον V   [206] λάχων Μ   [207] ἰχθύσι Μ
[208] ἐκλευκισμένος Μ: ἐκλελευκισμένος V: corr. Li.

wool, and only to pour in olive oil or bitter almond oil. Walking about, rising early, drinking white wine, refraining from garden vegetables. Bread and rock fishes to eat.[26]

64. Aspasius' wife had a dreadful pain of the tooth and jaw. She got relief by washing it with castorium and pepper, and holding the solution in her mouth. And she relieved her strangury. Barley meal with rose oil added heat. Castorium stops headaches from the womb. Most of the hysteric[27] affections: intestinal gas. An indication is belching and noise in the belly, swelling of the back, pains about the kidney and loins. Give also newly fermented sweet red wine, or a third part of aromatic herbs, two of fine meal: boil in fragrant white wine: pour it on a cloth, and after rubbing her with oil, apply the cloth as a poultice to treat her belly's hysteric pains.[28]

65. For Callimedon's son what also helped for the swelling on the neck, which was hard, large, unripe and painful, was opening the vein of the arm, putting on a plaster of saturated linen: wetting it in white wine and oil and binding it on, not hot or too much boiled. Or with melicrat boil meal of fenugreek, or barley, or wheat.[29]

66. For Melisander, when his gum was troublesome, painful and very swollen: opening the vein of the arm, Egyptian alum, for reducing it at its height.[30]

---

[26] Cf. *Epid.* 5.66.   [27] Hysteric probably means simply "uterine" in this passage, without the implications of the wandering womb.   [28] Cf. *Epid.* 5.67.
[29] Cf. *Epid.* 5.68.   [30] Cf. *Epid.* 5.69.

---

[209] ἐπαλείφοντα V   [210] Smith: πεφρυγμένον MV (cf. *Epid.* 5.68)   [211] ὅλου V

67. Εὐτυχίδει ἐκ χολερικῶν ἔπειτα τῶν σκελέων τετανώδεα· ἔληξεν ἅμα τῇ κάτω ὑποχωρήσει. κατακορέα χολὴν πολλὴν ἤμει ἐπὶ τρεῖς ἡμέρας καὶ νύκτας, σφόδρα ὑπέρυθρον· ὑπὸ δὲ τὸν ἔμετον ἔπινε, καὶ ἀκρατὴς ἦν καὶ ἀσώδης· οὐδὲν κατέχειν ἐδύνατο οὐδὲ τὸ ἐκ τῶν σιδίων· καὶ οὔρου σχέσις καὶ τῆς κάτω διόδου· διὰ τοὺς ἐμέτους τρὺξ μαλακὴ ἦλθε, καὶ κατέρρηξε[212] κάτω.

67b. Ὑδρωπιώδεα[213] ταλαιπωρεῖν, ἱδροῦν, ἄρτον ἐσθίειν, πίνειν μὴ πολύ, λούεσθαι[214] κατὰ κεφαλῆς πολλῷ μὴ θερμῷ, ἀλλὰ χλιηρῷ· οἶνος λευκός· ὕπνῳ μὴ πολλῷ χρῆσθαι.

68. Καλλιγένει, περὶ πέντε καὶ εἴκοσιν ἔτεα γεγενημένῳ, κατάρροος· ἡ βὴξ πολλὴ ἀνῆγε τὸ καταρρέον ὑπὸ βίης· οὐδὲν κάτω.[215] ἔτεα τέσσαρα διετέλεσεν· θέρμαι λεπταὶ ἐν ἀρχῇ ἐγένοντο. ἐλλέβορος οὐκ ὠφέλησεν, ἀλλὰ ὀλιγοσιτίη, ἀλλὰ τὸ ξυμπιασθῆναι· ἐσθίειν ἄρτον· οἶνος μέλας· ὄψα ὁποῖα ἐθέλοι· ἀπέχεσθαι δριμέων, ἁλυκῶν, λιπαρῶν, ὀποῦ σιλφίου, λαχάνων ὠμῶν· περιπατεῖν πολλά· γαλακτοποσίη οὐ ξυνήνεγκεν, ἀλλὰ σήσαμον πίνειν καθαρόν, ὠμόν, ὅσον ὀξύβαφον, σὺν οἴνῳ μαλακῷ.

69. Τιμοχάρει[216] χειμῶνος κατάρρους, μάλιστα ἐς τὰς ῥῖνας· ἀφροδισιάσαντι ἐξηράνθη πάντα· κόπος, θέρμη ἐπεγένετο· κεφαλὴ βαρέη· ἱδρὼς ἀπὸ κεφαλῆς

432

---

[212] κατέρυξε M   [213] Li.: ἰκτεριώδη MV (cf. *Epid.* 5.70)
[214] Corn.: λοῦσθαι MV (possibly correctly)

67. Eutychides, after biliousness, had cramping of the legs. He got better at the same time as the purgation through the bowel. He vomited much pure bile for three days and nights, extremely reddish. He took the drink for vomiting, and was weak and nauseated; he could hold nothing down, even the extract of pomegranate rind. The urine was stopped as well as the bowels. Because of the vomitings soft "wine lees" came out, and his bowels broke loose.[31]

67b. Hydropics should exercise, sweat, eat bread, drink little; wash around the head with much water, not hot but lukewarm. White wine. Use not too much sleep.[32]

68. Calligenes, when he was about twenty-five years, had a flow from the head. Much coughing brought up violently the matter that had flowed down. Nothing through the bowels. It persisted for four years. There were light fevers at the beginning. Hellebore did not help, but reducing food, yet fattening him. Eating bread. Wine, red. Meats, whichever he wanted. Refraining from acrid, salt, fat, juice of silphium, raw vegetables. Many walks. Drinking milk was of no benefit, but drinking, with mild wine, pure raw sesame, about one tenth of a pint.

69. Timochares, in winter, had catarrhs, mostly towards the nose. When he had sexual activity, all was dried up. Fatigue, fever came on. Head heavy. Much sweat from the

[31] Cf. *Epid.* 5.79.
[32] Cf. *Epid.* 5.70.

---

[215] κάτω ὑπέμενεν M

[216] τιμοχαρίη MH: τιμοχάριϊ V: (but V reads -ει in *Epid.* 5.72): τιμοχάριτι IR

πολύς· κατέρρει δὲ καὶ κατὰ τοῦ σώματος ὅλου· ἦν δὲ καὶ ὑγιαίνων ἱδρώδης· τριταῖος ὑγιάνθη.

70. Ὁ Κλεομένεος παῖς, χειμῶνος ἀρξάμενος, ἀπόσιτος, ἄνευ πυρετοῦ ἐτρύχετο, καὶ ἤμει τὰ σιτία καὶ φλέγμα. δύο μῆνας ἡ ἀσιτίη παρηκολούθησεν.

71. Τῷ μαγείρῳ ἐν Ἀκάνθῳ τὸ κύφωμα ἐκ φρενίτιδος ἐγένετο· φαρμακοποσίη οὐδεμία ξυνήνεγκεν· οἶνος μέλας, καὶ ἀρτοσιτίη, καὶ λουτρῶν ἀπέχεσθαι, χρίεσθαί τε καὶ ἀνατρίβεσθαι, μὴ σφόδρα θάλπειν, μὴ πολλῷ πυρί, ἀλλ᾽ ἠπίῳ.

72. Οἷσι ῥεύματα ἐς ὀφθαλμοὺς λεπτὰ καὶ χρόνια, ἢν πέπονα κατὰ ῥῖνας χωρήσῃ,[217] ὠφελέονται.

73. Ἧισιν ἐν κνήμασιν ἐκ πτώματος ἢ σπάσματος ἢ πληγῆς πόνοι, ἐπιεικῶς ἐν τῇσι τρίτῃσι δηλοῖ εἰ διαφθείρει.

74. Τῇ Σίμου[218] τὸ τριηκοσταῖον ἀπόφθαρμα πιούσῃ τι ἢ[219] αὐτόματον· ξυνέβη πόνος, ἔμετος χολωδέων πολλῶν ὠχρῶν, πρασοειδέων, μελάνων, ὅτε πίοι.[220] τριταίη, σπασμός· τὴν γλῶσσαν κατεμασᾶτο. πρὸς τεταρταίην εἰσῆλθεν· ἡ γλῶσσα μέλαινα, μεγάλη· τῶν ὀφθαλμῶν τὰ λευκὰ ἐρυθρά· ἄγρυπνος· τεταρταίη ἀπέθανεν ἐς νύκτα. |

434   76d.[221] Ὀρίγανος ὀφθαλμοῖσι κακὸν πινομένη καὶ ὀδοῦσιν.

---

[217] χωρήσει M   [218] τησίμου M
[219] τι ἢ Li. from *Epid.* 5.53: τῇ MV   [220] πίθι M
[221] From here to ch. 78 I restore the ms. order of the text, but indicate Littré's numeration

head, and it flowed down from the whole body; but he tended to sweat even when in health. On the third day he was healthy.[33]

70. Cleomenes' son, beginning in the winter, had no desire for food, wasted away without fever, and vomited his food and phlegm. The aversion to food persisted two months.[34]

71. The butcher in Acanthus developed a humpback after phrenitis. No drug helped. Red wine and eating bread, and refraining from bathing, being rubbed with oil, warming the back, not excessively, by a small and gentle fire.[35]

72. People who have thin chronic flows into their eyes are helped if the flows come ripened through the nostrils.

73. In the case of women in pregnancy who have distress from a fall or from a convulsion or from a blow: generally a woman shows clearly on the third day whether she is aborting.

74. Simus' wife had an abortion on the thirtieth day, from drinking something, or spontaneously. She had pain, vomiting of much bilious material, yellow, leek-colored, black, whenever she drank. On the third day convulsion. She kept biting her tongue. Towards the fourth day it invaded that: her tongue was black, swollen. The whites of her eyes were red. Sleepless. She died towards night on the fourth day.[36]

76d. Oregano, when drunk, is bad for eyes and for teeth.

[33] Cf. *Epid.* 5.72.
[34] Cf. *Epid.* 5.51.
[35] Cf. *Epid.* 5.52.
[36] Cf. *Epid.* 5.53.

77. Ἡ ἀπὸ τοῦ κρημνοῦ κόρη πεσοῦσα ἄφωνος· ῥιπτασμὸς εἶχεν· ἤμεσεν ἐς νύκτα αἷμα πολύ. κατὰ τὸ οὖς τὸ ἀριστερὸν πεσούσης αἷμα σύχνον ἐρρύη· μελίκρητον χαλεπῶς κατέπινεν· ῥέγχος, πνεῦμα πυκνόν, ὥσπερ τῶν ἀποθνησκόντων, φλέβες αἱ περὶ τὸ μέτωπον τεταμέναι, κλίσις[222] ὑπτίη, πόδες χλιηροί· πυρετὸς πολύς· ὁπότε ὀξύτατος μάλιστα κατεπλήσσετο. ἑβδόμῃ, φωνὴν ἔρρηξεν· αἱ θέρμαι λεπτότεραι εἶχον. περιεγένετο.[223]

75. Πυθοκλῆς τοῖσι κάμνουσιν ὕδωρ, γάλα πολλῷ τῷ ὕδατι μιγνὺς ἐδίδου, καὶ ἀνέτρεφον.[224]

76a. Χιμέτλων, κατασχᾶν, ἀλεαίνειν τοὺς πόδας, ὡς μάλιστα ἐκθερμαίνειν πυρὶ καὶ ὕδατι.

76b. Ὀφθαλμοῖσι πονηρόν, φακῆ, ὀπώρη γλυκέη, καὶ λάχανα.

76c. Τοῖσι περὶ ὀσφῦν ἢ ἰσχίον ἢ σκέλεα[225] ἀλγήμασιν ἐκ πόνων, θαλάσσῃ, ὄξει, θερμοῖσι καταιονᾶν, καὶ σπόγγους βάπτοντα πυριᾶν, ἐπικαταδεῖν δὲ εἰρίοισιν οἰσυπηροῖσιν.[226]

78. Ὀνισαντίδης τοῦ ὤμου τὸ ἄλγημα ἔσχεν ἐν θέρει γενόμενον ἐξ ἀποστάσιος· ἐν τῇ θαλάσσῃ βρέχειν τὸ σῶμα καὶ τὸν ὦμον ὡς πλεῖστον χρόνον ἐπὶ τρεῖς ἡμέρας· οἶνον λευκὸν ὑδαρέα ἅμα πίνειν ἐν τῇ θαλάσσῃ κατακείμενον, καὶ οὐρεῖν ἐν τῇ θαλάσσῃ.

---

[222] κληῖς MV (cf. Epid. 5.55)   [223] περιγένοντο MV (cf. Epid. 5.55)   [224] ἀνέτρεφαν M

[225] Τοῖσι περὶ ἴσχιον καὶ σκέλεα ἢ ἴσχιον V

77. The girl who fell from the cliff became speechless; restless tossing persisted. She vomited much blood towards night. She had fallen on her left ear, and there was a large flow of blood. She had trouble drinking melicrat. Rasping in the throat; rapid breathing, as of dying people; the blood vessels about the forehead stretched; lying on her back, feet warm. Much fever; when it was very acute she was especially prostrated. On the seventh day she broke her voice free. The fevers continued lighter. She survived.[37]

75. Pythocles gave ill people water, and milk which he had mixed with much water, which nourished them successfully.[38]

76a. Chilblains: scarifying, warming the feet, heating as much as possible with fire and water.[39]

76b. Bad for eyes: lentil soup, sweet fruit, and vegetables.[40]

76c. For pains of the lower back or hips or legs from exertion, pour over a mixture of sea water, vinegar, hot water, and give fomentations with sponges that have been dipped in it, and bind with wool that still has oil in it.[41]

78. Onisantides, in the summer, had the pain of the shoulder which comes from an apostasis. Soaking his body and his shoulder in the sea for as long as possible for three days; at the same time drinking watery white wine while lying in the sea, and urinating in the sea.

[37] Cf. *Epid.* 5.55.   [38] Cf. *Epid.* 5.56.   [39] Cf. *Epid.* 5.57.   [40] Cf. *Epid.* 5.58.   [41] Cf. *Epid.* 5.58.

226 ὑσυπυροῖσι ῥίνας V: οἰσυπηροισειν ῥῖνας M (cf. *Epid.* 5.58)

79. Ὁ γναφεὺς ὁ ἐν Σύρῳ, ὁ φρενιτικός· μετὰ δὲ καύσιος τρομώδης· σκελέων τὸ χρῶμα οἷον ὑπὸ κωνώπων ἐγκαταδήγμασιν·[227] | ὀφθαλμὸς μέγας, βραχείη κίνησις· φωνὴ κεκλασμένη, σαφὴς δέ· οὖρον καθαρόν, ὑπόστασιν οὐκ ἔχον· ἦρα διὰ τὴν ἀπὸ τῆς θαψίης ὑποχώρησιν; ὀκτωκαιδεκαταῖος ἀνῆκε μολυνθεὶς ἄνευ ἱδρῶτος.

80. Καὶ ὁ ἐν Ὀλύνθῳ Νικόξενος ὁμοίως· ἑβδομαῖος ἐδόκει χαλάσειν μεθ᾽ ἱδρῶτος· ῥυφήματα προσεφέρετο, οἶνον, βότρυν ἐξ ἡλίου. πρὸς ἑπτακαιδεκαταῖον εἰσῆλθον· καυστικός· τοιοῦτον, ἡ γλῶσσα· θέρμη ἔξωθεν οὐ πάνυ ἰσχυρή· ἔκλυσις σώματος δεινή· φωνὴ κεκλασμένη, ἔργον ἀκοῦσαι, σαφὴς[228] δέ· κρόταφοι ξυμπεπτωκότες· ὀφθαλμοὶ κοῖλοι· πόδες μαλακοὶ καὶ χλιηροί· κατὰ σπλῆνα ξύντασις· τὸ κλύσμα οὐ πάνυ τι ἐδέχετο, ἀλλ᾽ ἀνεπήδα· ἐς νύκτα[229] ἐπῆλθε κόπρος ξυνεστηκυῖα ὀλίγη, καὶ αἵματός τι, οἶμαι ἀπὸ τοῦ κλυσμοῦ· οὖρον καθαρόν, λαμπρόν· κλισίη ὑπτίη, σκέλεα διηνοιγμένα[230] διὰ τὴν ἔκλυσιν· παράπαν ἄγρυπνος. ἐντὸς τῶν εἴκοσιν ἡ θέρμη ἐμολύνθη ἀπὸ τῶν ἀπὸ κρίμνου, ἄλλοτε ἀπὸ μήλων[231] ὁμοῦ καὶ σίδης χυλὸς καὶ φακοῦ πεφωγμένου ψυχρός· καὶ ἀλεύρου πλύμα ἐφθόν, ψυχρόν· λεπτὸν ῥύφημα· περιεγένετο.

81. Τῶν κναφέων οἱ βουβῶνες ἐφυματοῦντο σκληροὶ καὶ ἀνώδυνοι καὶ περὶ ἥβην· ἐν τραχήλῳ ὅμοια

79. The fuller on Syros who was phrenitic: trembling, with burning fever. The color of his legs like that of gnat bites. Eye large, very little movement. Voice broken but intelligible. Urine clear, having no sediment. Was it because of the bowel movement from thapsia? On the eighteenth day the fever ceased, having come down without sweat.

80. The man in Olynthus, Nicoxenus, fared similarly. On the seventh day his fever seemed to slacken with sweat. He was given porridge, wine, raisins dried in the sun. Towards the seventeenth day I visited him. He had burning fever: his tongue was of that kind. The fever outside was not very strong. Dreadful disorganization of body; voice broken, a task to hear it, but intelligible; temples collapsed; eyes hollow; feet soft and warm. Tightness at the spleen. He would not take much enema, but it spurted out. Towards night came a small amount of feces, formed, and a little blood, I believe from the clyster. Urine clear, bright. He lay on his back, legs spread out because of disorientation, totally sleepless. Before the twentieth day the fever came down from the drink made from coarse barley meal, sometimes from apple and pomegranate juice and juice from toasted lentils, cold. Boiled water from soaking wheat meal, cold. Light porridge. He survived.

81. The glands of the fullers swelled up hard, painless, even in the groin; at the neck, similar large swellings. Fe-

---

228 σαφεῖς V
229 ἐς νύκτα om. V
230 διηνυγμένα M
231 μήδων MV: corr. HR

μεγάλα· πυρετός· ἔμπροσθεν δέκα[232] βηχώδεις ἀπὸ
ῥηγμάτων. τρίτῃ μὲν ἢ τετάρτῃ γαστὴρ ξυνετάκη,
θέρμαι ἐπεγένοντο· γλῶσσα ξηρή, δίψα· ὑποχωρήσιες[233] ἑκάστῳ χαλεπαί· ἀπέθανον.

82. Τὰ χολερικὰ ἐκ κρεηφαγίης, μάλιστα δὲ χοι-
438 ρείων ἐνωμοτέρων, καὶ ἐρεβίνθων,[234] καὶ μέθης εὐώ-
δεος παλαιοῦ, καὶ ἡλιώσιος, καὶ σηπίης, καὶ καράβων
τε καὶ ἀστακῶν, καὶ λαχανοφαγίης, μάλιστα δὲ πρά-
σων καὶ κρομμύων, ἔτι δὲ θριδάκων ἑφθῶν, καὶ κράμ-
βης, καὶ λαπάθων ἐνωμοτέρων, καὶ ἀπὸ πεμμάτων,
καὶ μελιτωμάτων, καὶ ὀπώρης, καὶ σικύου πέπονος,
καὶ οἴνου καὶ γάλακτος, καὶ ὀρόβων, καὶ ἀλφίτων
νέων. μᾶλλον ἐν θέρει τὰ χολερικά, καὶ οἱ[235] διαλείποντες[236] πυρετοὶ καὶ οἷσι φρῖκαι ἐπιγίνονται. οὗτοι
ἔστιν ὅτε κακοήθεις γίνονται καὶ ἐς νοσήματα ὀξέα
καθίστανται· ἀλλ’ εὐλαβεῖσθαι. μάλιστα δὲ ἡ πέμπτη
καὶ ἡ ἑβδόμη καὶ ἡ ἐνάτη δηλοῦσι, βέλτιον δὲ μέχρι
τῶν τεσσαρεσκαίδεκα φυλάσσεσθαι.

83. Κύδει[237] μεθ’ ἡλίου τροπὰς χειμερινάς, τῆς
νυκτὸς πλευροῦ δεξιοῦ ἄλγημα, καὶ πρότερον εἰθισμέ-
νον. ἔληξεν· ἠρίστησεν· ἐξελθών, ἔφριξε· πυρετὸς ἐς
νύκτα· ἀνώδυνος· βηχίον ῥέον τι ξηρόν. οὖρα οὐ
πολλά, ὑπόστασις πολλή, ξυσματώδεα μαδαρὰ δι-
εσπασμένα ἀπ’ ἀρχῆς, μετὰ δὲ τέσσαρας[238] θολερὰ
ὑφίστατο, καὶ τὸ οὖρον οὐκ ἄχροον, ὑπόστασιν ἔχον,
καὶ τὸ σύστημα οὐκ ἐφαίνετο ἐν τῷ χερνιβίῳ,[239] ψύ-

<hr />

232 δὲ καὶ coni. Li.   233 ὑποχώρησις M

ver. Ten days previously, coughing caused by fissures. In the third or fourth month the bowels liquified, fevers came on. Tongue dry, thirst. Painful bowel movements for all of them. They died.[42]

82. Choleric conditions, from meat-eating and especially undercooked pork, and from chickpeas and from fragrant old wine, exposure to sun, and cuttlefish, from crayfish and lobsters, from eating garden vegetables, especially leeks and onions, and also boiled lettuce, cabbage, and undercooked dock, and from pastry and honey cakes, and fruit and ripe cucumber, and wine with milk, and vetches, and fresh barley. Choleric conditions are more likely in summer, as also remittent fevers and those attended by shivering. These sometimes become of ill habit and are established as acute diseases. Beware. Especially do the fifth day and the seventh and ninth give indication, but it is better to keep watch up to the fourteenth day.[43]

83. Cydes, after the winter solstice, had in the night a pain in his right thorax, which had been common before. It went away. He ate a meal. When he went out he had shivering. Fever towards night, without pain. A little cough produced some dry flux. Not much urine, much sediment, pulpy, dispersed particles from the beginning, and after the fourth day it became turbid, the urine itself not of bad color, having suspended matter, and the deposit did not

---

[42] Cf. *Epid.* 5.59.   [43] Cf. *Epid.* 5.71.

---

[234] ἐνερεβίνθων M   [235] οὐ V   [236] διαλίποντες M
[237] Smith: Κύδη MV: Φερεκύδει recc.
[238] τὰς τεσσάρας V   [239] Galen writes this at *Epid.*
6.3.7 as καὶ σύναγμα ἐφαίνετο ἐν τῷ χερνίβῳ

χεος ἐόντος. τριταίῳ αὐτομάτη κοιλίη ὑπῆλθεν. τετάρτῃ, πρὸς βάλανον[240] εὖ κοπρώδεα καὶ χολώδεα· ὑγρὸς
περίρροος. ὕπνος ἐς νύκτα οὐ πάνυ, ἡμέρης δέ τι
ἐκοιμᾶτο· δίψα οὐ σφοδρή, τεταρταίῳ μάλιστα ἐς
νύκτα· δέρμα περὶ μέτωπον, καὶ τὸ ἄλλο μαλακόν,
ἐπέφερεν[241] αἰεί· πυρετὸς πρὸς χεῖρα ἐκρατεῖτο. ὑπενό-
τιζεν ἄδηλος· σφυγμὸς σφόδρα ἐν μετώπῳ φλεβῶν·
440 βάρος | ἐν τῇσιν ἐπιστροφῇσι καὶ ἐν τῇσιν ὑποχωρή-
σεσι πάσῃσιν ὀλίγον χρόνον· διὰ παντὸς ἀνώδυνος·
ἀπ’ ἀρχῆς ἀσώδης, καὶ κατὰ σμικρὸν ἤμει. ἑβδόμῃ
πρὸς βάλανον ὑπεχώρησε τρίς, χολῶδες[242] καὶ κοπρῶδες, ὑγρόν, σφόδρα ὠχρόν· καὶ ἤδη τι ὑπο-
παρέκρουε, καὶ νοτὶς ἐπ’ ὀλίγον ἔστιν ὅτε περὶ μέτω-
πον· τὸ ἱμάτιον ἐπὶ τὸ πρόσωπον· τὰ ὄμματα διὰ κενῆς
ὡς εἴ τις βλέπων ξυνέστρεφε,[243] καὶ πάλιν κατέμνεν·
τὸ ἱμάτιον ἀπέβαλλεν. ἐναταίῳ ἱδρὼς πρωῒ ἀρξάμενος
μέχρι στήθεος[244] διετέλει ἕως ἐτελεύτησε, καὶ ὁ πυρετὸς ἐπέτεινε καὶ ἡ παραλήρησις· περὶ μέτωπον πλεῖ-
στος ἱδρώς, καὶ οἷον δεινὸν μὲν ἢ ἔκλευκον τὸ χρῶμα,[245] τὸ ὑπὸ τὰς τρίχας ὡς ἐσμηγμένῳ· ὑποχόνδριον
δεξιὸν ἐπηρμένον· ὑφῆκεν ὑφ’[246] ἑωυτὸν χολῶδες.
ὀγδόῃ, ὡς ὑπὸ κωνώπων ἀναδήγματα. πρὸ τῆς τελευτῆς ἐνέβηξεν, οἷον ἐκ μύξης, μύκητα ξυνεστηκότα,
λευκῷ φλέγματι περιεχόμενον· ἀπεχρέμπτετο δὲ καὶ
ἔμπροσθεν σμικρά, λεπτά,[247] γαλακτώδεα.

240 βαλάνιον MV   241 ἐπέφερεειν M
242 χολῶδεες M   243 ξυνέτρεφε M

appear in the pan when the urine was cold. On the third day the bowel moved spontaneously. On the fourth in response to a suppository, a solid and bilious movement. Liquid flowed around it. Sleep towards nights, not much, but he had some sleep during the day. Thirst, not severe, on the fourth day, especially towards night. Skin about the forehead, otherwise soft, kept forming ridges. The fever was powerful to the hand. It became damp, but not obviously so. There was violent beating of blood vessels in the forehead. Heaviness for a brief time when he turned and (always) when his bowels moved. He was without pain throughout; nauseous from the first, and he vomited a little. On the seventh day, from a suppository, he had three bowel movements, bilious and formed, watery, very yellowish; at that time he became somewhat delirious, and had dampness briefly sometimes about the forehead. His covers over his face. He moved his eyes about emptily, like one glancing about, and again nodded off. He would throw the bedclothes off. On the ninth day sweat which extended to his chest began in the morning, and persisted until his death. And the fever intensified, and the delirium. Most of the sweat was about the forehead; his skin was frightening and whitish, that under his hair as on a person who has been scrubbed with soap. Distension of the right hypochondrium. Bilious bowel movement spontaneously. On the eighth day marks like gnat bites. Before the end he coughed up a mushroom-like object formed as if from mucus, surrounded by white phlegm. And he had previously brought up a little thin, milky matter.

---

244 στῆθος M   245 δέρμα M
246 ἐφ᾽ MV: corr. Li.   247 λευκά M

84.[248] . . . μετὰ τὸ δεῖπνον ἐν τῷ ὕπνῳ ἔφριξεν. πρωὶ
ἐξανέστη καρηβαρικός· ἔφριξεν, ἀπήμεσε, κεφαλὴν
ἐβαρύνετο· ἐς νύκτα ἀνῆκε μέχρι μέσου ἡμέρης ἐπι-
εικῶς·[249] ἔφριξε πάλιν· νύκτα χαλεπῶς. τὴν ἐπιοῦσαν
δὲ ἡμέρην, πυρετὸς ὀξύς· κεφαλῆς σφάκελος· ἔμετος
χολῆς πολλῆς, ἡ πλείστη πρασοειδής· ἔληξε πάντα·
ὕπνος ἐς νύκτα. πρωῒ περιέψυκτο· ἱδρώτιον, νοτὶς ἐπὶ
πολὺ τοῦ σώματος· κατὰ σπλῆνα τῇ χειρὶ ἐδείκνυεν
ὀλίγον χρόνον ξύστρεμμα ἀνώδυνον, παραχρῆμα |
442 ἐμαράνθη. ἐς νύκτα ἄγρυπνος· περὶ ἀγορὴν παρωξύνετο πυρετός· ἄση, σκοτόδινος, στρόφος, κεφαλῆς
ὀδύνη, ἔμετος πρασοειδής, λεῖος, γλίσχρος ὡς φλέγ-
μα· πρὸ δυσμῶν ἔληξε πάντα· ἱδρὼς κεφαλῆς, τραχή-
λου· ὑπεχώρησε μετὰ τὸν ἔμετον κοπρώδεα, ὑγρά,
χολώδεα, οὔτε μέλανα, οὔτε ἐπεοικότα. νύκτα μετρίως,
καὶ τὴν ἐπιοῦσαν ἡμέρην· ἐς νύκτα πάλιν ἄγρυπνος·
ἔμετος ὄρθρου ὁμοίως, καὶ τὴν ἐπιοῦσαν ἡμέρην ἄνευ
ἄσης· τῆς δὲ κεφαλῆς μετὰ τὸν ἱδρῶτα ἔληξαν αἱ
ὀδύναι· ἐς τὴν ἑσπέρην πάντα ἐχάλασεν. ἐνάτη, οὐκ
ἔτι ἤμεσεν· ἐθερμάνθη μᾶλλον· τοῖσιν ἄλλοισιν οὐκ
ἐδόκει πυρεταίνειν· αἱ ἐν κροτάφοισι ἐπήδων· ἀνώδυνος πάντων· δίψος ἦν αἰεί. ἐναταῖος ἐπὶ θῶκον ἀναστὰς ἐλειποψύχει σφόδρα· ὑπεχώρησε πρὸς βάλανον
ξυσμάτια μέλανα καὶ χολώδεα· χρῶμα κοπρῶδες,
ὅσον ἀπέσταξεν· ἡ φωνὴ κεκλασμένη· ἐν τῇσιν ἐπι-
στροφῇσι βαρύς· ὀφθαλμοὶ κοῖλοι· δέρμα μετώπου

---

[248] There is no break in the text of the mss.   [249] om. V

84. . . . [44] After the meal he shivered in his sleep. He arose in the morning heavy headed. He shivered, vomited, head weighed down. Towards night it relented somewhat until midday. He began to shiver again. During the night he was in a bad way, and on the subsequent day had acute fever. Violent headache, vomiting much bile, most of it leek-colored. All symptoms relented, there was sleep towards night. In the morning he was chilled. Sweat, dampness over much of his body. By the spleen he indicated with his hand, for a brief time, a painless gathering which suddenly withered away. Towards night, sleepless. At the filling of the marketplace the fever became acute. Nausea, vertigo, pain in the bowel, head pain, vomit of leek-colored matter, smooth, sticky, like phlegm. Before sunset all symptoms relented. Sweat of the head, neck. After the vomit he had a bowel movement, formed, watery, bilious, neither black nor proper-looking. He was easier in the night and the subsequent day. Towards night he was again sleepless. The morning's vomit was similar and through the day he was without nausea. The head pains stopped after the sweat. Towards evening all symptoms relaxed. On the ninth day he did not vomit again. He was warmer. He did not seem feverish in other respects.[45] The vessels at his temples leaped. No pain anywhere. There was continuous thirst. On the ninth day when he got up for the toilet he was very faint. In response to a suppository he passed black, bilious particles; what dribbled out was dung-colored. His voice was broken; he was heavy in his movements, his eyes hollow, the skin of his forehead tight.

[44] The beginning of this case history appears to have been lost.
[45] Or "to the other people."

περιτεταμένον· ἄλλως εὔπνοος, κόσμιος· πρὸς τοῖχον
τὰ πλεῖστα ἀπεστραμμένος, ὑγρὸς ἐν τῇσι κλίσεσιν,
ἐπικεκαμμένος, ἀτρεμίζων· γλῶσσα λευκή, λείη. περὶ
δὲ τὰς δέκα καὶ μετά, οὖρα ἐρυθρὰ τὰ κυκλώδεα, ἐν
μέσῳ²⁵⁰ σμικρὸν λευκόν. δωδεκαταίῳ, πρὸς βάλανον
ὅμοια ἀπέσταξε χολῷ²⁵¹ καὶ ξύσματι μετὰ τοῦ βαλάνου, λειποψυχίη· ἔπειτα τὸ στόμα ἐπεξηραίνετο, δι-
εκλύζετο αἰεί· καὶ εἰ μὴ σφόδρα ψυχρὸν ἦν,²⁵² θερμὸν
ἔφασκεν εἶναι· τῆς χιόνος, ἐπιεικῶς. εἶτα δίψα οὐκ
ἐνῆν· τὸ ἱμάτιον αἰεὶ ἀπὸ τῶν στηθέων ἀπεώθει· τὴν
ἀμίδα χλιαίνειν οὐκ εἴα· τὸ πῦρ πόρρω καὶ σμικρά·
γνάθων ἀμφοτέρων ἔρευθος· μετὰ ταῦτα ἄκροπις· ἀν-
εθερμάνθη μίαν ἢ δύο ἡμέρας, καὶ ἔληξεν. |

444   85. Ἀνδροθαλεῖ ἀφωνίη, ἄγνοια, παραλήρησις· λυ-
θέντων²⁵³ δὲ τούτων, περιῄει ἔτεα²⁵⁴ συχνά· καὶ ὑποστροφαὶ ἐγίνοντο. ἡ γλῶσσα διετέλει πάντα τὸν χρόνον ξηρή· εἰ μὴ διακλύσαιτο, διαλέγεσθαι οὐχ οἷός τε
ἦν· καὶ σφόδρα πικρὴ ἦν τὰ πολλά· ἔστι δ᾽ ὅτε καὶ
πρὸς καρδίην ὀδύνη. φλεβοτομίη ἔλυσε ταῦτα· ὑδρο-
ποσίη· μελίκρητον· ἐλλέβορον ἔπιε μέλανα, χολῶδες
οὐ διῄει, ἀλλ᾽ ὀλίγον. τέλος δέ, χειμῶνος κατακλιθείς,
ἐξ ἑωυτοῦ ἐγένετο· καὶ τὰ τῆς γλώσσης παθήματα
ὅμοια· θέρμη λεπτή· γλῶσσα ἄχρως· φωνὴ περιπλευ-
μονική· ἀπὸ τοῦ στήθεος τὸ ἱμάτιον ἀπέρριπτε, καὶ
ἐξάγειν ἑωυτὸν ὡς οὐρήσοντα ἐκέλευεν, οὐδὲν δυνάμενος σάφα εἰπεῖν, οὐδὲ ἐῶν παρ᾽ ἑωυτόν, ἐξῆγον αὐτόν·

²⁵⁰ ἐν μέσῳ] μέσου V

Otherwise he breathed easily and was composed. Mostly turned towards the wall, lying loosely on the bed, bent, not moving. His tongue white, smooth. Around the tenth day and afterwards there were globules in his urine, red with a little white at the center. On the twelfth day in response to a suppository he dribbled feces like bile and scrapings. At the time of the suppository, he was faint. Then his mouth became dry; he kept rinsing it. Unless the water was very cold he said it was hot, if like ice, moderately cold; then his thirst departed. He kept throwing the covers off his chest. He did not want the chamber pot warmed. The fire was distant and made little heat. There was redness on both cheeks. After that, speech slurred. He had fever for one or two days, and it left him.

85. Androthales suffered from voicelessness, unaware-ness, and delirium. When these things had stopped he went about many years, and there was a relapse. His tongue stayed dry all the time. If he did not rinse it he could not speak. And it was bitter most of the time. Some-times he also had pain at the heart. Phlebotomy stopped these symptoms, drinking water; melicrat. He drank black hellebore. No bilious excrement, save a little. At the end, in winter, he took to his bed; he lost his sense of himself. Similar affections of the tongue; light fever; tongue color-less; voice peripneumonic. He threw the covers from his chest and asked that someone get him up to urinate; he could not speak clearly, nor was he aware of himself. They

251 χυλῷ V
252 καὶ εἰ . . . ἦν om. V
253 αὐθέντων MV (cf. Epid. 5.80)
254 om. V

ἐς νύκτα ἐτελεύτησεν· ᾗσι κατεκλίνη,[255] δύο ἢ τρεῖς ἡμέραι ἐγένοντο.

86. Τὸ Νικάνορος πάθος, ὁπότε ἐς ποτὸν ὁρμῷτο, φόβος τῆς αὐλητρίδος· ὅτε ἀρχομένης[256] αὐλεῖν ἀκούσειεν ἐν τῷ ξυμποσίῳ ὑπὸ δειμάτων[257] ὄχλοι. μόγις ὑπομένειν ἔφη ὅτε νὺξ εἴη. ἡμέρης δὲ ἀκούων οὐδὲν διετρέπετο. καὶ ταῦτα παρείπετο χρόνον συχνόν.[258]

87. Δημοκλέης δὲ ὁ μετ᾽ ἐκείνου ἀμβλυώσσειν καὶ λυσισωματεῖν ἐδόκει, καὶ οὐκ ἂν κρημνὸν[259] ἔφη παρελθεῖν οὐδὲ ἐπὶ γεφύρης, οὐδὲ τοὐλάχιστον βάθος τάφρου τολμῆσαι ἄν, ὑπὸ δείματος μὴ πέσῃ, διελθεῖν, ἀλλὰ δι᾽ αὐτῆς τῆς τάφρου πορεύεσθαι· τοῦτο χρόνον τινὰ αὐτῷ ξυμβῆναι.

88. Τὸ Φοίνικος, ἐκ τοῦ ὀφθαλμοῦ τὰ πολλὰ ἐκ τοῦ δεξιοῦ ὥσπερ ἀστραπὴν ἐκλάμπειν ἐδόκει· οὐ πολὺ δὲ ἐπισχόντι, ὀδύνη ἐς κρόταφον τὸν δεξιὸν[260] ἐγένετο δεινή, ἔπειτα καὶ ἐς ὅλην τὴν κεφα|λὴν καὶ ἐς τράχηλον ᾗ δέδεται[261] ἡ κεφαλὴ ὄπισθεν τῶν σπονδύλων· καὶ ξύντασις καὶ σκληρότης ἀμφὶ τοὺς τένοντας· εἰ γοῦν ἐπειρᾶτο διακινεῖν τὴν κεφαλὴν ἢ διοίγειν τοὺς ὀδόντας, ἠδυνάτει,[262] ὡς σφόδρα ξυντεινόμενος. ἔμετοι, ὁπότε γενοίατο,[263] ἀπέτρεπον τὰς εἰρημένας ὀδύνας ἢ[264] ἠπιωτέρας ἐποίεον· καὶ φλεβοτομίη ὠφέλησε, καὶ ἐλλέβοροι ἦγον παντοδαπά, οὐχ ἥκιστα πρασοειδέα.

---

255 κατεκλίνει MV: corr. Li.   256 ἀρχόμενος V
257 ὑποδημάτων M   258 σύχνον χρόνον V

took him to urinate. He died towards night. He had been in bed two or three days.[46]

86. Nicanor's affection, whenever he went out drinking, was terror of the flute girl. Whenever he heard one starting to play at the symposium, masses of terrors rose up. He said that he could hardly bear it when it was at night. But when he heard it in the daytime he not affected. This continued over a long period of time.[47]

87. Democles, who was with him, seemed to lose his sight and become limp. He said he could not go along a cliff nor cross on a bridge over a ditch of the least depth for fear of falling, but would go through the ditch itself. This affected him for some time.[48]

88. Phoenix's problem: he seemed to see flashes like lightning in his eye, mostly the right. And when he had suffered that a short time, a terrible pain developed towards his right temple and then around his whole head and on to the part of the neck where the head is attached behind the vertebrae. And there was tension and hardening around the tendons. If he tried to move his head or open his teeth, he could not, as being violently stretched. Vomits, whenever they occurred, averted the pains I have described, or made them gentler. Phlebotomy helped, and hellebore produced variegated matter, much of it leek-colored.[49]

---

[46] Cf. *Epid.* 5.80.   [47] Cf. *Epid.* 5.81.
[48] Cf. *Epid.* 5.82.   [49] Cf. *Epid.* 5.83.

---

[259] παρὰ κρημνὸν M   [260] τὸν δεξιὸν om. V
[261] εἴδεται V: ἤδεται M (cf. *Epid.* 5.83)
[262] Smith: δύνατο MV: ὠδυνᾶτο Lind.
[263] R: γενοίαντο MV   [264] ἢ add. Lind.

89. Παρμενίσκῳ καὶ πρότερον ἐνέπιπτον ἀθυμίαι καὶ ἵμερος ἀπαλλαγῆς²⁶⁵ βίου, ὁτὲ δὲ πάλιν εὐθυμίη. ἐν Ὀλύνθῳ δέ ποτε φθινοπώρου ἄφωνος κατέκειτο ἡσυχίην ἔχων, βραχύ τι ὅσον ἄρχεσθαι ἐπιχειρέων προσειπεῖν· ἤδη δέ τι καὶ διελέχθη, καὶ πάλιν ἄφωνος. ὕπνοι ἐνῆσαν, ὁτὲ δὲ ἀγρυπνίη, καὶ ῥιπτασμὸς μετὰ σιγῆς, καὶ ἀλυσμός, καὶ χεὶρ πρὸς ὑποχόνδρια ὡς ὀδυνωμένῳ· ὁτὲ δὲ ἀποστραφεὶς ἔκειτο ἡσυχίην ἄγων. ὁ πυρετὸς δὲ διατέλεος, καὶ εὔπνοος· ἔφη δὲ ὕστερον ἐπιγινώσκειν τοὺς ἐσιόντας· πιεῖν ὁτὲ μὲν ἡμέρης ὅλης καὶ νυκτὸς διδόντων, οὐκ ἤθελεν, ὁτὲ δὲ ἐξαίφνης τὸν στάμνον ἁρπάσας τοῦ ὕδατος ἐξέπιεν· οὖρον παχὺ ὡς ὑποζυγίου. περὶ δὲ τὴν τεσσαρεσκαίδεκα ἀνῆκεν.

90. Ἡ δὲ Κόνωνος θεράπαινα,²⁶⁶ ἐκ κεφαλῆς ὀδύνης ἀρξαμένη, ἐκτὸς ἑωυτῆς καὶ βοὴ καὶ κλαυθμὸς πολύς, ὀλιγάκις δὲ ἡσυχίη. περὶ δὲ τὰς²⁶⁷ τεσσαράκοντα ἐτελεύτησεν· τὰς δὲ ἐπὶ τελευτῆς ὡς δέκα²⁶⁸ ἡμέρας ἄφωνος καὶ σπασμώδης ἐγίνετο. |

448   91. Καὶ ὁ τοῦ Τιμοχάριος θεράπων, ἐκ μελαγχολικῶν δοκεύντων εἶναι καὶ τοιούτων, ἐτελεύτησεν ὁμοίως, καὶ περὶ ἡμέρας τὰς αὐτάς.

92. Τῷ Νικολάου περὶ ἡλίου τροπὰς χειμερινὰς ἐκ πότων ἔφριξεν· ἐς νύκτα πυρετός. τῇ ὑστεραίῃ ἔμετος χολώδης, ἄκρητος, ὀλίγος. τρίτῃ ἀγορῆς πληθούσης ἱδρὼς δι᾽ ὅλου τοῦ σώματος· ἔληξε καὶ ταχὺ πάλιν

---

²⁶⁵ ἀπαλλαγὴ M   ²⁶⁶ θεραπαίνη M
²⁶⁷ δὲ τὰς] ἔτεα MV (cf. *Epid.* 5.85)

89. Parmeniscus had previously been affected by depression and desire for death, and then again by optimism. One time in Olynthus in the fall he took to his bed, voiceless. He kept still, hardly attempting to begin speaking. At times he said something, and again voiceless. Sleep came on, and periodically wakefulness, and tossing silently, and delirium, and his hand went to his hypochondria as though he was in pain. And at times he turned away and lay still. His fever was continuous and he was breathing easily. He later said that he recognized people who came in. At times for a whole day and night when they offered water to drink he did not want it, but at times he would suddenly seize the water cooler and drink it down. His urine thick like a mule's. He was cured about the fourteenth day.[50]

90. Conon's (female) servant, who began with a pain in the head, became delirious; shouting, much crying out, with some periods of quiet. She died about the fortieth day. For about ten days before her death she was speechless and convulsive.[51]

91. And Timochareus' (male) servant, from what seemed a similar melancholic affection, died similarly in about the same number of days.[52]

92. Nicolaus' son, about the winter solstice, had shivering after drinking. Towards night, fever. The following day, bilious vomit, unmixed, in small quantity. On the third day at the time when the agora was filled, sweat over his whole body; that stopped and he quickly again became feverish.

---

[50] Cf. *Epid.* 5.84.   [51] Cf. *Epid.* 5.85.
[52] Cf. *Epid.* 5.87.

---

268 δέκα] ἑξ V: δ' εχ (sic) M (cf. *Epid.* 5.85)

ἐθερμάνθη· περὶ μέσας νύκτας ῥῖγος, πυρετὸς ὀξύς.
ἡμέρης δὲ τὴν αὐτὴν ὥρην ἱδρώς· ταχὺ πάλιν ἐπεθερ-
μάνθη· ἔμετος ὅμοιος. τῇ τετάρτῃ, ἀπὸ ὕδατος λινοζώστιος ὑπεχώρησεν εὖ κοπρώδεα καὶ ὑγρά, ὑπομύ-
σαρα δέ, σποδοειδέα, οἷα ὕδωρ λινοζώστιος, οὐκ
ἀνόμοια, ὑπόστασις οὐκ ἦν· οὐδὲ πολὺ τὸ οὖρον,
ἐναιωρήματα σμικρά· ἀλγήματα ὑποχονδρίου ἀριστεροῦ καὶ ὀσφύος, ᾤετο ἐκ τοῦ ἐμέτου· ἐπανέπνει ἔστι δ'
ὅτε διπλόον· γλῶσσα λευκή, ἔχουσα ἐκ δεξιοῦ οἷον
θέρμου πρόσφυσιν ὑποβρυχίην·[269] διψώδης, ἄγρυπνος, ἔμφρων. ἑκταίῳ δεξιὸς ὀφθαλμὸς μέζων ἦν[270] τῷ
βλέπειν. ἑβδομαῖος ἐτελεύτησε· κοιλίη πρὸ τῇ τελευτῇ
ἐμετεωρίσθη· ἀποθανόντος[271] τὰ ὄπισθεν ἐφοινίχθη.

93. Μέτωνι μετὰ Πληϊάδος δύσιν πυρετός, πλευροῦ
ἀριστεροῦ ὀδύνη μέχρι ἐς κληΐδα οὕτω δεινὴ ὥστε
450 ἀτρεμίζειν οὐχ οἷός τ' ἦν, | καὶ τὸ φλέγμα κατεῖχεν·
ὑποχώρησις πολλὴ χολώδης. ἐν τρισὶν ἡμέρῃσι σχεδὸν ἔληξεν ἡ ὀδύνη, ἡ θέρμη δὲ περὶ τὰς ἑπτὰ ἢ ἐννέα.
βὴξ ἐνῆν· ἀποχρέμψιες ὑπόχολοι οὐδὲ ἐγένοντο, οὐδὲ
πολλαί, φλεγματώδεις δέ.[272] αἱ βῆχες παρηκολούθεον
σιτίων ἀπεγεύετο· ἔστιν ὅτε ἐξῄει ὡς ὑγιής· ὑπελάμ-
βανον ἐνίοτε θέρμαι λεπταὶ ὀλίγον χρόνον· ἱδρώτια
ἐγίνοντο ἐς νύκτα· πνεῦμα ἐν τῇ θέρμῃ πυκνότερον·
γνάθων ἔρευθος· περὶ τὸ πλευρὸν βάρος καὶ ὑπὸ
μασχάλην[273] καὶ ἐς ὦμον. αἱ βῆχες ἀπεῖχον·[274] φάρ-

---

[269] ἐπὴν βραχῇ MV: corr. Li.   [270] Smith: ἐν mss.
[271] ἀποθανόντας M   [272] δέ om. V

About the middle of the night, a chill, acute fever. In the day at the same hour, sweat; quickly again feverish; vomit similar. On the fourth day, from water of *linozostis* a good movement, formed and moist, but somewhat foul, cinder-like, such as water of *linozostis* produces, fairly uniform. There was no deposit, and not much urine; small suspended particles. He had pains in the left hypochondrium and lower back, from the vomit, he believed. He inhaled doubly sometimes. Tongue white, and with an underlying swelling on the right, as if from heat. Thirsty, sleepless, conscious. On the sixth day the right eye was greater to look at. He died on the seventh day. The belly before the end was elevated, and after death his back became bright red.[53]

93. Meton, after the setting of the Pleiades, had fever. Pain in the left side up to the collarbone, so severe that he could not keep still, and phlegm possessed him. There was much bilious excrement. Within three days the pain virtually disappeared, and the fever about the seventh or ninth. He had a cough, the expectorated material was not bilious nor was there much of it, but phlegmatic. The coughing persisted. He nibbled at his food. Sometimes he went out, as though healthy. But sometimes light fevers seized him for a brief time. There were sweats towards night. His breathing was more rapid in the fever. His cheeks were red. Around the thorax a heaviness, and beneath the armpits and up to the shoulder. The cough slackened. A drug

---

53 Cf. *Epid.* 5.88.

---

273 τὴν μασχάλην M
274 ἐπεῖχον Foës

μακον ἄνωθεν ἤγαγε χολώδεα· τρίτῃ ἀπὸ τοῦ φαρμάκου ἐρράγη τὸ πύον, ἀπὸ δὲ τῆς ἀρχῆς τοῦ ἀρρωστή-
ματος τεσσαρακοστῇ.[275] ἀποκαθῆραι[276] δὲ περὶ πέντε
καὶ τριήκοντα ἡμέρας ἄλλας, καὶ ὑγιής.

94. Τῇ Θεοτίμου ἐν ἡμιτριταίῳ ἄση καὶ ἔμετος καὶ
φρίκη ἅμα ἀρχομένῳ τῷ πυρετῷ, καὶ δίψα· προϊόντος
καὶ ἀρχομένῳ ἐξαίσιον τὸ θερμόν· μελίκρητον πιούσῃ
καὶ ἀπεμεύσῃ ἥ τε φρίκη καὶ ἡ ἄση ἐπαύσατο· καὶ τὸ
ἀπὸ τῆς σίδης ὕστερον.

95. Τῇ Διοπείθεος ἀδελφεῇ ἐν ἡμιτριταίῳ δεινὴ
καρδιαλγίη περὶ τὴν λῆψιν, καὶ παρείπετο[277] ὅλην τὴν
ἡμέρην· καὶ κεφαλαλγίη. καὶ τῇσιν ἄλλῃσιν ὡσαύτως
ὑπὸ Πληϊάδος δύσιν· ἀνδράσι σπανιαίτερα ἐγίνετο τὰ
τοιαῦτα.

96. Τῇ Ἀπομότου ἐν ἡμιτριταίῳ, περὶ Ἀρκτοῦρον,
δεινὴ καρδιαλγίη περὶ τὴν λῆψιν,[278] καὶ ἔμετοι, καὶ
πνῖγες προΐστανto ἅμα ὑστερικαί, καὶ ἐς τὸ μετάφρε-
νον ὀδύναι κατὰ ῥάχιν· ὅτε ἐνταῦθά οἱ ᾔει, ἔληγον αἱ
καρδιαλγίαι.

97. Τῇ Τερπίδεω μητρὶ τῇ ἀπὸ Δορίσκου[279] διαφθο-
452 ρῆς γενομένης μηνὶ πέμπτῳ διδύμων ἐκ πτώματος,[280]
τοῦ μὲν ἑτέρου αὐτίκα ὡς ἐν χιτῶνί τινι ἀπηλλάγη,
τοῦ δὲ ἑτέρου ἢ πρότερον ἢ ὕστερον τεσσαράκοντα
ἡμέρῃσιν ἀπηλλάγη. ὕστερον δὲ ἔλαβεν ἐν γαστρί.
ἔτει[281] δὲ ἐνάτῳ ὀδύναι δειναὶ κατὰ γαστέρα ἐπὶ πολὺν
χρόνον· ἤρχοντο ἔστι μὲν ὅτε[282] ἐκ τοῦ τραχήλου καὶ

---

[275] τεσσαρακοστήν M   [276] ἀπεκαθάρθη Foës

brought up bilious vomit. On the third day after the drug, the pus broke forth, the fortieth day after the beginning of the illness. With purging for thirty-five more days, he became healthy.

94. Theotimus' wife, in a semitertian, had nausea and vomiting; shivering at the time the fever began, and thirst. As time went on, and even at its beginning, the fever was extraordinary. When she drank melicrat and vomited, her shivering and nausea stopped. And later the solution from pomegranate skin.

95. Diopeithes' sister in a semitertian had terrible heartburn at its onset, and it continued all day. And head pain. So did other women about the time of the Pleiades' setting. Such symptoms were rarer for the men.[54]

96. Apomotus' wife, in a semitertian, at Arcturus' rising, terrible heartburn at the onset, and vomiting, and at the same time hysteric suffocation was obvious, and pains into the back around the spine. But when the pain was there the heartburn stopped.

97. Terpides' mother, who was from Doriscus, had an abortion of twins after a fall in the fifth month. She was immediately delivered of one that was in a sort of cloak, and of the other earlier or later by forty days. But she later conceived a child. But nine years later she had terrible pains in the belly for a long time. They sometimes started from the

---

[54] Cf. *Epid.* 5.89.

---

277 ξυμπαρείπετο M   278 λήμψιν M
279 Δωρίσκου M
280 πώματος M
281 ἔτι M   282 ὅτι M

ῥάχιος καθίσταντο δὲ ἐς ὑπογάστριον καὶ βουβῶ-
νας·[283] ἔστι δ᾿ ὅτε ἐκ γούνατος τοῦ δεξιοῦ καθίσταντο
δὲ ἐς τωὐτό· καὶ ὅτε μὲν κατὰ τὴν γαστέρα αἱ ὀδύναι
εἶεν ἐμετεωρίζετο ἡ κοιλίη· ὅτε δὲ ἐπαύσατο, ξυν-
έπιπτεν ἡ[284] καρδιαλγίη· πνιγμοὶ οὐκ ἐνῆσαν· ψύξις δὲ
τοῦ σώματος ὡς ὕδατι κειμένῳ κατὰ τὸν χρόνον ἡνίκα
ἡ ὀδύνη ἐγίνετο. ὑπέστρεφε δι᾿ ὅλου τὰ ἀλγήματα ἢ
ἠπιώτερα τῶν κατ᾿ ἀρχάς. σκόροδα, σίλφιον, τὰ δρι-
μέα ξύμπαντα οὐ ξυνέφερεν, οὐδὲ τὰ γλυκέα, οὐδὲ τὰ
ὀξέα, οὐδὲ οἱ λευκοὶ οἶνοι· οἱ μέλανες δὲ καὶ λουτρὰ
ὀλιγάκις. ἀρχομένη καὶ ἔμετοι δεινοὶ ἐπεγίνοντο, καὶ
τῶν σιτίων ἀποκλείσιες περὶ τὰς ὀδύνας. καὶ τὰ γυναικεῖα οὐκ ἐφαίνετο.

98. Τῇ Κλεομένεος, περὶ ζεφύρου πνοάς, ἐκ ναυσίης
καὶ κόπου[285] πλευροῦ ἀριστεροῦ ὀδύνη ἐκ τραχήλου
ἀρξαμένη καὶ τοῦ ὤμου· πυρετὸς καὶ φρίκη καὶ ἱδρώς.
ἤρξατο δὲ ὁ πυρετός· οὐκ ἔληγεν, ἀλλ᾿ ἐπέτεινε, καὶ ἡ
ὀδύνη δεινή, βήξ, ἀπόχρεμψις ὕφαιμος, ὠχρή, πολλή·
γλῶσσα λευκή· διαχωρήματα μέτρια, ὑγρά· οὖρα
χολώδεα. τετάρτῃ ἐς νύκτα τὰ γυναικεῖα ἦλθε πολλά·
καὶ ἔληξεν ἥ τε βὴξ καὶ ἡ ἀπόχρεμψις καὶ ἡ ὀδύνη,
καὶ ἡ θέρμη λεπτή.

99. Τῇ Ἐπιχάρμου πρὸ τόκου δυσεντερίη, πόνος,
ὑποχωρήματα ὕφαιμα, μυξώδεα· τεκοῦσα[286] παραχρῆμα ὑγιής.

283 R: βουβῶνες MV
284 ἡ om. M

neck and spine and settled in the lower belly and groin, sometimes from the right knee and settled in the same area. When the pains were in the belly, the intestines were elevated, and when it stopped heartburn came on. There was no choking. There was a chilling of her body, as for a person lying in water, during the time when she had the pain. The pains recurred totally or were more mild than those at the beginning. Garlic, silphium, acrid substances did not help, nor did the sweet nor the acid, nor did white wine. But the red wines and baths helped occasionally. At the beginning there were terrible vomits and exclusion of food at the time of the pains. And the menses did not appear.

98. Cleomenes' wife, when the zephyrs began to blow, after nausea and fatigue had a pain in her left thorax. It began from the neck and shoulder. There was fever and chills and sweat. The fever came first. It did not cease, but intensified. The pain was dreadful. Cough; matter brought up, bloody and yellow, in large quantity. Tongue white. Excrement moderate, watery. Urine bilious. On the fourth day towards night her menstrual flow came in great quantity. And the cough, and the expectoration, and the pain ceased, and the fever was mild.

99. Epicharmus' wife before delivery had dysentery, fatigue, mucous, bloody feces with phlegm. When she had given birth she was suddenly healthy.[55]

---

[55] Cf. *Epid.* 5.90.

---

100. Τῇ Πολεμάρχου ἐν ἀρθριτικοῖσιν[287] ἰσχίου
454 ἄλγημα ἐξαίφνης | γυναικείων οὐ γενομένων. τὸ ἐν τῷ
σεύτλῳ πιούσῃ ἔσχετο ἡ φωνὴ νύκτα καὶ ἐς ἡμέρης
μέσον·[288] ἤκουε[289] δὲ καὶ ἐφρόνει, καὶ τῇ χειρὶ ἐσήμαι-
νεν ἀμφὶ τὸ ἰσχίον εἶναι τὸ ἄλγημα.

101. Τῇ Κλεινίου ἀδελφῇ, τῇ ἀφηλικεστέρῃ, ἔμετος, ὅ τι προσδέξαιτο, τεσσαρεσκαίδεκα ἡμέρας, ἄνευ
πυρετοῦ αἱματωδέων· καὶ ἐρευγμοί· καὶ ξυνεστραμ-
μένον ἐχώρει πρὸς καρδίην πνῖγμα. καστόριον, σέσε-
λι πάντα[290] ἔπαυσε, καὶ τὸ ἀπὸ τῆς σίδης· ἀπέστη δὲ
ἄλγημα μέτριον ἐς κενεῶνα. βολβοῦ χυλός, καὶ οἶνος
γαλακτώδης αὐστηρός, καὶ ἄρτοι ὡς ἐλάχιστοι ξὺν
ἐλαίῳ.

102. Τῇ Παυσανίου κούρῃ μύκητα ὠμὸν φαγούσῃ
ἄση, πνιγμός, ὀδύνη γαστρός. μελίκρητον θερμὸν
πίνειν καὶ ἐμεῖν ξυνήνεγκε, καὶ λουτρὸν θερμόν· ἐν τῷ
λουτρῷ ἐξήμεσε τὸν μύκητα, καὶ ἐπεὶ λήξειν[291] ἔμελλεν ἐξίδρωσεν.[292]

103. Ἐπιχάρμῳ, περὶ Πληϊάδων δύσιν, ὤμου ὀδύνη
καὶ βάρος δεινὸν ἐς βραχίονα, ἄση. ἔμετοι συχνοί,
ὑδροποσίη.

104. Τῷ Εὔφρονος παιδὶ ἐξανθήματα, οἷα ἀπὸ κω-
νώπων. |

456 105. Μετὰ Ζέφυρον αὐχμοὶ ἐγένοντο μέχρις ἰσημερίης[293] φθινοπωρινῆς. ὑπὸ κύνα δεινὰ πνίγεα καὶ
πνεύματα θερμά, καὶ πυρετοὶ ἱδρώδεις, καὶ πάλιν

---

[287] ἀρθρητικοῖς Μ   [288] μέσον ἡμέρης Μ

100. Polemarchus' wife, in an arthritic condition, had a sudden pain in the hip joint at the time her menses failed to appear. When she had drunk the decoction in beet juice, her voice was checked during the night and up to midday. But she could hear, and her mind was clear; she indicated with her hand that the pain was around the hip joint.[56]

101. Cleinias' youthful sister vomited a bloody vomit for fourteen days, whatever she ate or drank. There was no fever. Belching, and with contractions a suffocation went to the heart. Castorium, seseli stopped all symptoms, and juice from pomegranate rind. The pain shifted, in moderate form, to her flanks. Onion juice, and acrid wine with milk, and minute amounts of bread with olive oil.

102. Pausanias' young daughter, when she ate a raw mushroom, had nausea, choking, pain in the stomach. Drinking warm melicrat and vomiting helped, and warm bathing. In the bath she vomited the mushroom and when she was about to recover she sweated.

103. Epicharmus, about the setting of the Pleiades, had pain in the shoulder and terrible heaviness into the arm, nausea. Frequent vomits, drinking water.[57]

104. Euphron's son had eruptions like gnat bites.[58]

105. After Zephyrus there were droughts until the fall equinox. At the Dogstar there were bad heat waves and hot winds, and there were fevers with sweats which imme-

---

[56] Cf. *Epid.* 5.91.   [57] Cf. *Epid.* 5.92.
[58] Cf. *Epid.* 5.93.

---

289 ἤκουσε M   290 πάντας MV: corr. recc.
291 ἐπει λήξειν] ἐπιλήξειν MV: corr. recc.
292 καὶ ἐξίδρωσεν V   293 Lind.: ἡμέρης mss.

εὐθὺς ἐπεθέρμαινον· φύματα παρ' οὓς συχνοῖσιν ἐγί-
νετο· τῇ γραίῃ τῇ βηχικῇ[294] περὶ ἐνάτην· τῷ μειρακίῳ
τῷ σπληνώδει τῷ τῆς παιδίσκης, κοιλίης ὑγρανθείσης, περὶ τὸν αὐτὸν χρόνον· Κτησιφῶντι ὑπ'
ἀρκτοῦρον σχεδὸν περὶ τὰς ἑπτά· τῷ παιδὶ μοῦνον
ἐπωθοῦντο·[295] Ἐρατύλλου ἀπεμωλύνθη παρὰ ἀμφότερα. ἀνιδρώτεις,[296] γλῶσσαν ὑπὸ ξηρότητος ψελλοί.
οἱ ὀρνιθίαι[297] ἔπνευσαν πολλοὶ καὶ ψυχροί, καὶ χιόνες
ἐξ εὐδίων ἔστιν ὅτε ἐγένοντο· καὶ μετὰ ἡμέρην ἰσημερίην νότια διαμίσγοντα βορείοισιν· ὕδατα συχνά·
ἐπεδήμησαν βῆχες, μᾶλλον[298] δὲ παιδίοισι· παρὰ
τὰ[299] ὦτα πολλοῖσιν οἷα τοῖσι σατύροισιν· ὁτὲ δὲ
χειμὼν ὁ[300] πρὸ τούτου τοῦ χρόνου σφόδρα χειμερινὸς
ἐγένετο καὶ χιόνι καὶ ὄμβροισι βορείοισιν.

106. Τῷ Τιμώνακτος παιδίῳ ὡς διμηνιαίῳ ἐξανθήματα ἐν σκέλεσι, καὶ ἐν ἰσχίοισι, ὀσφύι, ὑπογαστρίῳ,
καὶ οἰδήματα σφόδρα ἐνερευθῆ.[301] καταστάντων δὲ
τούτων σπασμοὶ καὶ ἐπιληπτικὰ ἐπεγίνοντο ἄνευ πυρετῶν ἡμέρας πολλάς, καὶ ἐτελεύτησεν.

107. Τῷ τοῦ Πολεμάρχου ξυνέβη μὲν τῷ ἔμ-
458 προσθεν χρόνῳ ἐμ|πυωθῆναί τε καὶ ἀποχρέμψασθαι·
ὕστερον δὲ θέρμαι, καὶ ὑδρωπιώδης τε ἦν, καὶ ἐπί-
σπληνος, καὶ ἀσθματώδης πρὸς αἶπος[302] εἴ ποτε
ἴοι,[303] καὶ ἀδύνατος, καὶ διψώδης, καὶ ἀπόσιτος ἐπιεικῶς ἔστιν ὅτε, καὶ βηχία ξηρὰ ἐπὶ πολὺν χρόνον

---

294 Βεσσιακῇ conj. Foës ex Gal. *Gloss.* s.v.
295 ἐπωηθυτο (sic) M   296 ἀνιδρώτεες MV: -σιες recc.

diately became again hot. Many people got swellings by
the ear. The old woman who had the cough, towards the
ninth day, the youth with the swollen spleen (son of the
serving girl), whose bowels liquified, around the same
time. Ctesiphon at about Arcturus' rising, around seven
days. They pushed outward only in the slave. Those of
Eratyllus went down on both sides. Patients were without
sweat, lisping from dryness of the tongue. The bird-winds
blew strong and cold, and there was snow at times out
of calm skies. And after the equinox there was southerly
weather mixed with northerly, much rain. Coughs were
epidemic, especially in children. Swellings by the ears in
many people, as in satyriasis. There had been very stormy
periods earlier in the winter, with snow and rain from the
north.[59]

106. The infant of Timonax at two months had eruptions on the legs and at the hips, on the lower back, the area
under the stomach. The swellings became quite red. When
they went down he had convulsions and seizures without
fever for many days, and he died.

107. Polemarchus' son had previously had empyema
and expectoration. He later had fever and was hydropic,
had swelling of the spleen and asthma; if he tried to climb a
hill he could not; he suffered thirst and sometimes rejected
food. For considerable periods he had dry coughs in this

[59] Cf. *Epid.* 5.94.

---

[297] Foës: ὀρθίαι MV   [298] μάλιστα cit. Gal. Comm. in
Epid. 6   [299] τὰ παρὰ τὰ cit. Gal.   [300] om. V
[301] ἐνευρέθη M   [302] ἔπος MV: corr. recc.
[303] ἴη V

οὕτως ἐγίνετο· ὑπεφέρετο, καὶ εἰ μή τι εὔλυτος γίνοιτο
κοιλίη κάτω, πλήρης ἐγένετο ἄνω, καὶ πνῖγμα καὶ
ἆσθμα ἐγίνετο μᾶλλον. τέλος[304] δὲ κατάρροος, καὶ
ἀπόχρεμψις ἐπικατῆλθε, καὶ βήξ, καὶ ἀπόχρεμψις,
παχέα καὶ ὠχρὰ πῦα· καὶ πυρετὸς σφοδρός· καὶ ἐδόκει
λῆξαι, καὶ ἡ[305] βὴξ πρηϋτέρη, καὶ ἡ ἀπόχρεμψις
καθαρή. ὑπέστρεψε ὀξὺς πυρετός, πνεῦμα πυκνὸν ἦν,
ἐτελεύτησε. προσερρίγωσε πόδας, καὶ κατεψύχθη·
μετὰ ταῦτα πνεῦμα μᾶλλον ἐγκατελήφθη· ἡ[306] οὔρησις ἔστη· τὰ ἄκρα κατεψύχθη· ἔμφρων ἐτελεύτησε
τριταῖος ἀπὸ τῆς ὑποστροφῆς.

108. Τῷ τοῦ Θυννοῦ σφόδρα ἐν πυρετῷ καυσώδει
ἐλιμοκτονήθη· ὑποχώρησις συχνὴ χολῆς ἐγένετο
μετὰ ἀψυχίης καὶ ἱδρῶτος πολλοῦ· κατεψύχθη σφό-
δρα· ἄφωνος[307] ἦν ἡμέρην ὅλην καὶ νύκτα· ἐγχεόμενος
χυλὸν πτισάνης, κατείχετο, ἐφρόνει εὔπνοος ἦν.

109. Τῷ Ἐπιχάρμου ξυνέβη ἐκ περιόδου καὶ ποτοῦ
ἀπεψίη. τῇ ὑστεραίῃ πρωῒ ἄσης γενομένης, πιὼν
ὕδωρ, ὄξος, ἅλας, ἐξήμεσε φλέγμα· μετὰ δέ, ῥῖγος
ἔλαβεν· ἐλούσατο πυρεταίνων· τὸ στῆθος ἤλγει. τῇ
τρίτῃ εὐθὺ πρωΐ, κῶμα ὀλίγον χρόνον ἐπεῖχε, καὶ
ὑπελήρει, καὶ πυρετὸς ὀξύς· βαρέως ἔφερε τὴν νοῦ-
σον. τῇ τετάρτῃ ἄγρυπνος· ἀπέθανεν. |

460   110. Ἀρίστωνι δακτύλου ποδὸς ἡλκωμένου[308] ξὺν
πυρετῷ ἀσάφεια· τὸ γαγγραινῶδες ἀνέδραμεν ἄχρι
πρὸς γόνυ· ἀπώλετο· ἦν δὲ μέλαν, ὑπόξηρον, δυσῶδες.

[304] τέλεος M   [305] ἡ om. M   [306] ἡ om. V

way. He would relapse, and if his bowels did not break loose below he became overfull above and the asthma and suffocation came on more. Finally he had a catarrh and expectoration, and a cough. And the expectoration was thick, yellow, purulent matter. Violent fever. It seemed to abate and the cough was milder and the expectoration clean. Acute fever recurred, his breathing was rapid, he died. He had cold feet and was chilled. After that his breath was even more constricted. The urine was stopped. His extremities cold. He died conscious on the third day after the relapse.

108. Thynus' son, in a burning fever, took no food or drink. He passed large amount of bilious excrement, with faintness and much sweat. He was severely chilled. He was voiceless a whole day and night. He was administered strained barley broth, and kept it down. He regained his senses, breathing became normal.

109. After walking about and drinking, Epicharmus' son got indigestion. The following day in the morning he was nauseous. After drinking water, vinegar, and salt, he vomited phlegm. Afterwards shivering seized him. He bathed while feverish. His chest was painful. First thing in the morning on the third day a coma came on for a brief time and he was delirious and there was acute fever. He had great distress from the disease. On the fourth day, sleepless. He died.

110. Ariston, with a lesion on his toe, had fever and mental confusion. Gangrene ran up to his knee. He died. It was black, rather dry, foul-smelling.

---

307 καὶ ἄφωνος M
308 εἰλκομένου (sic) M

111. Ὁ τὸ καρκίνωμα τὸ[309] ἐν τῇ φάρυγγι καυθεὶς ὑγιὴς ἐγένετο ὑφ᾽ ἡμέων.

112. Πολύφαντος ἐν Ἀβδήροισι κεφαλὴν ὠδυνᾶτο ἐν πυρετῷ[310] σφόδρα· οὖρα λεπτά, πολλά· ὑποστάσιες δασέαι καὶ ἀνατεταραγμέναι· οὐ παυομένου[311] δὲ τοῦ ἀλγήματος τῆς κεφαλῆς πταρμικὸν προσετέθη ἐόντι δεκαταίῳ. μετὰ δὲ ἐς τράχηλον ὀδύνη ἰσχυρή· οὖρον ἦλθεν ἐρυθρόν, ἀνατεταραγμένον, οἷον ὑποζυγίου· παρέκρουσε τρόπον φρενιτικόν· ἀπέθανεν σπασμοῖσιν ἰσχυροῖσιν. παραπλησίως δὲ καὶ ἡ Εὐαλκίδου οἰκέτις ἐν Θάσῳ,[312] ᾗ πολὺν χρόνον τὰ δασέα ἐχώρει οὖρα[313] καὶ κεφαλαλγίαι ἐνῆσαν· φρενιτικὴ γενομένη ἀπέθανεν σπασμοῖσιν ἰσχυροῖσι· πάνυ γὰρ τὰ δασέα οὖρα καὶ ἀνατετραμμένα σημεῖον ἀκριβὲς κεφαλαλγίης καὶ σπασμῶν καὶ θανάτου. ὁ δὲ Ἁλικαρνασσεὺς ὁ ἐν τῇ Ξανθίππου οἰκίῃ καταλύων οὓς ἐν χειμῶνι ὠδυνᾶτο καὶ κεφαλὴν οὐ μετρίως· ἦν περὶ ἔτεα πεντήκοντα· φλέβα ἐτμήθη κατὰ Μνησίμαχον· ἐβλάβη ἡ κεφαλὴ κενωθεῖσα καὶ ψυχθεῖσα, οὐ γὰρ ἐξεπύησε· φρενιτικὸς[314] ἐγένετο, ἀπέθανεν· οὖρα καὶ τούτῳ δασέα.

113. Ἐν Καρδίῃ τῷ Μητροδώρου παιδίῳ ἐξ ὀδόντων ὀδύνη | σφακελισμὸς σιηγόνος, καὶ οὔλων[315] ὑπερσάρκωσις δεινή· μετρίως ἐξεπύησεν· ἐξέπεσον οἱ γόμφιοι ἢ ἡ σιηγών.

462

309 τῷ M
310 V writes ἐν πυρετῷ twice

111. The one whose cancer in the pharanx was cauterized was cured by me.

112. Polyphantus, in Abdera, had pain in the head with severe fever. Urine thin, lots of it, deposits shaggy and disordered. When the pain in his head did not stop, a sternutatory was applied on the tenth day. Afterwards he had a severe pain into the neck. He produced urine that was red, turbid, like a mule's. His mind was unsound in a phrenitic manner. He died with powerful convulsions. Similarly the maidservant of Eualcides in Thasos, whose urine was shaggy for a long time and who had headaches. She became phrenitic and died with powerful convulsions. For, indeed, urine which is shaggy and turbid is a precise sign of headache and convulsion and death. And the man from Halicarnassus who lodged in Xanthippus' house had ear pain in the winter and headache to an extreme degree. He was about fifty years old. He had a vein cut by Mnesimachus. His head was harmed by the emptying and the chilling, since he was not purulent. He became phrenitic, died. His urine, too, was shaggy.

113. In Cardia, Metrodorus' son had pain from the teeth, mortification of the jaw, and dreadful overgrowth of flesh on the gums. He was moderately purulent. His molars collapsed, or else his jawbone.[60]

---

60 Cf. *Epid.* 5.100.

---

311 πατομένου M
312 ἐν Θάσῳ] ἔνθα M
313 οὖρα ἐχώρει M
314 ἐνφρενιτικὸς M
315 οὐράων MV (cf. *Epid.* 5.100)

114. Ἀναξήνωρ ἐν Ἀβδήροισιν ἦν μὲν σπληνώδης καὶ κακόχρους· ξυνέβη δὲ αὐτῷ οἰδήματος γενομένου περὶ μηρὸν τὸν ἀριστερὸν ἐξαπίνης τοῦτο ἀφανισθῆναι· οὐ πολλῇσι δὲ ὕστερον ἡμέρῃσιν ἐγενήθη αὐτῷ κατὰ τὸν σπλῆνα, οἷον ἐπινυκτὶς ἐξ ἀρχῆς, ἔτι[316] οἴδημα καὶ ἐρύθημα[317] σκληρόν· μετὰ δὲ ἡμέρην τετάρτην πυρετὸς ἐγένετο καυσώδης, καὶ ἐπελιδνώθη πάντα κύκλῳ καὶ σαπρά. ἐδόκει ἄμεινον· ἀπέθανεν· ὑπεκαθάρθη δὲ πρότερον καὶ κατενόει.

115. Κλόνηγος ἐν Ἀβδήροισιν ἦν μὲν νεφριτικός· οὔρει δὲ αἷμα κατὰ σμικρὸν πολὺ χαλεπῶς· ἠνώχλει δὲ καὶ κοιλίη ἐντερική. τούτῳ πρωὶ μὲν ἐδίδοτο γάλα αἴγειον καὶ ὕδατος πέμπτη μερίς, ἀνεζεσμένον, τὸ πᾶν κοτύλας τρεῖς, ἑσπέρην δὲ ἄρτος ἔξοπτος, ὄψα,[318] σεῦτλα ἢ σίκυος, οἶνος μέλας λεπτός· ἐδίδοτο δὲ καὶ σίκυος πέπων· οὕτω δὲ διαιτωμένῳ καὶ ἡ κοιλίη ξυνέστη, καὶ τὰ οὖρα καθαρὰ ἐφοίτα· ἐγαλακτοπότησε δὲ ἕως τὰ οὖρα ἀποκατέστη.[319]

116. Γυναικὶ ἐν Ἀβδήροισι καρκίνωμα ἐγένετο περὶ τὸ στῆθος· ἦν δὲ τοιοῦτο· διὰ τῆς θηλῆς ἰχὼρ ὕφαιμος ἔρρει· ἐπιληφθείσης δὲ τῆς ῥύσιος ἀπέθανεν.

117. Τῷ Δεινίου παιδίῳ ἐν Ἀβδήροισι μετρίως ὀμφαλὸν τμηθέντι συρίγγιον κατελείφθη καί ποτε καὶ ἕλμις δι᾽ αὐτοῦ[320] διῆλθεν ἁδρά· καὶ ἔφη, ὅτε πυρέξειε χολώδεα ὅτι καὶ αὐτὰ ταύτῃ διῄει. προσεπεπτώκει τούτῳ τὸ ἔντερον πρὸς τῷ συριγγίῳ, καὶ διεβέ|βρωτο

---

316 Smith: ἔτ᾽ MV: ἔτι δὲ recc.   317 ὀλίσθημα V

114. Anaxenor, in Abdera, had spleen problems and bad color. And it occurred that it suddenly disappeared after a swelling had developed around his left thigh. But not many days later he developed also about the spleen, like an *epinyktis*[61] at first, swelling and hard redness. After the fourth day a burning fever developed; it all became livid in a wide circle and rotten. He seemed to get better; died. But he was purged beforehand and was aware.

115. Clonegus, in Abdera, had a kidney disease. He urinated much blood, painfully, little by little. In addition a dysenteric bowel troubled him. He was given, early in the morning, goat's milk and a fifth part of water, boiled, to a total of three cotyls[62] in all. And in the evening roasted bread, a main dish, beets or cucumber, thin red wine. And he was given ripe cucumber. When he kept this regimen his intestines stabilized, and the urine became pure. But he did drink milk until the urine became restored.

116. A woman in Abdera had a cancer on the chest. It went this way: bloody serum flowed out through the nipple. When the flow was interrupted, she died.[63]

117. Deinias' child in Abdera: when a small incision was made at his navel a small fistula was left behind. Once a full-sized worm came through it. And he said that whenever he was feverish actual bilious material came out by it. The intestine had fallen into this fistula and it was eroded

[61] An *epinyktis* is a pustule that is most distressful at night.
[62] Ca. three cups.　　　[63] Cf. *Epid.* 5.101.

---

318 ὅσα V　　　319 ἀπεκατέστη M
320 διωτοῦ V: δι᾽ ἑωυτοῦ M: corr. Li.

464 ὡς τὸ συρίγγιον, καὶ ἐπανερρήγνυτο, καὶ βηχία ἐκώλυε διαμένειν.

118. Τῷ παιδίῳ τῷ Πύθωνος, ἐν Πέλλῃ, πυρετὸς
αὐτίκα ἤρξατο πολύς, καὶ καταφορὴ πολλὴ μετὰ ἀφω-
νίης· ὕπνοι ἐγένοντο, καὶ κοιλίη σκληρὴ παρὰ πάντα
τὸν χρόνον. προστιθεμένου δὲ τοῦ ἐκ τῆς χολῆς πολλὰ
διεχώρει καὶ αὐτίκα ἐνεδίδου· ταχὺ δὲ πάλιν ἡ κοιλίη
ἐπήρετο καὶ ὁ πυρετὸς παρωξύνετο, καὶ ἡ καταφορὴ[321]
διὰ τῶν αὐτῶν. τῆς δ' αὐτῆς ἀγωγῆς ἐούσης ἐδόθη τι
τῶν σὺν κνήκῳ, καὶ σικύου, καὶ μηκωνίου· χολώδεα
κατέρρηξε. καὶ αὐτίκα τὸ κῶμα ἐπέπαυτο, καὶ ὁ πυρετὸς ἐπεπρήυντο,[322] καὶ τὰ ὅλα ἐκουφίσθη. ἐκρίθη τεσ-
σαρεσκαιδεκαταῖος.

119. Εὔδημος σπλῆνα ἐπόνει ἰσχυρῶς· προσετάσ-
σετο ὑπὸ τῶν ἰητρῶν ἐσθίειν πολλά, πίνειν οἶνον μὴ
ὀλίγον λευκόν, περιπατεῖν συχνά· οὐ μετέβαλεν·[323]
ἐφλεβοτομήθη·[324] σιτία καὶ ποτὰ πεφεισμένως·[325]
περίπατοι ἐκ προσαγωγῆς· οἶνος μέλας λεπτός· ὑγιάνθη.[326]

120. Φιλιστίδι[327] τῇ Ἡρακλείδου γυναικὶ ἤρξατο
πυρετὸς ὀξύς, ἔρευθος προσώπου, ἐξ οὐδεμιῆς προ-
φάσιος· ὀλίγον δ' ὕστερον τῆς ἡμέρης ἐρρίγωσεν· οὐκ
ἀναθερμανθείσης σπασμὸς ἐγένετο ἐν τοῖσι δακτύ-
λοισι τῶν ποδῶν καὶ τῶν χειρῶν· σμικρὸν δὲ μετὰ
ταῦτα ἐπεθερμάνθη, οὔρησε τροφιώδεα, νεφελώδεα,
διεσπασμένα· νύκτα ἐκοιμήθη. δευτέρῃ ἐπερρίγωσεν
ἡμέρῃ, σμικρῷ μᾶλλον ἐθερμάνθη· καὶ τὸ ἔρευθος

like the fistula. And it was fissured. And coughing kept it
from staying still.

118. In Python's child, in Pella, there suddenly commenced a great fever, great lethargy, with loss of speech.
Sleep came on and his intestine was hard the whole time.
But when the suppository made from bile was applied,
there was a large bowel movement, and immediately the
disease improved. But soon his belly was elevated again
and the fever intensified and the lethargy in the same way.
When the same evacuation had been accomplished, he was
given a medicine with cardamum, cucumber, and opium.
The bilious matter broke loose, the coma was gone immediately, the fever had become mild, and he was relieved in
all respects. He had a crisis on the fourteenth day.

119. Eudemus had severe pain in the spleen. He was
told by the physicians to eat much, drink plenty of light
wine, walk much. There was no change. He was bled from
a vein. Food and drink sparingly. Walks increasing. Light
red wine. He recovered.

120. Philistis, the wife of Heraclides, began to have
acute fever, redness of face, from no clear cause. Shortly
afterwards she had shivering in the daytime. When she was
without fever, a spasm developed in her toes and fingers,
and shortly after that she became feverish. She produced
urine that was full of congealed matter, cloudy, separated.
She slept the night. On the second day she had shivering;
her fever increased a little. Her redness was less and her

---

321 καταφρόνη M
322 ἐπρηνετο V 323 μετέβαλλεν V
324 ἐφλοτομήθη V 325 πεφισμένως M
326 καὶ ὑγιάνθη V 327 Φιλιστίδη MV: corr. Asul.

ἧσσον, καὶ οἱ σπασμοὶ ἐγένοντο μετριώτεροι· οὖρα
διὰ τῶν αὐτῶν· νύκτα ἐκοιμήθη, σμικρὰ διαγρυπνή-
σασα,[328] οὐδεμιῆς δυσφορίης ἐούσης.[329] τριταίῃ οὔρησεν εὐχροώτερα, σμικρὰ ξυνεστηκότα·[330] τὴν δ' αὐτὴν
466 ὥρην ἐπερρίγωσε·[331] πυρετὸς ὀξύς· ἱδρὼς ἐς νύκτα |
δι' ὅλου· ὀψὲ δὲ τῆς ἡμέρης χρῶμα ἀνετράπη ἐς τὸν
ἰκτερώδεα τρόπον· νύκτα ὕπνωσε δι' ὅλου. τεταρταίῃ
αἷμα ἐκ τοῦ ἀριστεροῦ ἐρρύη καλῶς, καὶ γυναικεῖα
σμικρὰ ἐπεφάνη ἐν τάξει· πάλιν δὲ τὴν αὐτὴν ὥρην τὸ
πυρέτιον παρωξύνθη· οὖρα τροφιώδεα σμικρά· κοιλίη
δέ, φύσει μὲν σκληρή, πολὺ δέ τι μᾶλλον ξυνεστήκει,
καὶ οὐδὲν διῄει εἰ μὴ βάλανον πρόσθοιτο· νύκτα
ὕπνωσεν. πεμπταίῃ πυρέτιον[332] πρηΰτερον, καὶ πρὸς
τὴν ἑσπέρην δι' ὅλου ἵδρου· καὶ γυναικείων χώρησις·
καὶ νύκτα ὕπνωσεν. ἑκταίῃ οὔρησεν ἀθρόον πολὺ
τροφιῶδες, σμικρὴν ὑπόστασιν ἔχον ὁμόχροον· περὶ
δὲ μέσον ἡμέρης, σμικρὰ ἐρρίγωσεν, ὑπεθερμάνθη,
ἵδρωσε δι' ὅλου· νύκτα ἐκοιμήθη. ἑβδομαίῃ σμικρὰ
ἐπεθερμάνθη, εὐφόρως ἤνεγκεν· ἵδρωσε δι' ὅλου· οὖρα
εὔχροα· ἐκρίθη πάντα.

121. Τύχων ἐν τῇ πολιορκίῃ τῇ περὶ Δάτον ἐπλήγη

---

328 διαγρυπνεύσα M: διαγρυπνεύσασα V: corr. recc.
329 ἐνεούσης M   330 ὑφεστηκότα M
331 ὑπερρίγωσε M   332 πυρέτιον] τό τε πῦρ αἴτιον M
333 καταπέλτῃ . . . γέλως] καταπέλτην καὶ μετ' ὀλίγον ἐς
τὸ στῆθος καὶ μετ' ὀλίγον γέλως M

---

64 The end of *Epid.* 7, like the ends of a number of Hippocratic

spasms became more mild. Urine of the same kind. At night she slept, having lain awake briefly; there was no discomfort. On the third day she produced urine of better color, small suspended particles. At the same hour she had chills. Acute fever. Sweat towards night on the whole body. Late in the day her skin color changed toward the jaundiced type. At night she slept straight through. On the fourth day blood from the left nostril flowed freely and slight menstrual flow appeared on time. Again at the same hour the fever became acute. Small amount of urine with congealed matter. Her bowels, by nature obdurate, became much more compacted and nothing passed unless a suppository was applied. She slept at night. On the fifth day the fever was milder and towards evening she had sweat all over. And flow of menses. And she slept at night. On the sixth day she produced, in a flood, much urine with congealed matter, with a small sediment of uniform color. About midday, she had slight shivering, she became warm, sweat all over. She slept the night. On the seventh day she was slightly feverish, but comfortable. She sweated all over. Urine of good color. Total crisis.

121.[64] Tychon, at the siege of Datum, was struck in the

works, is a collection of chaff probably gathered in the process of its transmission. The meaning of many things in these last sections is unclear. The subject matter of chs. 121–124 may well be later in date than the rest of the work. In ch. 122, *hippouris* ("docked horse's tail") is an unknown complaint, *kedmata* is also dubious but may be an arthritic complaint. Lechery may cure dysentery because it was thought to be drying to the system (cf. *Epid.* 7.69 above). The allusive ch. 123 I take to be about problems of puberty.

καταπέλτῃ ἐς τὸ στῆθος, καὶ μετ᾽ ὀλίγον γέλως[333] ἦν
περὶ αὐτὸν θορυβώδης· ἐδόκει δέ μοι ὁ ἰητρὸς ἐξαι-
ρῶν[334] τὸ ξύλον ἐγκαταλιπεῖν τὸ σίδηρον κατὰ τὸ
διάφραγμα. ἀλγέοντος δὲ αὐτοῦ, περὶ τὴν ἑσπέρην
ἔκλυσέ τε καὶ ἐφαρμάκευσε κάτω· νύκτα διήγαγε τὴν
πρώτην δυσφόρως, ἅμα δὲ τῇ ἡμέρῃ ἐδόκει καὶ τῷ
ἰητρῷ καὶ τοῖσιν ἄλλοισι βέλτιον ἔχειν· ἦν γὰρ ἡσυ-
χαῖος. πρόρρησις, ὅτι σπασμοῦ γενομένου οὐ βραδέως ἀπολεῖται. ἐς τὴν ἐπιοῦσαν νύκτα δύσφορος,
ἄγρυπνος, ἐπὶ γαστέρα τὰ πολλὰ κείμενος. τρίτῃ,
πρωὶ ἐσπᾶτο· περὶ μέσον ἡμέρης ἐτελεύτησεν.

122. Εὐνοῦχος ἐκ κυνηγεσίης καὶ διαδρομῆς ὑδρα-
γωγὸς γίνεται ὁ παρὰ τὴν Ἐλεαλκέος κρήνην. Ὁ περὶ
τὰ ἐξ ἔτεα ἔσχεν | ἱππουρίν τε καὶ βουβῶνα κατ᾽[335]
ἴξιν καὶ κέδματα. Ὁ τὸν αἰῶνα φθινήσας ἑβδομαῖος
ἀπέθανεν. Πυοποιούντων ἄπεπτον ἁλμυρὰ μετὰ μέλιτος. Πορνείῃ ἄχρωμος[336] δυσεντερικοῖσιν ἄκος.

123. Τῇ Λεωνίδεω θυγατρὶ ἡ φύσις ὁρμήσασα
ἀπεστράφη,[337] ἀποστραφεῖσα, ἐμυκτήρισεν·[338] μυκτη-
ρίσασα διηλλάγη· ὁ ἰητρὸς οὐ ξυνεῖδεν· ἡ παῖς ἀπέθανεν.

124. Ὁ Φιλοτίμου παῖς ἔφηβος ἦλθε πρὸς ἐμὲ
εὑρὼν[339] κρανίον ὀστέον[340] ἔρημον ἰητρικῆς ἐς ἰηχίνα,
τὰ κρύφιμα οὐχ ὁρῶν ἑστῶτα.

334 Smith: ἐξαίρων mss.
335 καὶ MV: corr. M. Rosenbaum
336 ἄχρομος V

chest by a catapult, and shortly later there was around him raucous laughter. The physician who removed the wood seemed to me to leave the iron in at the diaphragm. He was in pain. Towards evening he administered a clyster and a purgative drug below. He passed the first night in distress, but at daybreak he seemed to the physician and the others to be better, since he was quiet. The prediction was that convulsions would be followed by death soon after. Towards the subsequent night, the patient was distressed, sleepless, lying on his belly for the most part. On the third day early he had convulsions. Towards midday he died.[65]

122. The water-carrier by the spring of Elealceus became a eunuch from hunting and running. The boy of about six years had *hippouris*, and glandular swelling in the groin on the same side, and *kedmata*. The one with lifelong consumption died in the seventh year. For people producing unconcocted pus, give salt water and honey. Unrestrained lechery is a cure for dysentery.

123. In the case of Leonidas' daughter, her nature, rushing forward, was repulsed, and, repulsed, she bled from the nose. Having bled from the nose she was relieved. But the doctor did not understand. The girl died.

124. Philotimus' son, when in military training, came to me for healing when he found the bone of his skull bare of medical treatment because he had not seen the hidden things established.

---

[65] Cf. *Epid.* 5.95.

---

337 ἀπετράφη V   338 ἐμυκτήριζε corr. to -ισε M
339 εὑρὸν M   340 ὀστέων V

INDEX

INDEX